KINETIC ATMOSPHERES

This book offers a sustained and deeply experiential pragmatic study of performance environments, here defined at unstable, emerging, and multisensational atmospheres, open to interactions and travels in augmented virtualities. Birringer's writings challenge common assumptions about embodiment and the digital, exploring and refining artistic research into physical movement behavior, gesture, sensing perception, cognition, and trans-sensory hallucination.

If landscapes are autobiographical, and atmospheres prompt us to enter blurred lines of a "forest knowledge," where light, shade, and darkness entangle us in foraging mediations of contaminated diversity, then such sensitization to elemental environments requires a focus on processual interaction. Provocative chapters probe various types of performance scenarios and immersive architectures of the real and the virtual. They break new ground in analyzing an extended choreographic – the building of hypersensorial scenographies that include a range of materialities as well as bodily and metabodily presences. Foregrounding his notion of kinetic atmospheres, the author intimates a technosomatic theory of dance, performance, and ritual processes, while engaging in a vivid cross-cultural dialog with some of the leading digital and theatrical artists worldwide.

This poetic meditation will be of great interest to students and scholars in theatre, performing arts as well as media arts practitioners, composers, programmers, and designers.

Johannes Birringer is a choreographer and media artist. Since 2004 he has co-directed the Design and Performance Lab and headed the Centre for Contemporary and Digital Performance at Brunel University London, where he teaches performance technologies. DAP-Lab's immersive dance installations, metakimospheres, began touring in 2015–19. He is author of *Theatre, Theory, Postmodernism*; *Media and Performance: Along the Border*; *Performance on the Edge: Transformations of Culture*; and *Performance, Science & Technology*.

KINETIC ATMOSPHERES

Performance and Immersion

Johannes Birringer

Routledge
Taylor & Francis Group

LONDON AND NEW YORK

First published 2022
by Routledge
2 Park Square, Milton Park, Abingdon, Oxon OX14 4RN

and by Routledge
605 Third Avenue, New York, NY 10158

Routledge is an imprint of the Taylor & Francis Group, an informa business

© 2022 Johannes Birringer

British Library Cataloguing-in-Publication Data
A catalogue record for this book is available from the British Library

Library of Congress Cataloging-in-Publication Data
A catalog record has been requested for this book

ISBN: 9780367632571 (hbk)
ISBN: 9780367632618 (pbk)
ISBN: 9781003114710 (ebk)

DOI: 10.4324/9781003114710

Typeset in Bembo
by codeMantra

Front cover: Elisabeth Efua Sutherland in *kimosphere no. 3*, created by DAP-Lab. NailFeathersDress designed by Michèle Danjoux. 2016. © DAP-Lab

An ancient pond.
A frog plunges.
The sound of water.

(haiku, by Matuso Bashô, 1686)

I am, because we are
We are, therefore I am

(Ubuntu wisdom)

CONTENTS

ACKNOWLEDGMENTS

Numerous artists, performance organizations and research institutions have helped to make this book a reality. I thank everyone who contributed to the performance practices and cleared the pathways described here. I am also indebted to all those who kindly provided images and illustrations of their work or exchanged their technical and artistic expertise during workshops and performances. Above all, my gratitude extends to fellow travelers and practitioners who collaborated with AlienNation Co. and the Design and Performance Lab (DAP-Lab) in many workshops and artistic productions over the past decades, who invited us to their performances spaces or visited our labs or residency programs at Interaktionslabor (2003–15) in the abandoned coal mine Göttelborn or who were students and postgraduate researchers in the directing, acting and digital multimedia performance classes at the university levels where I taught or at the community theatres where I was a guest artist.

Many of the original ideas for multimedia performance and research into composition developed in Houston during the 1990s; I am very grateful to many people and organizations who believed in the experimental visions I developed in Texas and participated in the performance and installation projects of Alien-Nation Co. (www.aliennationcompany.com), which became a non-profit arts organization in 1997. In the years after the turn of the century, my base of operation first shifted to Ohio State University (2000–03) and after that my time was divided between Houston and London. In 2004 I founded the Design and Performance Lab (DAP-Lab) together with Michèle Danjoux, fashion designer and co-director of the studio and ensemble productions. Over the past sixteen years we have produced innumerable workshops and a series of stage productions and immersive installations – kinetic atmospheres – that provide the ground on which I walk here.

I must acknowledge the formidable support received from individuals, arts councils and government agencies who supported the work. The feedback from artists and audience members at various venues where we presented performances and working techniques has been very helpful. These venues include the following: DiverseWorks, Houston; Interaktionslabor, Göttelborn; Second International Congress on Art and Technology, University of Brasilia; University of Brasilia Art Gallery; Media Center, Graduate School of Semiotics and Communication-PUC-University, São Paulo; Subtle Technologies Conference, Toronto; Association of Internet Researchers Conference, Toronto; ScreenPlay Festival, Nottingham; RePerCute: Reflexiones sobre Performance, Cultura y Tecnología, Hypermedia Studio, UCLA, Los Angeles; DAMP Lab, V2_ Institute of the Unstable Media, Rotterdam; Flesh Made Text Conference, Aristotle University, Thessaloniki; TransNet Performance and Science Laboratory, Simon Fraser University, Vancouver; Beijing Dance Academy; Real-time, Light, Video and Sound Space workshop series, Monaco Dance Forum; Synaesthesie und Multimedien Workshop, Universität Erfurt; MotionBound to Virtual Skeleton, Dance Technology workshop, Essexdance, Guildford; Çaty Dance Studio and TECHNE Platform, Istanbul; Wearable Futures: Hybrid Culture in the Design and Development of Soft Technology, University of Wales, Newport; Architectures of Interaction laboratory, Chisenhale Dance Space, London; In-Between Time Festival, Arnolfini Gallery, Bristol; Performance Space, Sydney; IN TRANSIT Festival, Haus der Kulturen der Welt, Berlin; Performance Studies International Conference, Queen Mary, University of London; Digital Resources in the Humanities and Arts, Dartington College of Arts; Anthropology Department, Rice University, Houston; Digital Dance workshop, Monaco Dance Festival, Monaco; Movement, Sound and Network Project, SARC, Queen's University, Belfast; Boston Cyberarts Festival, Boston; 3rd Annual Moves Festival of Screen Choreography, Manchester; Prague Quadrennial, Prague; a-m-b-e-r 07 Festival, Istanbul, Turkey; Oficina de idéias, Funarte – Casa do Conde, Belo Horizonte, Brasil; Post Me_New ID Forum, CYNEtart_08 Festival, Hellerau; Subtle Technologies, Arts and Science Conference on Networks 2009, Toronto, Canada; kedja/CODA dance festival, Oslo, Norway; Taller Teatro Multimedia, Talento Bilingüe de Houston; Choreolab, Krems Donau Universität, Austria; UKIYO Symposium, Keio University, Tokyo, Japan; Motion Bank/The Forsythe Company, Frankfurt; METABODY, Madrid; Orff Institute, Salzburg; International Symposium on Electronic Art (ISEA)2017/ Festival de la Imagen, Manizales; National Center for Dance, Bucharest; ISEA 2018, Durban; Moody Center for the Arts, Rice University, Houston; MECA, Houston; Artaud Performance Center, Brunel University; Telematic Encounters, CINETic, Bucharest; Vibra Workshop, Norwegian University of Science and Technology, Trondheim; Body IQ Somatics Festival, Berlin; Tate Modern, London.

Excerpts and passages from previously published essays and reviews are reprinted with kind permission of the editors of *Theatre Journal, PAJ: A Journal of Performance and Art, Performance Paradigm, Theatre and Performance Design,* and

Texto Digital. Excerpts from "Audible and Inaudible Choreography: Atmospheres of Choreographic Design," first published in *Klänge in Bewegung: Spurensuche in Choreografie und Performance*, ed. Sabine Kaross & Stephanie Schroedter, reprinted with kind permission of transcript Verlag. Excerpts from "Gestural Materialities and the Worn Dispositif," first published in *Digital Movement: Essays in Motion Technology and Performance*, ed. Nicolás Salazar Sutil/Sita Popat, and excerpts from "Metakimospheres," first published in *Digital Bodies: Creativity and Technology in the Arts and Humanities*, ed. Susan Broadhurst and Sara Price, reprinted with kind permission of Palgrave Macmillan.

I thank the reviewers of the manuscript for their close reading and helpful advice, and my editor and the publisher for the enthusiasm brought to the book project. It will be apparent throughout the book where personal relationships to artists have been particularly influential for the development of our ideas. The works of numerous companies and directors are cited to provide a comprehensive context for the further study of contemporary digital art and performance. I have gained much insight from observing these works, and I also offer sincere thanks to those who dedicated their creativity and passion to the continuous theatrical explorations carried out in Houston, the Design and Performance Lab in London and the Interaktionslabor in Germany. My special gratitude is owed to my close collaborators who accompanied me through years of passion; and foremost to my partner Michèle Danjoux whose love I hardly deserve but whom I admire with all my heart. My thinking and physical practices resonate artistically and philosophically with the beliefs and actions of many others, kindly provocative friends with whom I have celebrated artmaking, amazing human beings I have encountered throughout a life time: too many to name them all here, but without their shared passion and friendship none of this work would have been possible.

PREFACE

FIGURE 0.1 Inside the *Horst* forest area, Saarland, Germany, summer 2019. ©
Interaktionslabor.

It is early morning, with a soft tint of light falling through the leaves, as I walk in
the forest and imagine discovering a hidden place, barely visible bird nest or ant
hill, then climb over a fallen, decomposing tree to reach for a small rock, covered with *Nacktmundmoos* (naked mouth moss) and ivy, almost invisible as rock
but painted with the earthy colors of thick underbrush, tangled mess of browns,
greens, purples. There is moisture in the naked mouth moss, I feel it between my

fingers. I think of fossils and animic lines on the skin, blurred species fingerprints of time, and of covert spaces that are produced beneath surfaces of ordinary objects. I touch the un-manifested, the layered and thick strings, the light flutter of wings, as I listen to wayward winds up above me, sensing the air getting colder in the atmosphere.[1] The crickets are quiet but something is fluttering, in the upper branches. The moisture reaches the skin on my face and neck, as I lie down – or is it the humidity? – I am immersed and crave for more, a wellspring of being acting like high voltage, do you have any last words?

In my mind's eye, I am a bird that has landed on this small mossy ooloi rock, a precipice for rest, after a longer flight that brought me back, in the summer season, to the terrains of childhood, my youth where roaming in deep and darker areas of the woods meant performing discoveries about environment and body, about the land and my sexuality, perhaps even an early unconscious form of *body weather* training practiced without knowledge and foreboding, then. No one had prophesied to me then that immersive experience in atmospheric dance and multimedia architectures would become a possible subject for a book on performance art and design. But "atmosphere," I gather, has become a more popular subject recently: symposia on *staging atmospheres* pop up, even books are now translated from architecture, geography and philosophy into performance studies. Atmospheres are metaphorically compared to stage design, to choreographic objects.

Having just read the autobiographical seasons of Karl Ove Knausgaard's latest outpourings – *Autumn, Winter, Spring, Summer* – alongside Didier Eribon's memoir *Retour à Reims*, I sense it is not the right time for an elegy about growing up in the hinterland. Childhoods are of interest, and all my friends talk about autobiographies we need to write. A reviewer of a novel by Elena Ferrante recently confessed, in the *London Review of Books,* to being self-conscious about reading the book too much through her own body, her own history.[2] Yet, scattered departures and returns into a range of mediums seem closer now to the diverse preoccupations that inspire debate, and certainly inspired my work on choreographic objects and kinetic atmospheres that I intend to share here. Returning home may be as difficult as moving out, moving away. My homegoing *retour* here is complicated by a sense of foreboding, the fifth act of *Macbeth* rings in my ears, ever since as a young student I jumped up onto the platform of the old Roman Theatre in Orange (France), wanting to test the acoustic resonance. Out, out brief candle. Till Birnam forest come to Dunsinane.

I often feel drawn to geologists, cavers, deep sea divers, archeologists, climate scientists and botanists. Have you ever heard of the marvelous book, with pressed flowers and herbs, by Balthasar Johannes Buchwald: *Specimen medico-practico-botanicum, oder kurtze und deutliche erklärung derer in der medicin gebräuchlisten und in Dännemarck wachsenden erdbewächse, pflanzen und kräuter...*(Copenhagen, 1720)? It has wonderful illustrations that impressed me as completely intriguing and enchanting, in a tactile sense, seeing the actual species of herbs, twigs, petals, fronds, blades pressed onto the paper of the book's pages, with small vertical columns above, written in Latin, Dutch, Danish, German and French – commentaries on the herbs, their flavor or medicinal healing capacities. For example, the bittersweet nightshade: *Amara dulcis, Dulcimara, Solanum candens, solanum dulcimara*; *Bittersüss, Je länger je lieber, Hirschkraut….* There they are, a *mise en scène* of leaves and plant fibers, 230 specimens pressed onto writing, a performance of organic and linguistic, historical poetics, the book a consummate artifact – I think only a dozen copies have survived and been preserved in libraries.

Pressing flowers into pages of a diary, this is something we do as school children, when flowers mingle with love poems, reflections on the days in school or in the village when we were traumatized or overjoyed by the attention of others, those who impressed themselves on our young bodies and minds. Now, in my later years of life, the sense of foreboding has shifted, not to the life ahead, growing up and into the bittersweet world, *je länger je lieber,* but glancing over the shoulder, toward the stranger by the lake, and at the art, theatre, dance and music communities I have traveled with, collaborators and instigators of wild discovery, along the substations, energy fields, stages, abandoned seasons and disappearing acts.

Light falls through leaves and branches, conjuring a stage or a precipice, a relic of convictions and then one hears the percussion, distant beats. They call, and fading up are colors of cymbals and gongs, and passageways, the dance always moving as if movement were captured in the lesser moments of disrupted dreams. The dreams can act as prophecies too. This book I wish to write here is dedicated to movement, movement toward life (as Anna Halprin once called it) and away from it too, into small distances, aimed at perceiving more and more landscapes and terrains of nature (real and designed) as bearers of intelligence. In reverse, looking into the dark goggles of VR and virtual spectrums, hopefully not ghastly dystopias of bullet-riddled game worlds but sensual 3D environments where you touch things and others.

Kinetic art and atmospheres, in the book of time, may not be remembered, as so much or so little. And how do you record movement and atmosphere, dissolving shapes or contours, boundarylessness? Do you write phenomenological diaries? Describe moods, emotions and feelings inside atmospheric conditions, which are processes of affective relations that change, and perhaps are quickly forgotten? Do you aim for high theories and definitions? Why is there a body of techniques (somatechnics) and how do we shift body weather and sensor knowledge, microcosmic pressures, off onto scales and planes of sound that reverberate? I have not done the scientific measurements. I echo my thoughts on performance technologies, whenever I rehearse and write, and yet I do not plan to write or theorize much here. Rather, can I suggest you enter into a quiet meditation on immersion, through the chapters, photographs and paintings here, through rehearsal workshops that are affective, I believe, because they physically and cognitively explore interactivities and interfaces. May I invite you to detect matter and mediations, eavesdrop and pick up some of these notes toward an evolving or different understanding of the *extended choreographic,* the atmospheric in motion, the *kimospheric*? These chapters that follow are not organized strictly or coherently, they are spun, thrown and let go, after many rehearsals – resembling scores, suppositions, spectrograms.

FIGURE 0.2 *Eclipsemoon*, acrylic on canvas. 2018. © Johannes Birringer.

They aim to reflect on theatre and performance, dance and sound, architecture and movement, as if they were ephemeral atmospheric models: I call them *kinetic atmospheres*. Thus they also stake out a few ideas on technologies, choreographic objects, biomedial environments, microperformativities and metabodies/metakinespheres in which the body is not necessarily the only basis of perception and of the less visible world of ubiquitous computational data processing, the living "postnatural" natureculture environment. Having looked over the shoulder to see the digital ground grow, behind me, as I am catching up, it makes me smile to read about the post digital, post nature, post humanity, as if we really knew how to grasp evolving media processes and transmissions, getting ready again for the next turn, formulating a new physics or metaphysics for the time being, or what Mark Hansen at one point, optimistically, called a "new philosophy for new media."[3] Some new media quickly get old. The Japanese media artist Takahiko Iimura, in a recent post to a listserv discussion on digital curating and preservation, spoke about his early work in 8mm film, having gone out to the beach of Tokyo Bay where he discovered all kind of junk thrown everywhere, desks, shoes, tatami, clothes, along with animals, dead cats, dogs, etc. He said he was not only shooting those objects, but also trying to use them – mixed media in nature. A leg of a dead dog moved like a living one, motioned by a wave from the sea.

Other philosophers, such as Donna Haraway, Anna Lowenhaupt Tsing, Timothy Morton, Frank Wilderson, Saidya Hartman, Fred Moten, Wendy Hui Kyong Chun and Armen Avanessian, strive to delineate troubling entanglements, old habits remaining the same and new temporalities in the "current ubiquity" beckoning to us from the future pre-emptively. Wilderson's Afrofuturism is pessimistic. Avanessian's metaphysics speculate on the post-contemporary, or what we used to call future, the "post" marking how what is happening now is in relationship to what has happened but is no longer.[4] Are you following me?

For the time being, looking at the world of cultural production in the Anthropocene from an ecologically minded perspective – and thinking through performance and mediation – a critical light on the aestheticization of experience could draw attention to environments, complex systems and infrastructures, how they are narrated, how they begin to take shape in manifold forms and airy subterfuges, as engineered choreographies of organic machines, actors of information, matter, energy and transmissions, audible scenographies, conductions and effervescences. If I now choreograph live forms, or find ourselves collectively choreographed and driven to certain alignments to dominant or exploitative determinations and technical operations, then how does one conjure up practices (and this evokes an ethics of such practices) connecting us, perhaps also in a ritual sense, to a scrambled, less aligned poetics of the sacred and of creation, of prime matter and of distanced potentiality? How does one conjure up intra-action with fossils and industrial ruins, plants and animals, viruses, matsutake and the moon?

In what follows, the sensorial and atmospheric will stand out, with weathered bodies and machines at play, and sensual choreographies as articulations of the real and the virtual. Technical systems stretched a little, subtracting here and there, accessories, costumes and masks adding up. With surveillance equipment increasingly used everywhere, why not also wear it on the body, conspicuously, dialectically. And yet, whose bodies are trailed, what bodies that matter, in the sensory labyrinth, in the mossy gardens of Kokedera (Saihō-ji, a Rinzai Zen Buddhist temple located in Kyoto), the streets of Minneapolis or São Paulo or the old decaying forests of my childhood Saarland?

The pre-emptive potentialities may include accidents, and blindness is welcome here, in the vibrant performance interfaces, among friendly machines learning, promiscuous malfunctions, dirty electronics and cracked loops. If only I could continue to return to hidden places or Octavia Butler's outer spaces where the Oankali mingle with us, on this so-called computational planet, retro-engineering enchanting forests of the imagination, as if reading a few ideas about time and animacy, tended and untended nature, in the palms that are falling off from high.

NOTES

1 My psychogeographical and psychophysical experience in the forest was matched, if not outscored, by an experience I had inside a VR installation of "The Plank" during the Digital Materialism workshop at Tanzhaus NRW (Düsseldorf) in May 2019. I describe it in more detail in Chapter 5. My first evocation of "atmosphere" here is meant to set the tone, like listening to overtures of a string quartet or jazz concert – I am trying, I expect, to avoid a definition of my book's title, preferring to work with reverberating echoes of what is not easily communicable.
2 Lockwood 2021: 22.
3 See Hansen 2004. At a lecture for the 2011 transmediale in Berlin, Hansen offered new ideas on the panel for "Delimination of Life – Affective Bodies and Biomedia"

(https://vimeo.com/20753681), suggesting that his big claims about embodied tech-nesis and technological interfaces already were no longer quite accurate and needed to be revised. See also: https://vimeo.com/20753675. In 2020, the journal *Performance Research* 25(3) published a special issue on "Microperformativity," edited by Jens Hauser and Lucie Strecker.

4 Haraway 2016; Tsing 2015; Morton 2015; Chun and Keenan 2016, Wilderson, 2020; Moten 2003; Hartman 2019; Avanessian 2018. For Avenessian's ideas on the post-contemporary, see also his conversation with Suhail Malik on "The Time-Complex. Postcontemporary." Available at: http://dismagazine.com/discussion/81924/the-time-complex-postcontemporary/.

1

INTRODUCTION

Theatre, atmospheres, living systems

> Act so that you will be spared the necessity of deceiving anyone.
>
> (Gertrude Stein)

> Art must start work where something is defective.
>
> (Bertolt Brecht)

Resilient theatre

We live in a 21st century marked by waning resources and various crises in the social and natural worlds. Human-induced global climate change and the deterioration of nature now receive as much attention as the economic fallout and attending financial stresses of global capitalism or the impact of the new migrations on old notions of national sovereignty and border security. Increasingly, one senses naturecultural immunitary deficiency, with no *deus ex machina* in sight, and no paleocybernetic remedies. And yet it is not always apparent how environmental problems are intertwined with social and political problems, and how such intertwining inevitably affects artistic production and the creative industries. A crisis of the political imagination could have the opposite effect on the arts – indeed one might expect artistic positions to articulate diagnoses of the world or offer visions of resilient interaction and transformation: bold new visions of what could be, imaginative projections of how human beings might harmoniously relate to one another, to other species and the living earth.

Artistic positions or forms of expression reflect principles of organization, certain kinds of management of resources, and in this respect theatres, or the performing arts in the broadest sense, are quite naturally subject to economic realities, to limits of sponsorship, travel opportunities, access to work spaces and tools, various safety rules and risk protocols. Even in the countries

DOI: 10.4324/9781003114710-1

of overdeveloped capitalism, many under- or unemployed artists, actors and dancers find themselves, like the migrant workers, belonging to the precariat laboring under conditions of everyday vulnerability in our age of austerity. Is it not inevitable, therefore, that performing and directing, writing, composing and designing as shared theatrical labor engage their practices, forms and spaces of expression in the search for fresh viewpoints? In the search for lenses able to focus on the archive and the repertoire, as well as on the new content that so-called immaterial labor might shape through the use of old and renewed vocabularies? In the search for new spaces and protocols under the impact of unforeseen lock-downs and pandemics?

About five or six years ago I began to work with a large collective of artists and arts organizations on a shared project in Europe, titled METABODY, which led my London-based ensemble to the development of a series of installations. There was no hesitation to shift from stage or film-work – the physical-digital – to other digital or software-based and networked experiments, in the spirit of immaterial labor, for example, linking distant performers telematically through the internet, creating a composite virtual stage or learning how to work with Kinect camera interfaces, sensors or virtual reality (VR) technologies. The physical-digital intertwining was a premise for working with bodies, not beyond but with and about living bodies and living environments that mattered. Taka-hiko Iimura's media work on the beach: gone forward into the gallery, then back to the shore; in his email post he proposes that you choose this one or that one, whatever the media, you do select one of them at a time, the one you fit to this or to that.

The digital always retained physical materiality for me. I understood the shifts to be a part of the construction of complex built environments. In fact I had always imagined the site-specific in my work as a combination of engaging a particular place and adapting to it with all the tools and media (somatechnics) available. Sometimes one also succumbs to and receives from an environment, an off-site specific nature or urban commons, outside of theatre. A place understood as milieu, as habitat and as experiential atmosphere. The energies of bodies and our imaginations enter into material sites while drawing from the sites' affective materiality. If you perform in a quarry, the quarry will constitute your environmental atmosphere and rocky habitat. The tools, materials and media help to modify the site or be modified with the site's power – the site transforming into a constructed reality with a particular atmospherics or aesthetic energy. The kinetic atmospheres I explore are somatechnical, full of mutual referrals, real worlds.

Kinetic also, of course, means moving (from the Greek *kinesis*), motioning, moveable. In art historical terms, kinetic means "relating to motion." Since the early 20th century, artists have incorporated movement into their art, their objects, either to explore the possibilities of movement or to introduce a temporal and architectural dimension, duration in the experience of material textures or behaviors, and also to reflect on the importance of the machinic and the light, as well as technologies and projection/rotation techniques (especially involving electric light and motors) in our modern world. These artistic kinetic experiments have also very often been related to explorations of perception, optics and the nature of vision. Movement, in regard to kinetic art, has been produced mechanically by motors (e.g. in Naum Gabo's *Standing Wave*, 1919–20) or by utilizing natural movement of air within a space (e.g. Alexander Calder's mobiles that he began to work with in the 1930s). Kinetic art became a major phenomenon of the late 1950s and 1960s; it was an international exploration, as I became aware visiting a number of museum exhibitions, for example, *Lo(s) cinético(s)* at Reina Sofia Madrid in 2007, or *Kinesthesia: Latin American Kinetic Art, 1954–1969,* first shown at the Palms Springs Art Museum in 2017.[1] In Houston I had seen many more exhibitions of art from Latin America, Japan, and Korea that involved experiments in light and colored space art, projection art, technical experiments, moving "sound" (multichannel acoustic architectures) and visual projections that drew attention of synesthetic experience. I remember the exhibition *See this Sound: Versprechungen von Bild und Ton* (Lentos Kunstmuseum Linz) – its catalog with a whole range of provocative essays edited by Cosima Rainer et al. in 2009. At that moment, after having been well aware of visual (sound) artists like Brain Eno, William Kentridge, Bob Wilson, Meredith Monk, Hélio Oticica, or Steina and Woody Vasulka, I began to connect ambient sound and atmospheric visual design/choreographies in my thinking much more explicitly.

Later, during my years in London, I also became aware of an annual festival called Kinetica, initiated in 2009 by Dianne Harris and Tony Langford who founded the Kinetica Museum. Open to galleries, curators and artists from around the world who "focus on universal concepts and evolutionary processes though the convergence of kinetic, electronic, robotic, sound, light, time-based and multidisciplinary new media art, science and technology," as stated in the press release, the exhibitions are structured like an art fair, with a wild, unpredictable mix of elements in each showcase, combining futuristic fantasy sculptures, or animal-machine hybrids punning on evolutionary processes, with the latest gadgets and amusing gizmos brought there by inventors themselves. One also discovers stunning laser works or projections such as Venezuelan artist Carlos Cruz-Diez's *Chromosaturation*. Participating in each year's event between 2011 and 2015, I also noted a performance space hosting a range of digital, interactive shows and graphic projections; a key feature here was the 3D stage of Musion Academy, which offers its advanced projection system to other artists (the DAP-Lab was invited in 2013 to show "Tatlin Tower," a wearable electroacoustic instrument) as well as displaying its own 3D animation research. Judging from the exuberant atmosphere on the opening nights I attended, almost half a century after *Cybernetic Serendipity: The Computer and the Arts* (ICA, 1968, curated by Jasia Reichardt), the public's interest in inventive new contraptions and kinetic art objects has not subsided at all. This exhibit attracts up to 10,000 viewers each year.

As so much of my work since its beginnings in 1986–87 had to do with creating scenographies for dance-theatre, for movement and projected movement/light/visuals, drawing also from the synergies of the locale where I was working (while experimenting with choreographing abstractions, abstract virtualities), I assumed that scenographing an *atmosphere* also meant that the public would become immersed in it, move through it or be moved by it. Atmospheres in this sense are like the weather: they act, embrace us and slip into us or spread, move our sensations, vibrate in us, shape our presentiments. Except that in theatrical-architectural terms they are also engineered, constructed, actuated. They might deceive us. We become subjected to illusions. The sound of wind can be heard? Do we hear wind? Atmospheres are tricky, they are tricksters, technical actors, they are actionable and they are performed.

In this book I am about to write, weather is being linked to forest knowledge, a sensual understanding of elemental archives so to speak (if you accept nature as infrastructural, environmental media and biocycle), and thus reaching back deeply into an internal time sense, my times of growing up around forests, navigating, building models of the *Umwelt,* comparing, crafting and forecasting perceptions. These are also times of growing older, readapting the techniques with which I naturalize, readapting memories that have permeated my sensorimotor muscles. It is my way of sharing choreographic operations with you, which may also to an extent be resilient operations. They certainly resist commercial pressures or expectations, as well as the conventions that are associated with the traditional dramatic and text-based theatre. Choreographing atmospheres might be the simplest way of putting what I plan to do here.

Choreographing atmospheres, expanding kinesis

The prevailing atmospheres of the locales where we worked also of course affected me and my collaborators – those were not created but creating, affecting and in-forming. Atmospheres are reciprocal. In meditative terms, one could speak of communal designs, spiritual energies and chance operations (also in the sense in which John Cage's compositional process included random events and tacit exchanges with indeterminacy, or Pauline Oliveros's work relied on deep listening) that come together and provoke what is experienced as *vibrational atmosphere*, in a participatory embodied manner with affect-intensities, changing amplitudes of sensory phenomena. The climate – or the condition of the light and the temperature – where you work also has an immediate effect on how you experience the work. My two main home bases are Houston, Texas and London, UK, and their respective climate temperatures, humidities, the dryness or wetness and the biophysical transmissions of space one experiences are quite distinct from one another, thus affecting my metabodily being. My third base is the childhood forest valley where I grew up, in the Saarland (a tiny state in southwest Germany on the border with France and Luxembourg).

Languages, and lilting accents, spoken here and there, differ. Body languages differ. Round, flat, sharp edged, steep, curvy – physical contours are different, and so are the predominant colors and hues. Word machines differ too; when I travel from Texas to Europe, words, their contours and grammars, change and their phonetics require a different acoustic sensorium. But such vehicular shifts are relative, as we may be rehearsing with Chinese, Japanese, Italian and Brazilian performers, reminding our ears of very small nuances in the pronunciation of single syllables and homophones. Listening to atmospheres of the spoken is a feat of perception, and I have always considered it uplifting. During the METABODY project we traveled to work in various locations – Madrid, Amsterdam, Genova, Weimar, Paris, Dresden and back to London – and again the material world kept changing, and with it the reservoir of the imagination. I recall that the workshops in Madrid were most pleasurable as we woke early in the morning, every day, to sunshine and a walk to the nearby Retiro Park, to gather together and practice Qigong. A somatic preparation for the many hours that followed inside the Media Lab Prado. While in the park, I enjoyed listening to the slowly repeated guiding instructions for our movements, softly provided by Marcello Lussana. I also tried to take in the distant barking of dogs, the rustling of the wind in the trees, the urban noise of a city waking up outside of the perimeter of Retiro, on one side lined with many dozens of tiny wooden stalls displaying used books – the Cuesta de Moyano urban "library" dating back from 1925. Listening while moving, sensing felt presences and breathing the air above the grass under the sky – communication happening at elemental levels, transporting us.

I did not know at the time that atmospheres and immersion would gain a greater notoriety in the 21st century as terms relevant to theatre and performance. I also did not immediately think of these terms as pertinent to resilient theatre operations. And the writings that follow are not necessarily trying to do justice to new theories of immersion or of the Anthropocene. Rather, they are grounded in my experience of making installations, performances and films, of foraging into and exploring assemblages, responding to the world and spatial relations through the expanded choreographic sensibilities I intend to sketch for you. I hope these sketches are helpful. And moving along, step by step, you will enter into the worlds, above and beneath the feet, that conjure up choreographic objects, rhythms and dynamics that concern us here, along with a number of basic questions.

When I say beneath the feet, I also imply subterranean spaces and underground spatialities – the latest explorations I am able to add to my writings, inspired by a four-month laboratory carried out in early 2019 with anthropologists and artists in Houston, climbing down into tunnel systems, old cisterns, and even older paleolithic caves in West Texas. My subterranean metabodily being, i.e. the immersive experience of rock, natural limestone caverns, fabricated concrete dome structures, vast systems of steam and chilled water pipe lines, buried communication cables, vents and underground graffiti – all this has transduced me to new imaginings of bodily practice and abstracting choreographies of Virtual Environment (VE) mediation: inner kinaesthetic thinking flying off into outer planes of digital craft. The "outer planes" of VE are a contradiction in terms of course, as brains are tricked into thinking/sensing experiencing visual nature, proprioception hooked into a fascinating simulation. Prepare yourself wearing your goggles, even if this is not easy with writing or reading. An interactive website that I plan to design to accompany the book, however, comes along with this print medium. On the website you can listen to my ASMR whisperings or tune into my 3D films of the underground. I will also upload my *Gravel Maraboutage* and *Sisyphus of the Ear* films, shot in the Grand Canyon like quarry that lies north of my village. There I immersed myself into sliding down the gravel hills, waiting anxiously for the *maraboutage* effect.

The link between *atmosphere* and *immersion* needs to be clarified further (*maraboutage*, by the way, is a Senegalese term for spiritual and magic practices, for a certain form of preparation with a belief in the supernatural transitions). On a basic level, one evokes the other: we move through atmospheres and become immersed into them, for example, when we walk through a light rain during sun set, spotting a rainbow appearing in the sky. Or enter a cathedral or a discotheque and feel the echo and the vibrations, the spatial volume and the light. Take St Stephan's Church in Mainz, Germany, for example: it features astounding stained-glass windows designed by Marc Chagall, the Russian-French artist known for his vibrant paintings. The church's nine Chagall windows are distinctive for their indigo-blue hues, colors that render the portals brilliantly luminescent when backlit by the sun. Chagall began working in the medium late in his career; he had designed a window at the Metz Cathedral in France, a 12-window series at the Hadassah-Hebrew University Medical Center in Jerusalem, as well as the "Peace Window" for the United Nations building in New York City. In 1978, the first Chagall window by the then 91-year-old artist was fitted in St Stephan's. A further eight followed, six for the east chancel and three in the transept – they include "Vision of the God of the Fathers," "Vision of the History of Salvation," and the lateral windows of the East choir with the theme of "Praise of Creation." Many have described the experience of being inside the church's atmosphere as mystical, having witnessed a beauty reflected by the stunning blue light that envelops you as soon as you enter. Experiencing the windows, I imagine, is to experience an especially powerful aura, spiritual or aesthetic, that is generated by the luminance of the color of light. I have experienced this same sensation a number of times visiting the Gedächtniskirche, with its majestic cold blue stained-glass window in the memorial chapel next to the bombed-out cathedral at Ku'damm in West Berlin. The sensation is making me freeze.

We could find many other examples, just think of the Hagia Sophia, the pyramids of Teotihuacan or the Death Valley in the Mojave Desert. When we are trapped outdoors in a rain storm, we feel the wind and the moisture pressing against our skin and clothes, we get heavier as we get wetter, we feel colder, our whole infrastructure, if we understand the body as such (cf. Durham Peters 2015: 266), becomes saturated, all sensations of wetness and windiness coursing through our internal environments, neurological and musculoskeletal. This is not a new experience in human time, in time that connects climate and culture and the particular habitats in which we live (in Texas that includes forever repeating hurricanes; in other states it may include forever repeating earthquakes and tsunamis, droughts and endless dark winter nights).

Why then did immersive theatre or audience participation become so fashionably *au courant*? Did this fashion suddenly arise or has it been around for a much longer time, except not discussed in the same terms as it is now? Is *immersion* a term that relates largely to new media (such as VR technologies, optical media and games) or can it not be traced back to ritual and ceremonial performance, and to earlier historical epochs that deployed whatever technical means available to produce illusionary visual spaces and panoramic views, or any wide angle view/representation of a physical space, as Oliver Grau has persuasively argued, whether in *painting*, drawing, photography, film, seismic images or three-dimensional models (Grau 2003)?

And when did we become aware of an *aesthetics of atmospheres*, as if the weather and its variable conditions and meteorological predictions had moved to the fore and were examined by artists, designers and philosophers of spatiality and geography? At what point did producers and curators, in the performing and visual arts, shift their attention more closely to atmospherics, moods and affects, to interactive and participatory engagement, preservation, cultural heritage and immersive community projects (and re-enactments)? How would socially responsible artwork be related to (social) atmospheres? Why are cultural heritage and preservation now such hot research subjects that receive generous funding and state sponsorship?

I should think that such attention had existed before, surely throughout the 20th century and during the various experimental art movements which focused on materialities and since the 1960s increasingly spawned poly-disciplinary practices, installation, bodywork and intermedia work. I shall comment on some of these practices in the chapters that follow. Cultural heritage tends to point to older historical strata, down to the buried archeological ones. The movie *The Dig* (2021, dir. Simon Stone) is a charming reminder of this. But the 1960s are now our early live art historical horizon that we love to revisit. "Reperformance" is the name for such visits, first applied by Marina Abramović who is not just *present* (in her long-durational 2014 installation at MoMA) but also has taken to restaging, and presencing, older works by other performance artists, for reasons unknown. *Seven Easy Pieces,* performed at the Guggenheim in 2005, included Valie EXPORT's *Genital Panic* and, even more curiously, Joseph Beuys' *How to Explain Pictures to a Dead Hare.* If I am not mistaken, she has even founded an institute for long-durational performance projects (MAI – the Marina Abramović Institute) in upstate New York, buying up a former movie theatre and converting it to her archival domicil. Beuys, when I encountered his performances in the 1970s, struck me as someone who performed a dangerously self-mythologizing shaman role, in light of Germany's military history and the way he exploited his survival of the war. Although admiring his often quaint and queer acts of posturing, covering things with felt and fat or spending the nights locked up with a coyote, I would not have imagined a feminist body artist like Abramović needing to revisit such male shamanic pretensions, unless immersion beckons. The later concept inscribes itself into a genealogy and revisits it.

In early 2019, I met Iranian-Canadian performance artist Sahar Sajadieh, when she visited the DAP-Lab in London. She told me she also had created some "revisits," probably inspired by Abramović. When she showed me her version of Beuys' performance, now entitled *How to Explain Pictures to a Live Paro*, I was intrigued since it was quite different from the original. As I have not seen her piece, I quote from Sajadieh's website blog (Figure 1.1):

> *How to Explain Pictures to a Live Paro* is an interactive durational performance artwork about gun control policies in the US, and the safety and security of students. It is an invitation to a conversation about these issues and a tribute to the students who have lost their lives in shooting incidents at schools. The artist will be performing with Paro, a robotic baby harp seal (a therapeutic robot), which is scientifically proven to be helpful in hospitals, elderly houses, and trauma therapy. Paro has come to help facilitate the interaction, communication, and healing through a dialogue that has been long postponed in our community. This piece is a digital reenactment of Joseph Beuys' 'How to Explain Pictures to a Dead Hare.' The original performance was about Beuys' relationship with his artworks and his frustration with 'explaining art' to the people intellectually. He completely ignored the audience in his performance and talked to a dead

FIGURE 1.1 Sahar Sajadieh, *How to Explain Pictures to a Live Paro,* 2017. Photo courtesy of Sahar Sajadieh.

hare for 3 hours about his artworks. This digital reperformance, however, is an invitation to the audience to interact with the artwork and talk about the pictures on the wall, which are polemical and disturbing. The pictures are of the victims of every mass shooting that has happened in the last eight years in American schools. (http://www.saharsajadieh.com/portfolio/howtoexplainpicturestoaliveparo/)

Sajadieh proposes to "explain" pictures of the Real rather than her art. She is still and enclosed in the installation space, and yet the work is opened up to all who want to lean into it, face these faces of the dead, hold the Paro like a transitional object. This revisit also creates an acoustic space of commonality, as Sajadieh invites her visitors to the installation to leave their own explanations of the pictures to the Paro, speaking into a recorder, transmitting their feelings and sensations, confessing their reactions.

The curious and preposterous might gain a positive valuation today, in times of "grotesque immersion," as Nicolás Salazar Sutil points out so eloquently in his surprising new book on *Matter Transmission*, which landed on my desk just as I had returned from my first descent into a cave. It turns out Salazar Sutil is a caver, too, having traveled to Lascaux, Chauvet and Altamira to dive into the old Ice Age environments wanting to revisit natural, paleontological formations of material transmissions, geophysical memories and intra-active energies (Salazar Sutil 2018: 91). In the echoic resonances of dark caves, he also encounters the highly eroticized and the grotesque aspects of "landesque immersion" (92): the kind of immersion marked already by very ancient flutings, carvings, engravings or drawings – bodily gestures found on the limestone walls and rock surfaces, intestines of an underworld of traces. His explorations are stunning, as he experiments with an archeology of mediation that loops the Lascaux caves (ancient as well as VR-simulated approximations, constructed for tourists in the replicas they built) into "space" travel – a kind of sensory vision-reply to a cultural studies industry that is forgetful of all origins in "brute material," neglecting the *kinaesthetic* and the *kinetic* that I mentioned earlier. Salazar Sutil's paleocyber way of life connects with Tim Ingold's animic ontology (248) – proposing that we may need to reconnect to humanity's childhood and the sedentary conditions of our social and political organization: "the construction of a paleo ontology, based on this conjunction of indigenous worldviews and posthistorical philosophies, is a fundamental aspect of paleocybernetics" (219). And what he clearly proposes is an "integration of body and landscape" (225) (mentioned earlier in his writings on "Prehistories of Media I: Immersion"), very enticingly intimated as a way of surrendering one's body to the "cave's own erotic geophysicality" (153). This, I take it, is a visceral homegoing, a deeply sensory experience of "immersive transmission" of landesque space, as well as of an acoustic and mental space (144–46). Furthermore, he offers a quite remarkable deconstruction of Werner Herzog's film *The Cave of Forgotten Dreams* (2010), shot at Chauvet Cave, rejecting both the filmmakers and the prehistorians' and archeologists' (cf. André

Leroi-Gourhan, Henri Breuil, Marc Azéma) claims that caves are pre-cinematic, a kind of proto-cinema or "archi-cinema" (as media philosopher Bernad Stiegler calls it) – and here Salazar Sutil's writing is truly inspired. His chapter on "Pre-histories of Media II: The Screen" (175–99) offers a wealth of insight into what he calls the raw matter transmissions of geophysical cave experience, linked to our reclaiming of an "indigenous way of thinking" that could be teased from "the material thinking embedded in image-filled caves" (186).

The emphasis, I take it, is not on the *cave replica* – on distanced screen perception, of projections made to fit a screen or delude us into a surround-3D virtuality, but on body memory, imagination and intuitive sensory mediation. Materialist avant-garde movements often had political subtexts or more explicit activist agendas striving for physical-political transformation, trying to leave their own revolutionary marks and gestures, sometimes cutting into their own flesh, drawing blood and injuring themselves. Blooded thought, Herbert Blau called it, in his 1981 book, trying to make sense of performance art of the 1960s and 1970s. Immersion thus practiced as a kind inversion, corporeal acts against one's own body, one's own flesh, yet also making the body actionable – putting one's body on the line, I thought, assuming these acts to have a concrete political and existentialist *raison d'être*, though of course not going to the extent of Jan Palach's protest suicide in Prague. The materialist notion of construction (constructivism) as an explicit sculpting *gesture* already haunts the early Russian revolutionary art and architecture – Tatlin's Tower most prominent among the Russian utopian political projects. It was never built and thus remains a pure gesture, a provocation. Tatlin is another strange shaman figure that might deserve a revisiting. I will come back to this tower at a later point in the book, when I discuss some of the performances we created with the DAP-Lab, with some of the extravagant metallic wearables my design collaborator Michèle Danjoux created, including the TatlinTowerHeadDress – a vibrant, vibrating sculpture worn on the head by a dancer and actioned as a "radio tower" in performance. The radio announces Khlebnikov's poetic address to the Futurians, those who have been born and not yet died, in the prolog to *Victory over the Sun* (the DAP-Lab's version was titled *for the time being*).

Russian avant-garde art and social-political transformation certainly informed my thinking over the past decades, and such thoughts were also indebted to my early understanding of Bauhaus-inspired architecture and theatre (the theatre of Oskar Schlemmer, the photograms and kinetic light sculptures of Lázló Moholy-Nagy, the rayonism of Mikhail Larionov and Natalia Goncharova). What was imagined and tested in the various studios of the Bauhaus (in Weimar and Dessau of the 1920s) and the revolutionary art school Vkhutemas, founded in 1920 in Moscow, evolved in full force in various avant-garde movements that exploded in many parts of the world after the 1950s. The Bauhaus just celebrated its 100th anniversary, Germany's cultural policy of commemoration (*100 jahre bauhaus*) naturally encouraging reflection and sincere (self)critique. The Bauhaus did not create a universal utopian design language, it only tried to implement an influential, ubiquitous set of design principles, of course also encouraging the whole idea of radical interdisciplinary art education. But it remained ideologically self-enclosed in its Western-centric basis of thinking and instituting a dominant language of design vision (which after Gropius and Mies van der Rohe informed and defined the International Architecture style of the 1950s and 1960s during the Cold War). It also remained self-enclosed in its old gender-biased thinking, i.e. almost all of the masters in the Bauhaus workshops were men, practically none of the many women Bauhäuslers, such as Marianne Brandt, Gertrud Grunow, Gunta Stölzl or Lucia Moholy-Nagy, are remembered. It can now be expected, though, that they will be revisited (see Birringer 2020).

In the 1960s, most art forms expanded, their systems stretched out, and most of them intermingled and fruitfully contaminated each other. Art and science/technology inevitably went hand in hand; there was no division even if C. P. Snow, in his famous "Two Cultures" essay, lamented the great cultural divide that he imagined to have separated two great areas of human intellectual activity. The gender hierarchies collapsed as well. We are more confident now, undisturbed by any particular imaginary binary divide, and the expansion of our interdisciplinary laboratories and practices is ongoing. But the new vocabularies of atmospheres and entangled materialities may need some closer inspection, as they seem in part derived from installation and intermedia art, but also from other disciplines such as geography, philosophy, ethnography and legal studies. New materialist theories also draw from the sciences (e.g. Karen Barad's work). I will try to track some of these main resources and methods of art making and entangled thinking, along with other tools available to the human imagination and the craft of producing art. In the course of following chapters, I will also develop some of my reflections on interactional, inversive and immersive art works further, thus providing a wider historical and critical framework for the ideas on immersive kimospheres I want to advance here.

Low end theory, or notes on the weather

It is an underlying assumption of this book that performance vocabularies are affected by global changes, and thus respond to important cultural, technological and economic questions about how we "model" our worlds and how we communicate both real and virtual constraints/potentialities. This also implies asking how we train and increase our bodily awareness of interaction, how we read the coding in our bodies and how the environment invites or impels perceptual (re)orientations − or *recodings*, to use digital terminology (cf. Hansen 2006). The digital vocabulary is carefully suggested here. Although I have published on performance technologies for more than 20 years now, I tend to understand the technical as well as social components of our work as very closely sewn into fabrics and tissues of our bodies, and thus the technical (re)orientations are also always somatic (somatechnic). As John Durham Peters, in his book *The Marvelous Clouds,* notes very succinctly: "Both bodily organs and technical ὄργανα (organa;

tools' in ancient Greek) are a hodgepodge of different environments layered upon each other" (2015: 33). I would perhaps phrase it a little differently, thinking of *moving soma* and nervous system and permeating morphogenetic energies. The kinetic energies mean ongoing change – where environments are moving too, and therefore a "layering" is more like a diffusion or defraction. In fact, one could ask how transmissions and integrations, say, of body and landscape, of soma and weather, are both immanent and transhumant, always fluid, moving on, retooling each other, like fungal mosaics. What Peters also suggests, in his approach to a philosophy of elemental media, is a combination of elements and vessels of storage, transmission and processing, i.e. "soft" and "hard" infrastructures that can range from electrical power lines and railroads to clouds and naturalized background such as trees, daylight, moss, dirt, water and steam (2015: 30–32). This may well encompass, then, the post-digital and the pre-digital interfaces between nature and artifice, unsupported and supported bodies – the mycelia of all this animate life.

Already in the early 2000s, media philosopher Mark Hansen spoke of "bodies in code" and examined how human bodies experience and adapt to information-rich digital environments. Directing his theoretical argumentation against technoculture writers prone to overlooking biological reality when they speculate on virtual reality, Hansen's book, with its emphasis on a phenomenological body-schema (and thus the body-environment coupling derived from our unconscious awareness of our organism and what is outside of it), strongly advocates what has become a commonplace now, namely, the notion of *embodiment*. But he places Merleau-Ponty's body image/body schema in relation to interactive virtual environments, and thus posits VR to be anchored in the body (Hansen 2016: 20). Embodied life is "'essentially' technical," he argues, perhaps a bit too stridently, and inversely one could say "that technologies are always already embodied, that they are in their own way 'essentially' embodied, if by this we mean that they mediate – that they express – the primordial fission, the gap, within the being of the sensible" (59). It is the body that allows someone to feel like they are really "moving" through virtual reality or being "touched" by its technical implementation or evocation. Such movement as virtual experience also profoundly affects internal perspectives and our interactional understanding of what it means to live as embodied beings. On a more basic level, I would argue that this kind of thinking through technicity brings us back to an awareness of touch, to our flesh, bones and skin, and how we adopt various senses of the sensible, so to speak. Touching and being touched requires technique (and yet also relies on intuition and memory).

Some years ago I curated a workshop on adoptive and adaptive systems, and its location – in an abandoned coal mine in southwest Germany, near my childhood forest – in fact strongly affected, and to some extent produced, the conditions for performative artistic and scientific assemblages, for things to happen and for new techniques to emerge. Following Gertrude Stein's cue for actions that are not self-deceiving, it becomes ever more necessary to study how we are embedded in worlds saturated with complex infrastructures, in-forming architectures and in-formation technologies, data streams and too many images, how we engage our entrance into the virtual (or down into a dark mine shaft, for that matter), and how we explain this embeddedness to ourselves, or to others, or to the decaying industrial park of a pit with its postindustrial virtualities, the retired miners, the bees and the bee-keepers, the returning wolves, pine trees and mushrooms, and even, let's say, to the kobudai (also known as *Semicossyphus reticulatus*).

The kobudai, if I can make this sideways reference, made an appearance in *Blue Planet II,* a documentary TV series (2017–18) by David Attenborough in which we learn about dead sperm whales and faceless fishes, and also one fish, the tusk, living in the deep, which has invented a tool-using mechanism to crack open a tasty shell, namely, banging it vigorously against some deep sea rock, until it relents and opens up. Impressive. Then there was also a fish (female) that made love to the ugly bulbous older male (who seemingly had very bad teeth), and then, surprisingly, began to develop a hormonal mutation or gender meta-morphosis, becoming male. This is the bulbous kobudai fish in the Indo-Pacific ocean, belonging to the species of "sequential hermaphrodites," also known as "protogynous." Scientists who study species evolution and behavior ascertain that such sex changes are not uncommon, and for both female-first (protogy-nous) and male-first (protandrous) species, the advantage of size appears to mo-tivate the change. For some fish it is an advantage to be of a different sex once having attained a certain size; thus in protogynous fish like the kobudai, where a dominant male controls a group of smaller females, it is of advantage to become male later in life. The leaders of the sex changing game in fish – known as serial or sequential hermaphrodites – can change back and forth in either direction, depending on the environmental circumstances in which they are embedded (cf. Cormier 2017). In a more recent TV series, *Seven Worlds One Planet* (2019), At-tenborough, inspired by underwater jellyfish in the Antarctic and their coupling behavior, offers quaint trans advice: "If it is hard to find a partner, it pays not to have a gender."

It is these mellifluent circumstances that lie at the heart of the book, trans circumstances that to some extent are contingent while also interrelated to de-sign, scenography and the tools of fabrication. I will tell stories about combined circumstance, hidden metamorphosis, uncanny transitions, floating islands, re-verberations and assemblages, and about methods of how to look at the embed-ded, the immersed and the entranced and possibly hallucinatory experiences in performance and installation art, and more widely in participatory theatrical en-counters. I hope you, the reader, have experienced entrancement or exhilaration in performance as well and are drawn to it.

Examples and case studies are largely chosen from works I have physically encountered or studied in the archives of cultural memory or overheard in the recollections of friends who were there. I am also moving and thinking through

a range of performance and design experiments I helped to produce with my ensembles and collaborators, some of these entering the virtual – in my most recent
kimospheres that include VR – but always also returning to the real, or rather,
remaining nourished by the elemental, the spirit of place and of the immanent.

The choices I will make of the former set of examples will also be motivated
by a sense of historical depth – not attempting an archeology of immersive media
performance, but at least probing some of the layers, or undergrounds, of exchange between systems as they evolved over the past century, and especially, of
course, during the post-1960s decades I lived through myself, those past four or
nearly five decades that drove and affected my work with the journeys, constant
relocations and writings. The more recent stages of my biography involve the
international community of dance and technology artists, creators and tinkerers,
composers, anthropologists, software writers, fashion designers and engineers.
This community of experimenters has been nothing but astonishing. In my last
book, *Performance, Science & Technology*, I acknowledged them to be formative for
my understanding of many of the interrelationships that producers and cultural
workers explored when connecting artistic and scientific ideas in their generative
methods and visceral entanglements with living systems.

The technological processes that came to the fore remained puzzling to many
who assumed dance to be the most physically embodied performing art, and it
is also true that many dancers felt suspicious of or alienated by computers and
software. It would be incongruous, however, to imagine a somatic (re)turn for
the digital technologies and wearables. Or would it not? The evolution of sensors and other interaction systems or devices, motion capture, gesture recognition and gesture-sound mapping techniques, machine-learning models allowing
real-time human pose estimation, etc., involved so-called human factors, among
other industrial, engineering and geological applications. Merce Cunningham
never had a problem looking at Life Forms (a software) and its animated figures, drawing inspiration for his choreography and new forms of disalignment.
The human factors that interest me here are physical aptitude and kinaesthetic
awareness, on the one hand, and imaginary anchorages to the environment on
the other. The latter clearly also have cultural and psychological, tangible and intangible connotations. The somatic has always been there (in the human factors).
The digital, one might imagine, had to adjust as well, to the old physical bodies
that Stelarc, our leading experimenter with robotic extrusions, third arms and
third ears, artificial intelligence extensions, virtual avatar limbs and sculptural
introversions, once smilingly declared obsolete. The exoskeletons he wore in
performance intrigued me. His laughter, witnessed in many direct encounters,
also encouraged me, naturally. I imagined he was not entirely serious about our
demise. After we became friends, I was in fact quite sure that this believer in organic obsolescence had a perfectly wonderful sense of humor. This does not, of
course, invalidate a concern about corporeal corruption or decrepitude.

I am also gathering wisdom from my African American brothers, who taught
me the trick about the low end. Robert Hodge, a Houston-based painter and hip

hop music producer, alerted me to the deep confluences of rhythm and color, the beats of black art and the syncopation of a kind of blue, a kind of "so what" that hovers in everything and also of course in (jazz) improvisation, and this meant for me to relax about theoretical pretensions and academic high theory. This goes along with the decrepitude bit. I frankly do not care about the theory side that much, I try to be exacting in other ways, behind the wind and a constant angle, dancing under the crows and their shadows in the air. I enjoy it when musicians like drummer Makaya McCraven (who recently released *Universal Beings* and discusses the music in a documentary by Mark Pullman) speak of moving into the 5th or 6th dimension during their live recording jam sessions, when they all enter into the vibe of energetic ambience, shape-shifting with each other.

We learn to adapt to the weather and the seasons, we wear certain clothes under certain conditions, our skin responds to temperatures, we change, much as our climate and weather conditions change too. We respond to constant generative challenges, and also to physical-corporeal challenges that may restrict or disable our bodily abilities. With age, we may also become more careful, apprehensive, and suspicious. We do not want to fall. We should trust our bodies more; yet as they do fail us, or even become incoherent, when we lose balance we get scared. I gave a talk about underground spatialities today, to an audience of anthropologists, showing them, in the dance studio where my audiovisual lecture took place, how to fall (in order to look at the moon from a different angle) and how much I enjoy falling. I noticed how the room fell silent, and I enjoyed it even more. I continued falling down in front of my audience, for a while, to enjoy the "loop" (McCraven mentions this too, in reference to jazz improvisation, when he says that you want to remain immersed in that sequence when you hit upon it and it lifts you up, entrances you).

I believe interaction with a highly technologized prosthetic and data-rich environment, with all its generative processes, complicates intuitive learning processes, retentions and protensions, getting into the vibe at various stages of our ageing process – at least I make the assumption that training is ongoing and can never rest – and thus one might wonder how physiological experience, tacit knowledge and cognitive capacities are continuously reshaped by our contemporary tools. This of course also implies that locally informed cultural/material practices – specific older generation tacit knowledge – are in danger of being lost. This is perhaps what Salazar Sutil means by "indigenous way of thinking." And that is part of the low end of the theory, the end that is ending. I imagine, on every bright morning in fact, when I return to writing, that this is my last book, *der letzte Satz*.

was denn
sucht noch die stimme
im kurzen windstoss unter krähen
schwarmhaft und scharfkantik
um den flügelschlag

auch diese
wissen nicht rat in den schwingen
sitzt eine not und der gerechte
himmel versagt ihnen den eigenen schatten

what then
can the voice still search for
in the short burst of wind under the crows
that are like flocks, with sharp edges
at the flap of the wings
they, too, don't have any advice on their wings
sits a need and the just
heavens deny them their own shadows.

This is a strophe from Esther Kinsky's poetry collection *kö növény kökény* (2018: 46), with my poor translation into English, meant here as an allegory: I cannot voice a strong theory or coherent narrative of these processes and how I have come to understand interactivity, kinetic atmospheres and immersion. Working on sites and with choreographic installations happened in fits and starts, along with breaks and faltering syncopes. But I am interested in the low end, for various reasons (Hodge's paintings being one of them: his work in music and visual art was tremendously inspiring as it opened up new horizons onto the hip hop culture in the South while also signaling toward the Black Lives Matter movement, and the push for freedom and justice for all). Hodge's visible music, in fact, taught me to re-imagine my experience of jiving with the rave culture of Houston in the 1990s, just after the bleak end, the dancing on the graves and the devastation of the AIDS crisis. I fumbled in the dark at the time, creating concerts and performances that were homage to death. AD MORTEM, I titled one of my concerts, this is how bad it was. The end is the beginning, self-destruction as creation: stop seeking without what's already back in. Immersion, looked at from a political perspective, is also always a matter of cultural atmospheres of critical interactivity, liaisons with dreams that are colder than death, and with shadows that escape and proliferate, even without much advice hanging from the wings.

The chapters that follow will be organized in a non-linear, improvised manner, as if a series of rehearsals were to take place and you can tune into some of the reflections that have accumulated over the past years. They reflect and build a kind of poetic project collage – if that is the right term – and a sculptural text. I wish to suggest that such a poetic project is much closer to the truth of the work I am sharing here, and therefore I hesitate to delve into too much high theory and discourse reliant on other sources from other fields. My own field is the forest.

As you imagined, the question of a field is not even so clear. I am a choreographer and media artist: I build scenographies for dance, alongside collaborators who help pushing the envelopes. Scenographies for dance are choreographies of architectural space. So I am quite convinced that the envelope of theory, if I

were to sample the sound of many other writers on performance and art, architecture, scenography, geography, neo-materialism, geology and anthropology,
would stretch indeed to a very colorful patchwork quilt. The quilt is a good
metaphor, in fact, of fabrics and politics (if we recall the AIDS Quilt of the early
1990s). It is also spiritual. Fabrics will make frequent appearances in my writings;
the memory of loss and the more brutal and suppressed ordeals are perhaps not
always apparent in the rhythms of the chapters in this book. The fabrics – in the
costumes and architectures of the *kimospheres* I describe – are often soft and for-
giving, sometimes they are also tantalizing, sturdy, nettling and provoking. They
provoke sensorial reactions, and thus they touch on our corporeal realities and
visceral being in the world including all the cracks and kaleidoscopic mechanics
of our existences and remembrances. Our past has never been a foreign country.

The past of Futurisms

A century ago, passionate manifestos appeared and addressed themselves to the
future. What Futurism promised then, namely, the dynamic vitality of technological revolution, today's Bio-Art with its various artistic reframings of bioscientific method tends to question and interrogate, subtly refocusing cultural
reception of critical tissues of the *bios* and the ethical-political debates on genetic evolution, reproductive technologies and interspecies relations. The artistic
and activist reframing also refocused the body as a troubling, confusing space
where assigned genders had been limiting, reductive and overly simplistic, so
that contemporary disruptions of the coercive binary system by transsexual and
transgender people in the LGBT movement are also historically path-breaking
and decolonizing. Expansive, non-binary fluid possibilities for life on our planet
could in fact draw on alternative (and sci-fi) narratives, the history of other gender markers being available and operative in virtual spaces and imaginative liter-
atures of earlier days predating Donna Haraway's Cyborg Manifesto or Octavia
Butler's *Exogenesis* books and tomorrow's cyber/video games.

Marinetti's Futurism with its celebration of war and of human triumph over
nature of course now looks rather abhorrent. Its collusion with totalitarian political and ideological movements in the early 20th century had been readily
noticed, if not sufficiently condemned. The appeal of techno-fantasies has kept its
hold on mass-mediated culture as we see in each new and improved product that
enters the market, in each new cellular device that promises better connectivity
and faster Googling of the vast sea of data. Radical or provocative articulations,
alongside Dadaist eccentricities, had been a prerogative of the avant-garde, at
least in the modern age in the West, even as one might fruitfully compare such
radicalism with different – humanist and softer ecophilosophical – art movements
that happened around the same time, for example, with the foundation of Émile
Jacques-Dalcroze's Hellerau Institute for the teaching of eurhythmics. This art
school, perhaps the first significant experimental theatre laboratory in Europe
dedicated to training and education through a synthesis of the arts of theatre,

music, dance, stage design and lighting (an early somatechnics lab) was housed in the midst of the newly created Garden City of Hellerau, built in 1908 on the meadows overlooking the city of Dresden, Germany, and carefully planned according to utopian principles of fostering well-being and social harmony. The community was to enjoy an environment of finely designed houses and crafts workshops, public areas and various recreational and educational facilities. On top of the hill one can still see today the grand Festspielhaus with its studios and large open hall – now renovated and repaired after years of neglect and decay following its occupation by the Soviet army after World War II. The theatre was built by architect Heinrich Tessenow in the spirit of the remarkable design visions of Adolphe Appia who had drawn his own futurist sketches for the use of light and space – the exploration of "living art."[2]

This, in short, is the philosophical focus of the book you are reading: light and space, living art, architecture and vibrant interactions with the world around its internal system, which is in itself intertwined with or constitutive of atmospheres. Breathing, living and changing – what I have learned from the previous books I wrote on performance media, dance and science, now allows me to venture into a larger, expanded landscape, a kind of *forest* that I will repeatedly evoke in the following chapters. The forest is the environment of my childhood. I grew up foraging and playing in forests – they symbolize everything I associate with alluring and tangible atmospheres of experience, seductive, dangerous yet also calming, open-ended, full of secrets and surprises, soothing and troubling at the same time. Here is a pertinent citation from anthropologist Anna Lowenhaupt Tsing:

> To walk attentively through a forest, even a damaged one, is to be caught by the abundance of life: ancient and new; underfoot and reaching into the light. But how does one tell the life of the forest? We might begin by looking for drama and adventure beyond the activities of humans. This is the puzzle [...]: can I show landscape as the protagonist of an adventure in which humans are only one kind of participant?
>
> (Tsing 2015: 155)

I will return again to this idea of forest landscape and the dance it stimulates as one wanders through its bewildering lines of life, as only one kind of participant, pursuing them through all the senses, with movements and orientations. This is a dance signifying a form of "forest knowledge – but not that codified in reports" (Tsing 2015: 204) – it is a wilder dance lived and enacted and thus it is also knowledge growing up, unthinkingly rummaging through the thick, with childhood imagination, dreams and fantasies. Then there is the drama beyond humans, of vegetation and animal life, of fungal spores, flammable woods, diseases and pests. Such forest knowledge then will be the ground on which I move, metaphorically but also concretely, having grown up. No doubt forest knowledge also still includes myths and fairy tales, alongside its empirical side with the botanical and ecological ways of knowing.[3]

The past of Futurisms then also branches out, from Marinetti's synthetic the-atre and Russolo's noise music to Afrofuturism, Sun Ra's Arkestra and George Clinton's Parliament, to Robert Glasper's *Black Radio* and Makaya McCraven's organic beat music of *Universal Beings*. How artistic practices learn from history, or pretend to make history, is an uncomfortable question, not least with respect to the example of the Hellerau Institute's brief existence (1911–14), cut short by the outbreak of World War I. Its Futurism was already past before it had really started. In this regard the relationship of utopian ideas of Futurism to science fiction or to the messianic (as described by Walter Benjamin in his "Theses on the Philosophy of History") deserves to be examined if we place our faith in the constructivist side – and the communal ethos – of the performing arts. How pessimistic do we have to be? An awareness of past utopias, ill-begotten con-quests, ritual promises and failed triumphs of the will could guide us to a more poignant sensitivity toward the environment in which we live and breathe, with which we move and become. In his "Theses on the Philosophy of History," Benjamin makes an analogy between historical time and messianic time.[4] The historical time, being infused with the messianic, opens windows through which at any moment a messiah may enter, and such expectations, in whatever form, are recurrent in our cultures. On the political level, the election of the first black president in the United States in 2008 raised enormous hopes in many people, while it was quite unlikely, on the other hand, that this black messiah would utterly transform US politics or perform any miracles, before another president would take over and, as it were, reverse the course, polarize and charge up a new unhealthy climate of incivility, populism, racism and nationalism.

Over the past 20 years, throughout the political crises, the landscape of con-temporary art has also shifted in many ways; new windows have opened up. I prefer to think of them as small tectonic shifts to avoid touching upon the overexcited hyperboles of the earlier pronouncements on the *digital revolution* or the information age, which littered the side of the highways like so many reli-gious billboards promising salvation. The digital age and its mobilities have not revolutionized our political culture in the way that techno-futurism (or Afrofu-turism, for that matter) imagined. And not only since 09/11, the attack on New York's World Trade Center and its aftermath, has it become clear that we live in times of war and the production of bare life, not utopian liberation. Globali-zation has raised many more ethical demands and obligations than capitalism would ever be prepared to address. Today's bio-politics are marked not only by proliferating concerns about security, immunity and migration, but about infor-mation gathering/reprocessing itself, as the relationship of power to knowledge is transformed when secrecy fosters anxiety over things that cannot be known and things that cannot be seen. How, then, do we avoid the darker pessimism of the critics of technological progress who deplore the impact of the cybernetic and prosthetic imagination on our culture, or who predict that computers will exceed human intelligence in the *Age of Spiritual Machines* (Ray Kurzweil), when digital technologies become invisibly embedded in everyday things anywhere in our neighborhoods?

This is a good moment, then, to revive the question whether the living theatre, or what Appia in 1919 called *l'oeuvre d'art vivant* (the work of living art), can make a difference in the world by means of social intervention in matters aesthetic, political and ethical. Has the theatre found ways, perhaps a natural, energetic force, to reflect on its resources as well as its embeddedness in a changed cultural paradigm of global information processing networks?

More notes on the weather

In a recent interview, video artist Bill Viola addresses this embeddedness in response to questions about the synergies between painting, film and the performing arts within the "hybrid and liminal territories of video and multimedia installations." The interviewer wonders what made the characteristics of each discipline disappear, causing theatre to resemble real-time film, film to resemble installation and production processes of installation to acquire the methods of performance and film. Viola responds (Figure 1.2):

> We are living in extraordinary times. Cinema, and its extensions, has become not only a new way of making art but a part of daily life. People's behaviour is influenced by the characters they have seen in the movies – what clothes they wear, what cars they drive, how they talk, how they behave, how they think. Television extended these luminous apparitions right into the home every week, amplifying their familiarity and authority. I see social networking websites such as Facebook and Twitter as contemporary extensions of this phenomenon of identity-modelling and social shape-shifting. The influence of the moving image now continues into the digital age where cinema is no longer cinema but a fantasy that is played out on individual screens. When kids today play video games or communicate online with friends, they are usually alone in their rooms, and their friends are usually miles away, alone as well. Their minds and movements are extended into the virtual space of the Internet where they are communicating in a virtual landscape with virtual characters. And all this is originally derived from electronic media; from the telegraph, radio, telephone and, in terms of visual images, from cinema. But of course cinema itself is derived from theatre…
>
> (Valentini 2009: 59–60)

Viola adds that we now take the digital tools in our world for granted. He has not even anticipated the current lives on ZOOM, or redefinitions of "liveness."[5] The widespread and easily accessible means of producing live images in real-time

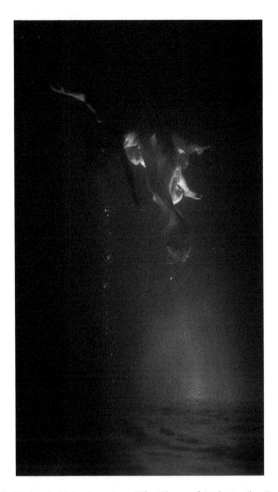

FIGURE 1.2 Bill Viola, *Isolde's Ascension (The Shape of Light in the Space after Death)*, 2005. Color high-definition video on flat panel display mounted vertically on wall; stereo sound 61 1/4 × 36 3/8 × 5 in. (155.5 × 92.5 × I 2.7 cm) 10:30 minutes. Performer: Sarah Steben. Photo: Kira Perov © Bill Viola Studio.

do not merely continue older techniques of recording and communication, they constitute a remarkable transformation of the representational apparatus that existed since the invention of the *camera obscura* in ancient China and its descendent techniques of creating three-dimensional illusions in painting and cinema. A new technology has appeared, Viola suggests, that allows a live, artificial image to co-exist with living human beings in real-time. A theatre of doubles and avatars, so to speak, has become second nature.

How does our relationship to the changing environment affect our psychology and our physical practices? This book seeks to posit theatre and performance as a living system, or ecosystem, a smaller-scale environment interrelated to a larger-scale world and its climates, an organism which is subject to the same pressures of technological engineering that are continuously transforming our concepts of, and relationships to, life forms, flows, movements, becomings and flourishings of cultures. The industrialized civilizations depended on engineering and socioeconomic progress inextricably linked to psychological and philosophical assumptions about human self-realization, subjectivity, agency, health, procreation, access to work and means of survival, housing, education, changing standards of living, etc. A whole range of ideologies thus determines how environments and space/times are experienced, how relationships to the world outside the self are identified, and what values regarding diversity of life forms and organisms are constructed.

As soon as we think an outside of the self and the body-mind, a relationship between subjects/objects and world moves into the picture, and this moving into the picture of course needs to be understood also through transformations. The environment is not still but living, always emerging and unstable, like climate and weather. This instability, if we remember the beginnings of photography in the late 19th century, led scientists and artists such as Étienne-Jules Marey, Georges Méliès or Eadweard Muybridge to devise new techniques for perception which, in turn, became new sensing techniques. We can think of Marey as an engineer of perception machines, intended to measure movement processes, to capture attitudes of the fleeting life of movement. In the early 20th century, Rodchenko, Larionov, Goncharova and Moholy-Nagy experimented with photography and light in ways that were equally challenging to perceptual processes.

The theatre has been a perception machine of particular complexity: its life force cannot be stilled, as we might see individual images in Muybridge's series of chronophotographs of, say, a "woman turning around in surprise and running away," but actors, dancers and scenes in the theatre are in continuous unfolding of present into future, through intensification of presences, blurred passages. Infinitely, they are also repeating themselves forward, anticipating the recurrences of action and mimesis, and the changes of weather, seasons, venues, circumstances, reception by audiences. How do actors train for this infinite repetition? And more recently, have we not seen the strangest and most affecting stage scenographies engineered by designers experimenting with fog, water, soil, mushrooms, sand, wood and steel (on the organic side) or digital projections, moving lights, onstage camera crews and foley artists (on the electrical and electronic side)?

"Theatre and Engineering" – this conjunction takes us back to another exhilaratingly complex historical period, after World War II, when artificial intelligence and cybernetics were still in their infancy, but social and artistic unrest exploded onto the cultural scenes of many nations, from Europe and North America to Central America, Venezuela, Brazil and Japan, shifting attention to new experimental happenings and collaborations. The 1960s saw the rise of the Living Theatre and performance art, Happenings and Fluxus events, informalism and assemblage art, Situationist interventions into the psychogeography of the Society of the Spectacle, as well as the concerted efforts of a few artists and scientists to support technological arts experiments and what we would now call "research and development."

The inspiration for a festival of such research or laboratory performances arrived in 1965–66, when Robert Rauschenberg's friendship and collaboration with Billy Klüver, a senior physicist and communications engineer at Bell Telephone Laboratories, led to planning a series of projects by artists and engineers from New York City to be presented at an "art and technology" event in Stockholm. When the Stockholm hosts could no longer endorse the process-oriented plans for artist-engineer collaborations submitted by Klüver and Rauschenberg, the New Yorkers went ahead on their own and started fundraising and location scouting. They secured the 69th Regiment Armory on Lexington Avenue and Twenty-Fifth Street (the place where the famous 1913 Armory Show had been held) and some corporate support; then Klüver and Rauschenberg proceeded to solicit the teams for *Nine Evenings: Theatre and Engineering,* the now legendary event of ten experimental performances presented in October 1966.[6]

Nine Evenings did not revolutionize the performing arts nor convince the theatre critics present at the event. In retrospect, it is surprising that any presumption of conventional theatre dramaturgy was brought to the event at all. The critics noted a lack of direction or hesitated to embrace the hybrid systems of physical and electronic organisms/circuits built by the artists' teams. Yet the dancers, choreographers, visual artists and musicians were not directing or staging plays but experimented with systems architecture. They created entire environments inside the Armory, live plastic sculptures, kinetic choreographies, and tactile, auditory and immersive sensorial experiences that we would now consider fundamental for an understanding of interactivity in so-called "intelligent environments" – spaces that detect/sense or capture behaviors and movements, "open scores" so to speak occurring within them, and that respond to data generated within the electronic circuits of interfaces. Those environments also behave, they have climatic qualities, they create humidity, tense temperatures, heat and energy, storms, cool winds and freezing affects, slowing down or speeding up the cyclical experience of temporal duration, for example, in the waning of the night.

Performers perform, machines perform, and the electronic circuitry and computational architecture generate reactions in the environment. The environment, in this sense, is alive and performative too. Within the chosen and prepared/designed parameters, performers and digital architecture together produce the event, through an exfoliating experience of mutual interconnection and synthesis, enveloping the audience or actively involving them in the occurrences, the movement of story and behavior, flows of images and sounds, unfolding of fictions and social behaviors. Rauschenberg's *Open Score* was perhaps the most exciting, unusual and yet paradigmatic performance architecture of the *Nine Evenings,* and it also "employed" several hundred people, extras brought in from the streets of New York City to enact a small social choreography proposed by

Rauschenberg who had asked them to gather on the tennis court built onto the space and allow themselves to be observed, in the darkness, by infrared cameras, then perform small instructions: touch someone who is not touching you; touch two places of your body; hug someone quickly then move on to someone else; draw a rectangle in the air; take off jacket, etc.

Why consider such a constellation an ecosystem? I propose to think of such installation-performance as an immersive weather system that functions like an organism, and use the term organism in both its biological and technological sense. It is crucial to imagine this system to be dynamic: any ecosystem encompasses both the interacting organisms and the environment in which they interact. To study a strong ecology of experience in performance is to study all the physical, psychological and technological dynamics in an environment created to foster experiential transformation, an emotional and aesthetic affect on living things, materials, forces. We will enhance our understanding of such performance ecologies by taking a journey through some of the main elements: action, movement, interaction, collaboration, design processes, and ecological and atmospheric principles.

The system I evoke is an intelligent system, by which I mean to imply that performance creativity, including acting, directing, ensemble work, technical operations, external circumstances and audience participation, takes place in an atmosphere – one might also call it working conditions – reflecting different levels of responsive behavior. An information-processing environment, along the lines of the live real-time systems mentioned by Viola, would be supported by common vocabularies of computer-based techniques, for example, the augmentations and reconfigurations in the software and hardware, the re-mixing of data, the "scripting" of "actor" or object behaviors (it is hardly a coincidence that patches in software environments like Adobe Director or Isadora are called "stage," and objects are called "actors"). The terms "digital performance" and "immersive installation" encompass a wide range of practices of composition and production, of sites and inter-spaces, of dynamic crossings between dramatic and postdramatic theatre, choreographic theatre, performance art and dance, installation, video, networked and media art, as well as mixed dramaturgies influenced by visual art, music theatre and sound art, games and television – a complex and ever-evolving world of hybridizations.

While I cannot address all dimensions of contemporary modes of performance, this introduction and the chapters that follow offer a practical and critical guide to an expanding field of artistic skills, ideas and working methods, drawing from observed examples and my own experience (in the labs and companies I direct) while emphasizing the ecological principles of the interrelations between the important elements of action, energy, movement, gesture, image and sound projection, costume design, scenography, light, spatial relationships/contexts, and the positions or roles of the spectator as protagonist. Such interrelations concern the making of experiences, the telling of stories, the modeling of worlds. My particular interest lies in parsing the synergies that exist in the ecosystem of performance, and thus I will pay close attention to the network of relations among organisms at different scales of organization, from micro-elements to larger architectures and combinatory flows, from cellular mechanisms to wider dramaturgical syntheses characteristic of hybrid aesthetics.

Digital ecosystems

As with all ecosystems, a philosophical understanding of deep ecology suggests that the organization of hybrid elements, choices and viewpoints includes diversity and heterogeneity, and the shared space of a "collective presence" in performance therefore also includes active contradictions between machining form and meaning in the relations of *mise en scène* and spectators. The term "machining" is here derived from provocative architectural theories that posit open, operative zones in which architectural materials perform or conjoin compositional systems, structuring forces and materialization processes that also include human interactions (Spuybroek 2004: 360–69). Contradictory presences are inevitable, as materials may also pose conflictual or resistant properties. Spuybroek's notion of "machining" can be considered an active artistic strategy or animating material energy that can trigger the materialization processes of structures, built environs and organic ecosystems. The contradictions are the necessary and elemental defect in these relations which Brecht proposed to examine in his critique of identification with mimetic illusion. For Brecht, contradictions are positive.

Brecht's critique of theatrical illusion, which he developed for his Epic Theatre in the 1930s, led him to develop specific techniques, for example, the *Verfremdungseffekt* (alienation effect), and a whole set of dramaturgical glitch principles which might be fruitfully revisited in regard to contemporary ideologies of acting/inter-acting, spectating and the kinds of visibility *and* defamiliarization that the technologically augmented theatre materializes. Digital machining in the theatre over the past two or three decades sometimes startled, if not alienated, their audiences and the critics (the latter often comment by asking why video projections were necessary or desirable for the scenography), bringing attention to what we now call *digital performance*. It also brought to the foreground new questions about virtuality – the virtual worlds, avatars and abstract motion graphics the audience would perceive and need to integrate into their perceptional habits. The growth of computer-based art is an accepted phenomenon in today's art worlds, yet there are few established theatrical conventions of receiving virtual art or interacting with non-mimetic code and a radically abstracted aesthetic logic. To my knowledge, it is not yet an accepted convention to regard the camera as a performer or a creature, and enjoy the expressive alien qualities of avatars manipulated by actors (or vice versa) and "drawn" by algorithms. The genre of digital performance and of technologized dance is still a very young one, barely defined and thus in need of historical and conceptual underpinnings.[7]

This is especially true in regard to dramatic theatre and its text-based strands, but also to the libretto-based sister arts: opera, music theatre, and sonic art (if we take coding as a form of libretto-writing), or the new wave of game design, networked performative coding and interactive story-telling. For all genres of embodied digital production, the emergence of new hybrid trans-theatrical forms inevitably requires a rethinking of a whole range of conceptions about the writing or staging of drama and its literary, oral, live art, documentary and cultural performance repertoires. The term *hybrid* here stands for assemblages that entwine the real and the synthetic, the physical and the virtual, the organic and the artificial.

The theatre has never been pure, medium-specific, as painting was perhaps understood at some point in modern art history. The theatre used painting, as it had used ritual, music, dance, writing, choreography, costume and set design. It had used machining, or dirty electronics, in the sense that theatre patches things together and always welcomes technologies of change. It has reacted to shifts in media history ever since the beginnings of notation systems in Antiquity when memorizing and pronouncing a written text on stage became a method of producing a vocal copy in the amphitheatre. *Kothornoi* elevated the actor, so audiences could see them better, and perhaps also added an interesting clacking sound on the stone floors. The theatre roughed it up, it grated its media and plugged in. The Latin word "auditorium" still reminds us of this cultural background of the history of theatre literacy: spectators came to listen to spoken writing amplified (alphabetical language slowly replacing the oral culture of the epic), watching and hearing actors reciting the dramatic dialog in action in the round, voices echoing in the amphitheatres and temples. I take it that Oliver Grau believes immersion is nothing new; it all goes back to these echo chambers, and perhaps to the idea of a vast expanse (panorama) of sound, in both its potential for illusion and anti-illusion – in the way that a modern piece like Steve Reich's *Pendulum Music* could be considered an audible sculpture that denies any illusion of musical time but focuses attention merely on the material of sounds. Audible sculptures have existed for millennia, one does not have to trace them to installation sound works in galleries, as Caleb Kelly does (Kelly 2017), but will have found them in aural architectures of many kinds, in nature and built space.

When film became a major new cultural medium and art form in the 20th century, theatre artists like Piscator or Svoboda sought to integrate it into the theatrical *auditorium*. Brecht had developed his "radio theory" in the early 1930s, linking the analysis of the new medium to utopian expectations. "A film must be the work of a collective," Brecht believed, assuming that the modes of technological production in the film imply the end of the bourgeois notion of art: "These apparatuses are predestined to be used for the *surmounting* of the old untechnological, anti-technological 'auratic' art, which was closely related to religious practices. The socialization of these means of production is a vital matter for art."[8] Today, in the world of social media and an instagramatic culture of broadcasting, publishing, distributing and consuming of digital images, it is hardly counterintuitive to recall radio and film as the first major mass media of the modern age, and to think of their "biomechanics" or social energies. The significance of film for the social, according to Brecht, lies not only in its collective production, but also in the new modes of reception. Only a collective, he explains, can create works of art, which transform the audience into a collective as well. It cannot be forgotten that Brecht had a radical socialist revolution in mind when he spoke of the collective and hoped for the mobilization of political response in the theatre.

In the absence of such a radical revolution, the social significance of theatre, in our current era of the internet, social media networks, streaming media and Reality TV, is surely subject to debate. Such a debate might raise doubts about the kind of collectivity the theatre can create today, using all the channels of its augmented space, patching them together hard. But the history of performance over the last century surely does reflect various kinds of reorganization of its means of production of such channels as well as its means of accommodation of related media such as photography, film, video and sound recording/amplification. Recording technologies also have become conventional forms of documentation and of shifting the modes of dissemination, for example, through the translations from stage to screen in public television adaptations of drama or in public live streaming. None of it is revolutionary. Film, television streaming platforms and social media have been a wider, much more distributed arena for actors (just as for musicians, when audiences are posting video clips of a concert online while the concert is going on): the sustainability of acting and the actor was one of my initial concerns, and one of the more depressing pragmatic and philosophical aspects of this book.

In a workshop at Brunel University, theatre director Geoffrey Colman (Central School of Speech and Drama London) argued that the conservatoire must strain for a new form of authenticity. He claimed that however much British actor training continued to aspire to a notion of a tradition and craft being passed down through the hearts and minds of successive generations, another perhaps more uncomfortable analysis must be considered. Actor training, Colman suggested, is now actually founded on a vocational rhetoric that hardly addresses anything other than the trainee's need for employment within a "camera-real" economy. For the actor's art to be manifest in such a context, it will be necessarily measured against a set of assumptions that are, in essence, about the successful realization of a general performance grammar – a camera-real acting style that somehow equates to a notion of good acting, and a non-camera-real or abstract acting style equating to that of bad acting. In order for this to work, the camera-real fictional nurses or cops that nightly haunt the television screens must appear to be performed in the same way, irrespective of style or period, and similar to other film or theatre worlds created elsewhere. In his discomforting summary of the reality of actor training, all has become neutered by the lens.[9]

This is a surprising indictment, coming from an acting teacher who works at one of the leading theatre conservatoires in Great Britain. On the other hand, the notion of authenticity is certainly a contested one, and Benjamin, perhaps glancing at Brecht's "theatre for the new scientific age," as the German playwright-director called it back in the 1930s, already observed that different techniques are indeed required for the "optical tests" of the film actor performing to a camera without any audience present. The optical test actor actually works in an anti-illusionistic space, performing to a lens. The presence of on-camera-acting in live theatre, however, is a phenomenon that Brecht and Benjamin did not anticipate. Actors are watching the optical test actor performing to a camera lens while an audience in a theatre is collectively witnessing this – and "this" means double (camera acting and simultaneous film projection of the captured shot, with all its zoom functions). I examine such changed optical tests in the work of various theatre companies who use the presence of the actor and the presence of the camera (recording and transmitting) to play with such doubling of the frame.[10]

The contemporary stage has generated a new intermedial liveness, and many forms of intricate and variable interplays characterize the mediations that present fascinating challenges to performers to become aware of their physical abilities, attention, balance and concentration, from the most intimate close up studies of emotional expression to outwardly expanding interactive and immersive real-time projection environments, as they have been used by Frank Castorf and stage designer Bert Neumann at the Berlin Volksbühne, for example, and by directors deploying massive projectors for cinemascopic effect, as Marianne Weems has done for The Builders Association, or Katie Mitchell in her many London and continental European productions. Opera directors often use projection mapping as *de rigeur* scenographic strategy – William Kentridge's extraordinary animated videographics for Alban Berg's *Lulu* (English National Opera at the London Coliseum in 2016) being just one example.

In large metropolitan theatres like the Berlin Volksbühne, where stage designers have sufficient budgets to build enormous "film sets," the self-conscious commentary on mediation of live (or pre-recorded) materials within the performance can be trenchant when Castorf, for instance, in his adaptations of Dostojevski's *The Idiot* and *The Humiliated and Offended*, not only uses continuous splitting and roaming of view-points (via camera live feeds) to create multiple scenic co-presences, but also treats the intimate close-up ironically as a kind of humiliation device, rendering the actors highly vulnerable, exposed to a particular style of low-tech, low-res video associated with "Trash TV," soap operas, Big Brother and porn movies. The Catalan performance troupe La Fura dels Baus also deploys such abject video imaging and simulated immediacy in their production of *XXX* (2003), a physical-digital phantasmagoria of sexual and pornographic projections inspired by de Sade's novel *Philosophy in the Bedroom*. Similarly, Christoph Schlingensief's direction of Wagner's *Parsifal* for the Bayreuth Festival (2004) uses a dramaturgy of aggressively overpopulating the stage with flickering videos, trash and kitsch objects, which turn the sensory overload of the mediums onstage into a kind of voodoo-TV-laboratory.

Some of these productions in the first decade of the 21st century have pushed the interactive use of hand-held cameras or microphones to a delirious edge, and I encountered many examples over the years, ranging from postdramatic multimedia theatre works (by the Wooster Group, The Builders Association, Complicité, John Jesurun, Hotel Modern, Ivo van Hove, Julie Taymor, Signa Köstler, Milo

Rau and others), music theatre productions (Teatro de Ciertos Habitantes) and digital dance works (Chunky Move, Dumb Type, Merce Cunningham, Trisha Brown, Bill T. Jones, Wim Vandekeybus, Wayne McGregor) to hybrid installations, telepresence performances and poetic stage adaptations of lyrical and narrative writing of the kind that theatre director Katie Mitchell has created over the past years. Mitchell is an interesting case of a practitioner who – like the well-known auteur of the US-American theatre scene, Anne Bogart, or the Japanese master teachers Tadashi Suzuki and Hironobu Oikawa – has not only directed her own work but written about actor training. After creating a critically acclaimed adaptation of Virginia Woolf's *The Waves* at London's National Theatre in 2006, she published her book *The Director's Craft* in 2009, surprisingly omitting any detailed reference to her dramaturgical process for using camera-actors and live recording technologies in the performance.[11] One chapter of this book is dedicated to a thick description of her vocabularies of intermedial staging.

If plays and dramatic productions have been adapted to the screen for some time, we have recently also seen the reverse angle, namely, the adaptation of movies and film scripts to the stage. The industrial and parasitical relationship of film and theatre is not the main focus of this book; rather, I examine the theatre's necessary confrontation of digital behavior and information environments in order to examine the experiential liveness of multimedia theatre and the directing challenges posed by the new "writing" and designing for the intelligent stage and hypersensorial ecoscenographies. The already existing lineage of adaptations (e.g. videodance/choreography for the camera) and multimedia performances, which incorporate projections of screen images, offers a broad background for understanding the compatibility between live performance and projected light/images, between the polyrhythms of movement, gesture, voices, and the digital behaviors of images, sound and virtual space.

Digital performance, however, is not primarily a screen-based medium, along the lines just suggested. Rather, it is an ecosystem characterized by an interface structure and collaborative compositional process in which the performing bodies of objects are the key to the interface. Digital performance can be said to include all performance works in which computational processes or electronic instruments are integral for the composition and content, the aesthetic techniques, interactive configurations, materials and delivery forms. The presence of actors is not required, though that of an audience (and thus participants) seems still essential and implies an alternative (delegated) modus of actuation. Thus, the notion of digital performance can be expanded to the sense of the infrastructural or confluential, to amplifying or nervous systems that necessarily reposition or redefine *corporeality* in the theatre, and also complicate the idea of immersion.

In many instances, the integration of human-machine interfaces implies the design of interactive systems and real-time synthesis of digital objects. Machining operations and artificial intelligence (AI) become part of the organism, and the materiality of machines and robotic devices can take on an agency of their own (as self-organizing systems), irrespective of human consciousness or intention within the interface. Installation architectures can replace the dominance of the proscenium stage; thus contextual design of programmable systems also becomes a new form of architecture, protocol and bio-informatic space. On the other hand, "theatre" still serves as an umbrella term for the place where performances take place or place themselves into becoming present: the presence of actors facing audiences and audiences facing actors. We have not yet gotten accustomed to speaking of audiences facing systems or recognizing themselves as functioning within protocol systems, aware of their interconnectedness to forecasting, to the programming of *techné*, so to speak. Many spatial configurations for staging interfacial performance are of course possible, and the notion of interfaciality will allow us to draw out ideas about social interaction, relationality and the presence/distance – real and ghostly – of digital objects and images, taking us perhaps closer to an object-oriented ontology too.

Kinetic Atmospheres: Performance and Immersion is written primarily from the point of view of artistic practice. It is less an analysis of the made, than a poetics of performance making and thus a passionate and often subjective engagement of compositional process and the living repertoire of the theatre's transformative powers. In the charged atmosphere of the rehearsal studio and the performance event, acting and design choices are being made, and propositions are worked into the choreographic ecology of relations – communicative, cognitive and corporeal-sensorial. Remembering that Brecht's theatre for the scientific age anticipated new modes of reception, the era of the digital has already transformed many of our habits and continues to do so at accelerated speed. In this book I give special attention to the transformation of conventions in both the machining architecture or performative machine of the theatre – its practical and metaphorical functioning as an organic "weather" or combinatory system – and the affective responses to the emergent digital aesthetics.

When widely known choreographers Merce Cunningham and Bill T. Jones collaborated with digital artists and computer scientists to create a series of dance works and installations exploring the artistic potential of motion capture technology – *Hand-drawn Spaces* (1998), *Biped* (1999), *Ghostcatching* (1999), *Loops* (2001–04) – the reviews in *Time* magazine spoke of "hypnotic groundbreaking performances" bringing dance, the most physical of the arts, into the digital age. Such a view assumes the physical to be different from the digital, and at the same time compliments it for having caught up with the advancements of technology. In light of hundreds of years of theatre machines and special effects, since the baroque era, this compliment is quite likely to be misleading. The theatre, at many points of its complex and differentiated cultural histories, has always negotiated – and is able to rupture or stretch – the available representational techniques and organizational methods. In the online Wikipedia, we read the following, for example, about European baroque theatre:

> In theatre, the elaborate conceits, multiplicity of plot turns, and variety of situations characteristic of Mannerism (Shakespeare's tragedies, for instance) were superseded by opera, which drew together all the arts into

a unified whole. Theatre evolved in the Baroque era and became a multimedia experience, starting with the actual architectural space. In fact, much of the technology used in current Broadway or commercial plays was invented and developed during this era. The stage could change from a romantic garden to the interior of a palace in a matter of seconds. The entire space became a framed selected area that only allows the users to see a specific action, hiding all the machinery and technology – mostly ropes and pulleys. This technology affected the content of the narrated or performed pieces, practicing at its best the *Deus ex Machina* solution. Gods were finally able to come down – literally – from the heavens and rescue the hero in the most extreme and dangerous, even absurd situations. (http://en.wikipedia. org/wiki/Baroque)

Today, the multimedia experience in the filmic use of motion capture-based digital graphics has already been widely seen in Hollywood and Hong-Kong martial arts movies; digital animation is a staple of television advertising, MTV, club-house VJ'ing, and the wide range of anime (especially from Japan). Tight choreographic systems with live and pre-recorded video projections are commonplace at many rock concerts, and musicians tune their shows with the ropes and pulleys of digital images. I remember Madonna's "Sticky & Sweet Tour" as a typical example of precisely cued and hyper-psychedelic interaction with video monitors or screen projections. All pop concerts now use digital screen doubling. The Japanese pop idol Hatsune Miku is performed as a hologram, backed in concert by a quartet of flesh-and-blood musicians; she is also an anime character, a video-game avatar, an array of sophisticated vocaloid code and an unnerving experiment in crowd-sourced pop art. Miku was released by Crypton Future Media in 2007, the vocaloid engines written by Yamaha, her figure drawn by manga artist Kei Garo. Her producers think of "her" (Miku) largely as a *software* for making music (and the old gendered pronoun here is of course absurd). The implications of this are far-reaching.

On a smaller scale, we have seen the hyperactive mixing of live actors and video actors in the many productions of the New York-based Wooster Group. Their work over the past three and a half decades reflects a continuous examination of the theatrical/representational apparatus and the performers' interaction with the technologies of dramatic and cinematic (re)production. It is hardly coincidental that in their production of *Hamlet* (2007), the Wooster Group re-plays *Hamlet* interpretations through a complicated meta-theatrical remediation – and re-editing – of Richard Burton's "Theatrofilm" of a 1964 Broadway production in which Burton played Hamlet under the direction of another well-known Hamlet, Sir John Gielgud. In the Wooster production, the actor Scott Shepherd, performing in a contemporary "camera-real" but also post-dramatic style, re-embodies Richard Burton's film acting in a mindboggling phantasmagoria of analog reconstructions of the now available DVD version of the older film/live broadcast of Burton-Hamlet. Such ghosting processes are compelling evidence of the contemporary theatre's paratactical effect of mixing analog and digital techniques of referencing realities and unrealities of play, of cultural memories, of repertoires and archives.

In Hollywood cinema, James Cameron's *Avatar* (2009) turned out to be an environmentalist romance of massive proportions – its planet Pandora probably the most vivid and convincing creation of a 3D graphic fantasy world yet seen in the history of moving pictures, at that point. Shot with the Cameron-devised Fusion Camera System (a single camera that shoots live action in stereoscopic 3D), it is a highly immersive sensory and sensual experience when viewed with stereoscopic glasses. Appropriate for the movie industry's science fiction extravaganzas, *Avatar* was produced by sophisticated digital magic. Cameron's "performance capture" technologies basically refined already existing motion capture involving the filming of actors wearing sensors and then rendering the resulting data through CGI computers, yet *Avatar* could be the first in a new generation of 3D movies (he has most recently produced and written the screenplay for *Alita: Battle Angel* [2019], based on Yukito Kishiro's famous manga *Gunnm*). Not an unappealing allegory for the digital spectrum here, Kishiro's protagonist is the young cyborg Gally, found on a scrapheap by the Cyberdoctor Daisuke Ido who rebuilds and redraws "her," becoming a surrogate father.

As I revise this introduction, a few years after seeing *Avatar* on a Christmas Eve in Texas, I realize that watching movies with 3D glasses has become already commonplace. Even a minor fantasy, such as Matt Reeves's *Dawn of the Planet of the Apes* (2014), which I watched in a small village cinema in Germany, included glasses for everyone in the ticket. I was also handed 3D glasses for *Atomos,* a dance concert by Wayne McGregor in London (2013). Curiously, I was not sure why and when I should wear them, as the need for the glasses was not really explained to the audience, and so most of us watching the dance, which has seven small flat screen monitors descending from the ceiling part way through the performance and displaying the same image on each, ended up just treating them as a gimmick souvenir.

The question of what is groundbreaking or vital in the coupling of live dance/ theatre and technology must be examined carefully in order to make any claims for a new ecology, a successful marriage of performance and interactive image or a sustained impact of new media (the digital data processing now having extended the photographic and the cinematic) on theatrical art. The history of film, breathtaking and fast as it seems, is barely older than a century, while theatre has existed in the world for millennia. The development of cinema is closely associated with the advances in reproduction and media technologies, and the diverse dissemination platforms and apparatuses of course tend to strengthen the industrial and economic power of the medium, whereas theatre and dance are bound to stay behind. This staying has to do with the fact that live theatre performance cannot be "taken along" (on a videotape, disc, power book, cell phone, iPad, etc.) and cannot be downloaded, copied and pasted. This may be contested by those who believe in cybertheatre or the participatory charm of

Twitter messaging, a live audience perhaps invited to messaging their tweets to the live actors, or even, as choreographer Joumana Mourad suggested at the V&A in London during her production of *IN_FINITE* (2013), contributing to the content of the emerging work.

Ropes and pulleys/somatechnics

When I heard of Mourad's invitation, I did not yet have a Twitter account nor realized she meant it as a method of community building. The theatre has often needed to acknowledge its groundedness, for example, in its role as community theatre or as preserver of national dramatic canons or national ballet and opera companies, as an agent of translation and adaptation (of works imported from outside of the *polis*), and a forum for fresh ideas in the relationships between actors and scripts, between directors, dramaturges and scenic designers often consigned to working at a particular theatrical institution. Yet the theatre also endeavored to expand its space of imagined scenes, ever since the beginning of the rise of the technological arts of projection (light, film, slide photography), as lighting became a crucial stage design technique when electricity entered theatre architectures. As lighting design moved from analog to digital control operation, from still lighting instruments to moving lights, so did sound in theatre become part of more elaborate amplification and diffusion systems, usually installed on stage or in the house. The more recent experiments with mobile sound and dislocations of the voice need to be addressed, along with the more comprehensive use of video/projection in theatre, which has inevitably affected methodologies of stage design. It remains to be seen whether the use of mobile media and social media is indeed not finite but open to constant development that touches upon, say, the theatre's interfacial relations to YouTube's proliferating interfaces with users. It is not yet obvious that social media relations build audience relations or even constitute a community that has any impact on theatrical content or form. But the global reach of music is a case in point: its mobility on wearable cellular devices accessible now to almost everyone everywhere, may of course build next streaming waves of displaced transmissions, way beyond what the old theatre meant to us.

It is a fact that video cameras already play a significant dramaturgical role in theatrical productions: the use of monitors and projections has become almost commonplace in many performances over the past decades. The so-called avant-garde theatre and dance companies started to work with the filmic medium early on, when professional three-quarter-inch tape became available with the Sony Portapak as the first portable video recording device (in 1967). Nam June Paik, trained as a musician and performing La Monte Young and Cage's graphic scores during the early 1960s, switched to video, TV and electronic technologies almost instantly, pioneering the most unusual concoctions, for example, the *TV Bra for Living Sculpture* (1969) for cellist Charlotte Moorman. By the 1980s, VHS tapes and hand held cameras were more widespread, soon to be displaced by the even smaller digital camcorders which were affordable and could be easily integrated in a performance making (and documenting) process. Today tiny GoPro cameras (that can be worn on the body) or smart phones make digital film clips instantly available for transmission. The relationship between production (making video) and documentation (of performance) is a fascinating one, which I will address when writing on the camera actor and some contemporary modes of video-theatrical *mise en scène*. Apart from well-known performance companies such as The Wooster Group, The Builders Association, Theatre Complicité or Societas Raffaello Sanzio, and dance companies choreographing with interactive digital film and animation, mainstream theatre directors have begun to embrace video production, though not quite to the same extent as the visual arts. Museums and galleries now emphatically include video installations, media sculptures, VR and other mixed reality arts works alongside the old traditional media (painting and sculpture). Van Gogh paintings can now be walked through (inside VR).[12]

Berlin Volksbühne director Frank Castorf and British director Katie Mitchell, among others, have perfected the use of onstage camera crews for their ropes-and-pulleys dramaturgy, mixing the real-theatrical and the filmic-theatrical (the screen doubles) on the fly (Figure 1.3). Many smaller companies and performance artists use video, Kinect cameras, microphones and electronic sound as if they were second nature. Yet the role of the physical body, vis-à-vis camera and computer software, and the limits of physical presence, can become significant theoretical

FIGURE 1.3 *Erniedrigte und Beleidigte*, dir. Frank Castorf, Berlin Volksbühne, 2002. Photo: Thomas Aurin.

issues for the discussion of technological embodiment. The currency of the term *embodiment* would be hard to explain if it indeed had not become an issue raised by the increasingly ubiquitous presence of technical transmissions. The role of the *subjects* of theatre is also crucial for the actors, directors and designers, in the most fundamental and practical sense. The role of the ropes and pulleys in generating and sustaining kinetic atmospheres and augmented realities is also critical, and a material theory of such generative machining is timely.

The development of new techniques which can address the modified and virtual presence of the actor should be a major artistic concern. I shall pay close attention to the potentialities and constraints of contemporary multimedia performance. It is no coincidence that a few years ago a new journal, *Somatechnics*, was founded with the objective of encouraging multidisciplinary research into the body and a critical engagement with the ethical and political implications of techniques – "somatechnics" referring to an understanding of corporealities always already bound up with a variety of technologies and techniques. A similar approach underlies my writing here, as I argue for the intermingling of *dispositifs* and physical enactment, and – in the sense of *body weather* and training on the edges body and environment – a deep interplay between bodily being and the technological context in which it occurs.

Technological environments and techniques have inevitable effects on embodiment and sociality, and thus also on transductive performance. New training becomes necessary, new possibilities and limitations arise, frustration occurs often. As our 21st century moves forward, it is important for performance practitioners to ask what kind of living system we like the theatre to be, and how we identify theatricality's capacity to organize space, time, technologically extended bodies and material objects, to integrate disparate elements and construct versions of reality environments. Since the intermedial stage does not grant the actor a privileged position in the interdependencies of performance/scenic elements, the question of the posthuman subject inevitably arises within the contextures of the bio-virtual.

The notion of theatre-as-living-system implies that theatrical environments are not only marked as varying kinds of artistic productions (frontal stagings, multi-spatial stagings with mobile audiences, indoor/outdoor site-specific stagings, promenade theatre, headphone theatre, mixed reality performances, etc.). They are also learning environments, where directors, designers, composers and performers acquire sensitivities and cognitive maps for understanding how the cellular organisms and materials of the content of a particular performance can be arranged, how resonances and affordances between phenomena are generated, and how a common vocabulary emerges through the interaction between agents and environment in a multiagent system. In the chapters that follow, I try to imagine such a vocabulary for multimedia composition or disposition in intelligent environments, recognizing the need to learn about systems by – metaphorically speaking – using an "engineering" approach: understanding by building ropes and pulleys. We gain basic insights about agents, action and interaction by asking how digital performance art scaffolds its environment.

Regarding the theatre's relationship to artificial systems, the idea of the *environment* includes of course various levels and dimensions. Apart from projections of pre-recorded material, and inclusions of screen-based worlds within physical architectures, the relations between screen-based spaces or VR environments and physical spaces need to be addressed. Furthermore, real-time interaction with a camera-vision, sensor or artificial intelligence system requires attention to the nervous system creation and to larger issues of the spatial *dispositif* and the sensory affordance of the digital, as well as to the constraints. It will become apparent that I am generally much more interested in constraints, as I believe the expansions of the real, through technologies, are very easily overestimated. In the theatre this should be obvious. The theatre as a place of multiple mediations necessarily interconnects different sensory processes. Old and new modes of perception and consciousness emerge from such experiences. When actors and dancers act with digitized moving images and sound, as well as with sensual, wearable techno-garments or devices that are audible and tactile, what are the outcomes of the interaction, and how do we understand the relations between physical action, the digitally embodied and the digitally extended? The digital is now being perceived more clearly as our virtual condition, our contemporary phenomenological dimension, full of technically mediated animate interfaces with "mixed reality."[13] Future performance artists and dance makers will be "born digital" and have grown up with pervasive computing and social media, feeding off multiple playing fields, drifting across offline/online platforms.

The author of this book did not grow up digital but playing in the forest. Then came training in the theatre, writing for the theatre and performing on stages or in galleries and other settings. Gradually, I learned to use image projections, electronic sound scores, film and then computation and the internet as complementary instruments of composition – writing *Raumpartituren* (spatial scores) for performance. In the 1990s, I began to look at performance as a real-time medium in the larger visual and electronic art field, and investigated the way human beings can experience space, narrative and movement and how we, as perceivers, reconstruct the internal and external worlds by means of our sensorial system. I became interested in various forms of making art in order to visualize things that are perhaps not in themselves visual but connect visual, aural, kinaesthetic/tactile and olfactory experience. This approach emphasizes embodied experiences within media-rich and sentient environments, presuming that sustainability requires a deeper listening between performer and world. My goal in the following chapters is to present methods of composing and directing such work that highlights both the concrete and the virtual ways in which performers interact with one another and with their environment, above all indicating the fundamental relationships between our internal and external worlds. I will begin, in the next section, to address the importance of collaboration.

Notes

1 The exhibition *Lo(s) cinético(s)*, curated by Osbel Suárez at Reina Sofia (March 27–August 20, 2007), was impressive and featured a wide selection of kinetic and op art (80 works by 45 artists). The catalog of the same title, edited by Osbel Suárez and Eugenio Fontaneda, included thought-provoking essays that inspired me to reflect on the relations between kinetic objects/projections and movement-based interactive or participatory installations, wondering of course also how such art dreams its audience, to use the apt title of a book by Theresa Hak Kyung Cha: *The Dream of the Audience*. Another exhibition that attracted by interest was *Vibration, Vibraçao, Vibracion: Latin American Kinetic Art of the 1960s and 70s in the Power Collection*, staged at University of Sydney Art Museum in 2012.

2 Cf. Birringer 1998: 33–34, 2000: 55–58. For a detailed discussion of the communal project and its communitarian philosophy, as well as Appia's significance for modern performance, see Beacham 2006. Selections of Appia's writings have been brought together in Beacham 2011.

3 Anna Tsing's ethnography includes a critique of capitalist destruction, and the ruination of forests, while also carefully probing the resilience of fungal ecologies and the salvage laborers of pericapitalist economies. Botany itself, a recent editorial in *The Guardian* (April 3, 2021) argued, in reference to London's Kew Gardens, needs to be "decolonized" and the plant collections, as cultural spaces, understood in their imperial-historical association with the "cabinets of curiosity."

4 Cf. Benjamin 1999: 245–55.

5 In a swift response of the COVID pandemic, the editors James R. Ball III, Weiling He and Louis G. Tassinary introduce a special issue of *International Journal of Performance Arts and Digital Media* (16:3, 2020) with an unexpected reference to the kinetic dimensions of the "zoom function" under lockdown social living regimes. The first article in the journal, by Andrew Starner, reviews the history of zoom technology and telematic connection: "Remote Viewing: a brief historical inquiry into theatre and social distance" (226–44). Early in 2021, the Australian media artist Suzon Fuks released a short "archival" video that also recapitulates telematic performance prior to the Zoom era: https://youtu.be/7JAyd8TBMwU.

6 For a more detailed description of the event and its context, see Tomkins 2005: 214–27. Rauschenberg's score and instructions for *Open Score* are reprinted in the catalog, edited by curator Catherine Morris, for the MIT List Visual Arts Center's 2006 exhibition *9 Evenings Reconsidered: Art, Theatre, and Engineering, 1966*, Cambridge: MIT LVAC, 2006, pp. 36–37. For an excellent archeology of significant 20th-century models for interdisciplinary art and technology research laboratories, see Michael Century's 1999 survey "Pathways to Innovation in Digital Culture," retrieved January 6, 2010, from http://www.nextcentury.ca/PI/PI.html. See also, Garwood 2007. For my earlier analysis of the event, see Birringer 2008.

7 Steve Dixon's *Digital Performance: A History of New Media in Theater, Dance, Performance Art and Installation* (2007) is perhaps the most comprehensive study to date of the use of new technologies in the performing arts. It is part of the MIT Press "Leonardo" series of books, edited by Roger Malina, focusing on new media arts, and the series includes other important studies by Kozel 2008, Popper 2007, Grau 2003, Manovich 2001, all the way back to Moser and McLeod 1996 whose book *Immersed in Technology: Art and Virtual Environments* offered case studies of several large-scale research projects conducted at the BANFF Center in Canada. Giannachi 2004 offers a theoretical account of hypertextuality and the theatricality of interactive technologies but rarely engages with performance practices. Paul 2003 provides an illustrated guide of artists and artworks since the 1980s, distinguishing between work that uses digital technology as a tool to produce traditional forms and work that uses it as a medium to create new types of art (net art, software art, digital installation, virtual reality, etc.). A critical and aesthetic engagement with these practices can be found in Birringer

2009. See also: Leeker 2001; Dinkla and Leeker 2002; Broadhurst and Machon 2006; and Chapple and Kattenbelt 2006; Salter 2010; Benford and Giannachi 2011. Carver and Beardon 2004 collect essays from both critics and practitioners, and their focus on design stands out among most of the other books. An overview of new media art movements and their gradual institutionalizations can be found in Tribe and Jana 2007. A useful book of case studies, investigating interdisciplinary arts and science research, is Brouwer, Mulder, Nigten, Martz 2005.

8 Brecht 1964: 158, 173. Brecht, along with Benjamin and later Marxist theorists understood the technical opportunities of radio and film to connect a multitude of people in non-hierarchic, interactive networks; the idea of the public microphone (as loudspeaker) was still prevalent in the recent OCCUPY movement.

9 Geoffrey Colman, 2010, "Actors Angels and Avatars: Contemporary Acting in a Flat Landscape," Lecture presented at The Performance Research Seminar Series, Brunel University, January 27, 2010.

10 See Benjamin 1999: 211–44. In the chapter on the "Screen Double" I will pay close attention to this matter of the optical tests.

11 Mitchell's *The Director's Craft: A Handbook for the Theatre* contains only two pages (pp. 90–92) on the use of video in theatre, claiming, astonishingly, that the use of video in mainstream theatre is still in its infancy.

12 The promotion of the "Original Immersive Van Gogh exhibit" coming to Houston in the summer of 2021 (https://www.houstonvangogh.com/) is just too precious not to mention here, even if it will eventually be merely an anecdote in the rise of impact museums:

> The original, blockbuster digital art experience that has received rave reviews from critics worldwide and entertained more than 200,000 guests since its North American debut last July is coming to Houston this summer. Official tickets are now available with prices starting at $39.99 per person. The Houston experience, which kicks off on August 12, joins already sold out and extended runs in Toronto, Chicago, San Francisco, Los Angeles, and New York. Designed by Creative Director and Italian film producer Massimiliano Siccardi – a 30-year pioneer in immersive content whose work was recently spotlighted in the Netflix series *Emily in Paris* – with music by Italian multimedia composer Luca Longobardi, the production harnesses 60,600 frames of captivating video totaling 90,000,000 pixels and 500,000+ cubic feet of projections to create a spectacularly breathtaking creative encounter like no other. Merging state-of-the-art technology, theatrical storytelling, and world-class animation, *Immersive Van Gogh* vividly brings the celebrated painter's dreams to life with unprecedented movements in his masterpieces. What's more, *Immersive Van Gogh* also stands apart from other recently announced experiences (many with similar brand names) in that each *Immersive Van Gogh* production is uniquely and safely designed for the specific venue in which it runs and does not include VR headset experiences – making this Van Gogh presentation an arrestingly visual happening for all participants. In Houston, the soon-to-be announced environment...will also become a multi-year, next-gen art, culture, and entertainment showcase.

13 I use the term "mixed reality" to indicate the combination of physical spaces/actions and cinematographic or computer-generated spaces and projected (virtual) actions. Thinking of animate and dynamic interfaces today also implies a vocabulary of navigation and immersion in extensive networked space of asynchronous real-time, a space that media artist/writer Mark Amerika sometimes refers to as "unrealtime." See Amerika 2007: 54. The discussion of the "posthuman" dimension of hybrid art practices was initiated by Hayles 1999.

2
COLLABORATION

Sustainable subjects

The introduction to this book meant to reveal a wider palette of what I want to refer to as the extended choreographic. The intro was a bit long, because I tried to cruise through some of the complex intermedial scenarios that underlie my experience as a choreographer working with multiple media. Rather than sketching the content of the book's chapters, I meant to create a flavor for such multiple media-choreographic manifestations – flows of echoic resonances among natural and technical environments. This meteorological palette already encompasses such a range of contemporary media practices that it seems to have left the frame of the theatre for good.

You also have guessed that I do not work in mainstream theatre. But the expansion of the notion of performance, and especially choreographic theatre with all the transmedial dimensions we have come to recognize throughout postmodernity, does not supersede one of the bases of theatrical production as a living system, namely, the process of embodiment that includes the material conditions of the performance to be experienced by selves and others – the audience. Embodiment and materiality are conjoined here in my outlook on atmospheric immersion: the specific affordances of material interactive manifestations will slowly move into focus. A focus on embodiment, which we can later re-view and question, would inevitably raise the issue of the actor being present in front of an audience – and to my knowledge actors have been *present with audiences* in all performance art traditions of all the world's cultures.

In a dance work presented in London in 2018–19, Chinese choreographer/dancer Zhi Xu opens with a solo under a full moon. The moon's image (projected) is overwritten with the word *Yuan* (Origin). The title of the choreography for five dancers, *X-Body*, appears to be an ironic allusion to the *X-Files*, one of the longest-running science fiction series in TV history, where special agents investigate unexplained and mind-bending cases. I like the idea of an unexplained body

DOI: 10.4324/9781003114710-2

of a performer (or of several), slowly offering itself as if under a microscope at the outset of a dance – it has intrigued me immensely.[1] I understand Xu's questioning to work into at least two, possibly contrasting, directions, namely, toward a sense of "origins" and traditions of cultural techniques of the body – and thus in his case embodiment trained in Chinese dance schools with both precise technical and also cosmology-based transmissions of breathing techniques (t'ai chi and other martial arts-based techniques, Peking Opera, and Chinese classical and Yangge folk dance combined with classical and contemporary ballet). But at the same time also into another direction, namely, an experimental technological dance path that uses interfaces and mediatization to expand its range of expression, and possibly alter its course. When I saw Xu's work in London, where he had moved after completing his degree at the Beijing Dance Academy, he was intent on combining his classical training with software-supported dance technologies and methods of interactive performance encountered in European and North American experimental dance contexts and Western training academies. He also wished to sustain his Chinese-trained body, and his dancing identity.

One of my initial examples in the previous chapter referred to a modern European school for the training of dancers and actors (Hellerau), and Émile Jaques-Dalcroze's eurhythmics system of musical and movement instruction, taught at his newly founded institute in synchronicity with Adolphe Appia's challenging experiments in scenographic philosophies and stage lighting (revolutionized by electricity). If I had chosen to begin with a reference to the historically persistent theatrical traditions in the East, I would have had to mention different philosophies, for example, the in-body actor training in Indian kutiyattam or kathakali, in Peking opera and Japanese Noh theatre, where mastery of the performance vocabularies and codified expressions require years of rigorous transmission processes between master and pupil (e.g. movement and gymnastic full-body training, exercises for eyes, eyebrows, facial muscles, hands and wrists, dance steps, coordination of face and hand gestures, choreography and roles in the repertory). A similarly rigorous training undergirds Balinese ritual drama and the shadow-puppet theatre where the puppeteer is often also considered a healer/shaman who must garner the intricate methods of manipulating the puppets as well as knowing how to summon protective spiritual powers when the shadows on the screen call out the local demons.[2]

Somatic demons

The ancient yet living shadow-puppet theatre tradition bears an uncanny relation to some of the intermodal performance practices I shall look at when focusing both on actor-centered and non-actor installations and machines, as well as interface-oriented strategies of play, performance and improvisation that can reveal particular forms of attendance and engagement of the structure/environment. These also include kinetic objects that choreograph: *choreographing objects*, to slightly modify William Forsythe's use of the term "choreographic objects"[3] and render it transitive. Objects that choreograph will slowly expand into the atmospheric, larger scale architectures that form the primary focus here.

Such expansions are metamorphoses. Zhi Xu's intent to maintain his "Chinese dancing body" strikes me as a particularly fascinating example of metamorphosis. He is of course aware that after 15 or 20 years of training in traditional dance (Yangge) he will have fully "embodied" this tradition. It lives on inside his muscular-skeletal memory. But he has also trained in modern dance and now works in the West where he is exposed to multiple influences not really compliant with ideologies of identity, national origin or cultural heritage. He recently took Ohad Naharin's Gaga classes and was excited by its energetic focus. In fact, I assume he experiences the full impact of globalization, diasporic confusion, nostalgia and curiosity, and inevitable amalgamations of styles, behaviors and living responsiveness – which Renaissance scholar Stephen Greenblatt once referred to as "self-fashioning" (Greenblatt 1980). Xu, in this sense, is also a "renaissance" person; he is dislocalized-embodied, experiences transcultural re-mix and collaborates. He recently also joined the DAP-Lab ensemble, which I co-founded in 2004 – a group of diverse, international artists practicing a wide variety of disciplines. He has expressed an eagerness to experiment with new body-worn technologies.[4] He will no longer be/live only in a Chinese dancing body.

Transmission processes are obviously equally relevant, though different, in Western theatre where performers might be trained by teachers influenced by the "methods" of Stanislavski, Vakhtangov, Meyerhold, Brecht, Grotowski, Barba, Strasberg, Adler, Bogart and others. Herbert Blau used to speak of "ghosting" (not quite the same as demons but close) when referring to actor training, with Shakespeare's *Hamlet* and the protagonist's ghostly father on his mind (Blau 1982a), and stimulated by Kafka's short story *The Burrow*. The ghostly demons, I contend, all convey physical techniques, gestural codes, as well as repressive or self-regulatory modalities for coping with the anxiety of influence. Today's range of "techniques" seems infinite – if we now also include the widely diverse and exciting somatics field with its embodied ways of creating knowledge, somatic mixed-abled epistemologies, and reflections on somatic means of teaching and researching artistic creative processes. These include inter-artistic performative tendencies, therapeutic practices, eco-performances, social work and practices with persons with special needs and disabilities.

Actor-centered approaches to theatre in the literate tradition has been dominated, however, by the dramaturgy for plays and the enactment of roles (characters) in the staging of fictional dramatic worlds rather than embodied practices directed at healing or social and political engagement. Today's testimonial and socially engaged theatre often prefers to deploy non-actors (amateurs or people who present their self-experienced plight), which in fact undermines the artistic logic discussed here. A refugee does not train to be a refugee, but Santiago Sierra or Tino Sehgal may cast a refugee into a performance. It might yet be presumptuous to think of actors as wounded healers, but our demons are burrowing through, and we act, perhaps, to feel with the body of another, the victimized, oppressed or exiled, to practice somatic ghosting, confusing the thresholds. In representational art, ghosting is necessary. In most cases of the Western theatrical repertoire the actor is trained how to embody dramatic roles, to take on a *persona* – "We always someone else," in the words of comedienne Marsha Warfield.[5] Yet this "We always someone else" reaches deeper (surely so with complex racialized and gendered overtones in reference to black actresses and stand-up comediennes), as we now have legacies of so many schools or methods of actor training, alongside various ballet, modern dance, tanztheater, physical theatre, somatic improvisation techniques, hip hop and multi-diasporic styles (capoeira in Berlin, tango in Tokyo, flamenco in Vancouver, Gaga in Beirut) – all of which have prepared the contemporary practitioners.[6] What do actors need to know and be capable of today when they are asked to perform in a postdramatic and multimediated environment? How do dancers put on wearable sensors or move into the lenses of Kinect cameras or motion capture systems? How do they offer to put the VR head-mounted displays (goggles) on their audience-visitors' heads? Or their own? (Figure 2.1)

In other words, what constitutes the "role" of acting in digital performance? Does the digital turn the physical-moving and acting body into something else? How do actors train with software, VR and Artificial Intelligence? Is such training accessible and sustainable? Who or what does the programming of the input and output environments, deciding on filters and other parameters, on scaling, or upside-downing? How do actors tangle with granular synthesis? How do we "train" audiences to wear a VR headset or head-mounted display (HMD), for

FIGURE 2.1   Visitor seated on floor wearing Vive VR headset, with cables held by actor guide. *VR Lost & Found* exhibition, Trondheim Verkstedhallen, 2019 © Johannes Birringer.

example, and then "move" inside a virtual atmosphere or do things? Can these questions be generalized at all or do we have to look at a large variety of practices which are particular to directors, performers, programmers and ensembles working with digital technologies, in different cultural contexts? Do media technologies care to work with actors? Do not intelligent machines now teach machines to act and self-organize? What are the somatic proxies of intelligent machine demons?

Perhaps we need to rephrase these queries and expand them even further, since my findings suggest there are no clear agreements on skills and practices in digital creation (and preservation) across the sectors, especially as there are so many overlaps between practices of playwrights, directors, stage performers, producers, dancers, choreographers, performance artists, singers, movie actresses, models, filmmakers, musicians, screenwriters, songwriters, video artists, installation artists, designers, composers, network artists, programmers, etc. Not to forget the outsider artists and outliers, the visual artists who direct performance, the interaction designers who paint and create VR worlds, like Mr Siccardi who now recreates van Gogh paintings in 3D, harnessing 60,600 frames of video totaling 90,000,000 pixels and 500,000+ cubic feet of projections!

How necessary is it to raise questions about roles if some approaches to digital performance environments assume the primacy of system design, the overriding significance of computational programming for the realization of the event score? I have had to explain this apparent primacy many times, face to face with performers who had thought the opposite – that they were in the spotlight and the systems cooperated with them. What if it were the other way round? And yet, how do digital ecosystems provide a safe place for actors while requiring specific and sustainable performer skills not associated with dramatic repertory but with adapting to different modes of real-time data controllers, *actuation techniques*?

In my experience, digital environments require an acute awareness and knowledge of working methods within such infrastructural atmospheres, the ropes and pulleys of the environments regardless of how controlled or how open they are. In the first instance, we could argue that any participants in a production of multimedia digital work need to know the kinds of operations set in motion: how they behave in a digital performance space, how they act upon objects or steer a feedback mechanism and how they negotiate a capturing *dispositif* that follows their actions and "translates" them. But who follows whom? Does the *dispositif* behave following its programming and internal operational logic, determining what can happen? Or is it dependent on outside input? Furthermore, how do actor and system negotiate the presence of camera sensing apparatuses or the invisible/hidden communication between smart devices and algorithmic computation? What are the actionable futures, if indeed most of the processes inherent to "the digital" are microperformative, taking place outside of the phenomenal field of human perception? If the intermedial and intelligent architecture of a performance environment harbors a certain autonomy, i.e. if it acts according to its own logic (or *daimon*), how can we continue to think of the actor as a sustainable subject empowered to control or influence the system through *somatechnics*, techniques of physical relationship with the system? In the following, I offer a few thoughts on the sustainability of acting in multimedia environments and the intrinsic necessity of collaboration. My starting point will be an example from physical heatre and the presumed predominance of the corporeal role of acting.

Physical real-time acting

Acting embodies the live and ephemeral manifestation of the role of "character" in a living system that represents the universe of a play or a devised dramaturgy of action. If the universe of the play or performance event envelops a computational system of mediation, any actor who takes part in the operational system needs to be trained in what we can call "system requirements" – a different set of rules for enacting a role. The dramaturgy of action, thus, presupposes an actor whose physical expertise is complemented by an understanding of working in an interactive environment that contains specific technical elements or dimensions augmenting the physical space (the so-called Augmented Reality). The physical actions are still elemental. But they are now correlated with technical and computational processes, to system details that are based on computational physics and algorithms, sometimes referred to as "physical computing," which are programmed to enable generative events (data outputs).

Physical theatre can be considered a phenomenon that came to prominence in the late 1980s and early 1990s, largely influenced by the impact Pina Bausch's *tanztheater* had on the theatre and dance communities worldwide (after Bausch's Wuppertaler Tanztheater started touring massively in Europe, Latin America, Asia and North America following their initial US concerts in 1984 and 1985). It had a clear focus on bodies and choreographies of physical relations, it also from its beginnings revealed a strong awareness of the role of cameras, transformation in the editing of the live work and its iterations, and documentation. This is certainly evidenced by the fact that physical theatre companies, such as DV8 and Ultima Vez (in Europe), Robert Lepage's company Ex Machina (Canada), Chunky Move (Australia), the artists collective Dumb Type (Japan) or The Builders Association (USA), were all known to have produced filmic versions of their stage production or included film projection inside their physical stage environments. They all work multi-modally.

The performers in these physical theatre groups not only practice awareness of spatial compositions with projections and projected light, with a sensorial environment in which cameras or microphones might be anywhere. They need to learn to trust their multifocal perceptions and anticipations, as well as their coming to terms with limitations and encumbrances. The latter is indeed a fascinating aspect of performing with technologies. The notion of virtuosity, therefore, entwines training and physical theatre practice with a further complication, namely pairing weightiness, strong and (sometimes) aggressive physicality as well as rhythmic and improvisatory skill with caution, proprioceptive self-knowledge and a butoh-like inner consciousness of an exerted body, or a body that is "worn down" (cf. my writing on Schlemmer's Bauhaus figurines).

Body-worn technologies require a heightened sensitivity toward the functionalities and behavior of sensors and accoutrements. The exerted body might be the dancing body that wears sensors on arms and legs, shoulders and head, or that has experienced motion capture sessions and rehearsals with Kinect cameras, thus knowing limited space extension and even more limited bodily extension – when the wearable exoskeleton (with wire cables), for example, hinders the movement and limits the range of possibilities. Acting, then, becomes a form of re-acting, too, coupled with a willingness to constantly adapt – in the physical-digital entanglement – to the vicissitudes of the exoskeleton. Wearing sensors also requires an understanding of the different types of sensors (accelerometers, muscle sensors, orientation sensors, heat sensors, bend sensors, biosignal sensors such as GSR, etc.) – all of which have their own distinct functionalities. The behaviors of these embedding sensorial systems are part and parcel of the moving world – the kinetic atmosphere – of a performance scenography. One could go so far as to claim that the sustainability of the physical-computational ecology depends on the functioning feedback connectivities between sensor technologies and actors. This could be the crocodile paradox or catch 22.[7]

The exerted body might also be a performer body that needs to know where the camera sits on stage and how to move in order to remain in its capture frame or to approach it and interact with it (without worrying about the doppelgänger). Cinematographic and motion capture vocabulary (as it is also used in animation) combines with formative work of physical acting, presentness and material fullness, yet "material" here is continuously reconfigured in the human exchange with tools and instruments within computational environments. Therefore, technological vocabulary, in fact, complements and at times supersedes the acting and choreographic languages inherent to theatre practice, and this is quite important to internalize for actors. Technology, John Durham Peters suggests in *The Marvelous Clouds,* provides a means of channeling the irreversibility of the flow of time. With digital media technology, time itself can become manipulable:

> Spatialized time allows us to spin optical, acoustic, and linguistic signs recording serial events, and even to send them in a crabwalk against its original time. New recording media prolong the afterlife of the dead by giving new homes for their faces, voices and movements.
>
> (Durham Peters 2015: 309)

The ensemble that has surely been most creative and influential in experimenting with recording/recorded material – and with prolonging afterlives – is the New York City-based Wooster Group, which has worked with microphones, video monitors, cameras and filmic projections for many years. In my opening chapter I mentioned their re-production of *Hamlet* (2007), and briefly sketched their dizzying use of meta-acting against/with the filmic recording of Richard Burton's 1964 Broadway theatre role, with the Wooster actor Scott Shepherd re-embodying and ghosting Burton's Hamlet. "Ghosting" is a term often used by Herbert Blau in his writings on performance and the techniques he tested with actors in his KRAKEN group, facing the challenges to actors (in the text and on the stage) within the total infrastructure of "appearances" which constitute a performance environment.

The process of ghosting, for Blau, is not only a reconstruction of, or a *burrowing* into, acting possibilities inscribed in the textual apparatus of the old drama. It is also a confrontation with postdramatic worlds and intersections with performance and physical languages (Blau 1982a: 204ff), thus the notion of tracing could now be linked to the digital, which was not yet a trace structure available at the time of Blau's reflections. Customary appearances, in Blau's excavations of theatre and theatrical thinking, were aligned with physical behaviors, purposive action, narrative, and the roles and shadows of those roles, which he often pondered in his comments on theatre vis-à-vis performance art, i.e. the wilder, troubled methods of body artists who cared little about illusions or narrative space (cf. Birringer 1991: 96–97, 2000: 240–47).

Wilder somatechnics: real-time interaction laboratory 1

Interactive performance is a kind of ghosting that demands a heightened awareness of how the framing (capturing apparatus, microphones, sensing environment) and how the atmosphere brushes up against the actor, i.e. how the system informs or determines the range of actions/reactions that emerge. This leads to a first major recognition, namely, that the primary agency of the actor is no longer the given of physical theatre, since the unity of form of the individual, of the acting physical agency and speaking linguistic subject, becomes relational and dependent. The human actor, therefore, is no longer the focus of the scenographic architecture, nor the spotlit subject of action and dramaturgy. Rather, the *inter-actor* joins the atmospheric, and the choreographing objects and forces in the programmed yet nervous multisensory environment exceed the old human subject. They exceed the director as well. One could think of it (following Octavia Butler) as polyamory blurred consent.

I want to summarize some of my findings from laboratory rehearsals, in order to explicate this further. Already in the early 2000s I had created an *Environments Lab* with a specially designed Interactive Performance Series at Ohio State University. This series, in turn, was motivated by the earlier LBLM (Live Bodies Lively Machines) workshops I conducted in the 1990s. In a later chapter, I will sketch the most recent experiments in VE (Virtual Environment) design and augmented virtuality. I offer these summaries to all of those readers who wish to gain a closer in-depth look at working methods, at questions that are still asked and will continue to be asked. Rehearsal workshops I lead in 2019 are different from the ones I led 20 years ago; yet they in fact introduce very similar concerns. Thus I imagine the following pages offer a helpful praxis guide.

A workshop I organized in 2002 with Scott deLahunta brought together some of the most exciting international interaction designers and artists working in digital performance. They have been influential in the field ever since. As we were identifying pathways for further artistic strategies that have not changed, the following impressions can prove useful. Our Think Tank focused on the practical and conceptual implications of working with interactive tools, instruments and computer-controlled systems within atmospheric environments – both on the level of artistic composition of embodied interfaces and on the level of interface design for users/audiences.[8] An awareness of atmospheres as acting/being outside of artistic control had not emerged yet at this point, although the following pages attest to the implications of how immersive conditions destabilize human engineering.

While taking theatrical, concert performance and exhibition parameters as starting points, the laboratory supported a wider range of conceptions of physical interaction within responsive environments, and interaction between physical and virtual worlds (e.g. VR installations, networked space, telepresence). It also preconceived the notion of an atmospheric constellation as I am using it now in the latest series of *kimospheres* created with the DAP-Lab. The characteristics of a *responsive environment* were becoming so affectively clear at the time that we never even questioned (i.e. bothered to theorize) the vast industrial warehouse space, on the Columbus campus of OSU, and its sensory, morphic and echoic quality. I had moved into Hadley Hall the moment I arrived at OSU, finding the former engineering building (used for a while by the Film Department) much more conducive to experimental work than any of the theatre stages and studios in the dance department. The presentations of the invited artists, discussions and lab demonstrations focused on the particular use of tools and interactive designs, and on interface design for performance spaces and intermedia art installations that function as a shared, collective, social, and playful space. Through exploring improvisation technologies, wearables, movement-sensing design, interface architectures and mapping processes, participants exchanged ideas about the physical, cognitive and transformative possibilities inherent in emerging technologies.

This is precisely the core question of this chapter – namely whether such possibilities allow us to imagine a sustainable subject role for actor and dancers, or whether the computational environments alter our understanding of acting, and actionable images, altogether.

Specific discussions in our Think Tank are organized around issues derived from three subject areas:

Level One: The technological and infrastructural;
Level Two: The artistic;
Level Three: The social/ behavioral.

Discussions of the Think Tank are structured as a series of cluster conversations. Participants form three workgroups of six or seven discussants each. These workgroups are invited one at a time to engage in a themed discussion in a closed circle. This inner cluster tries to define and discuss the key questions, while an outer circle of seated participants carefully listens and takes notes, preparing feedback responses.

Level one: the technological and infrastructural

Mark Coniglio facilitates, and the following is a list of keywords and phrases shaping this first cluster conversation.

> Tools versus Systems: What is 'toolness'? How is the tool an extension of the body incorporating the intention of the operator (hammerer using hammer, violinist using violin, dancer using dance technique, bodily knowledge with movement-sensing devices, etc.). What is 'systemness': the interactive system and its protocols incorporate the user but not the user-intentions. Design practices which make the interface as invisible as possible or seek to make the navigation/use easy and intuitive; the performance-user (performer, choreographer, composer) as extended instrument of interactive system; design of 'characters,' sonic display/instrument design, composed instrument, frame of animation; painting – drawing – dancing (motion capture); the creation, implementation and saturation of media protocol; gesture/speech recognition; data mapping and inframedia.

The discussion circulates primarily around the notion of tool, instrument or system, always in regard to the actor, of course. Coniglio comments that no agreement on terminology seems attainable, but it is interesting to look back and see that initially the tools/system relationship seemed to harbor a richness of possibility for discussion – but in fact it resolves into a set of clichés before our very eyes and by the end of the three days "tool" or "system" could only be referred to in quotation marks. This certainly underlines the slipperiness of language and discourse, and our lack of clear agreement on the *infrastructural atmosphere*.

David Tinnaple's concept of a system is something that begins to dictate content – where (linear) causality dissolves. Coniglio proposes that a tool could be an instrument; medical and precision instruments are mentioned. The concept of "taking a hammer and using it like an instrument" is mentioned as well – an object being able to achieve a dual sense of purpose, or many. Coniglio argues that his concept of system is more rigid/restrictive in the sense that the system comprises something that is reliable, taking input and giving output. Then cybernetics is brought up – the concept of the oarsman who steers within an environment within which s/he has some agency but is dependent on getting and understanding feedback from a multiplicity of factors. Axel Roesler reminds us that the primary informational "input" component of contemporary sensor-based heating/cooling system in houses are not the fuel and the electricity but the temperature air outside the house, and only then the inside temperature.

Tinapple's definition of his process of making entails starting on a MAX patch on the day of performance, working intensely all day and performing at night – this way he retains in short term memory the specifics of how the patch will perform and how he can throw it away when finished. This is reminiscent of conversations with programmers who build something and cannot remember how they did it a year or a month later. The specifics of the underlying code are only retained during the periods when the code is actually being built. Tinnaple's experience returns in the final public session when it is remarked that his "making a patch and using it the same day" process seems remarkably physical.

Coniglio asks David Tonnesen if thinks of himself as improvising when he programs. Computer scientist and graphics specialist Tonnesen's presentation featured a system with forms of "artificial intelligence" built into it: FoAM Lab's *T-Garden* (http://f0.am/tgarden/). This project, developed with Maja Kuzmanovic and other co-workers, is conceived of as a responsive play-space where visitors can "converse" with sound, perform with images and socially, that is collectively, shape media, constructing musical and visual worlds "on the fly." The performance aims to dissolve the traditional lines between performer and spectator by creating a computational and media architecture allowing the visitors-players to shape their overall multisensorial atmosphere through their own movements, as well as their social encounters with each other. Tonnesen, describing his expertise as "mapping physics onto mathematics," notes that the work had a component defined as the ROOM LOGIC, one that would evolve as it learned from the behaviors of the visitor-players.

The *T-Garden* develops the concept of the visitor-player further by proposing very clear pedagogical phases or thresholds of experience – the waiting room, the changing room, followed by three levels of learning in the system. These levels of learning start from the simple individual and slowly build to emergent social/ collective interaction at which point the system is meant to begin to recognize behavior patterns and to "learn" from the visitor-players (the ROOM LOGIC). At the end of Tonnesen's provocative presentation, a short discussion ensues regarding the efforts to measure and evaluate audience responses in these environments, with the implication that the measurements are not perhaps looped back directly to the artist/maker, as one might assume they should, but could also prove to be valuable contributions to the overall evolving cultural context we find ourselves in – trying to explain ourselves, maintain artistic integrity, raising necessary funds and finding access to the expertise (in the computer science and engineering field) to pursue the work.

The *T-Garden* is interesting to compare to other implementations of technologies used in performance and immersive installation art, especially regarding the attempts made in our discussion to distinguish between tools and systems. Tonnesen mentions that this is the first time someone has asked him this question.

As regards sensor or interactive systems, the *T-Garden* team aims to develop a system that learns and develops some form of understanding the participants, able to store this as history and feed it back into the ROOM LOGIC. Similar to the cybernetic oarsman, you can steer within the system, not "controlling" it but noting some sort of "feeling" of how the system works. Where does this become an intuition? And when would this become a form of full embodiment?

Roesler brings up the principle in systems engineering of viewing the "model" as a cognitive mechanism whereby the viewer or participant builds up an understanding through "decomposition" of the object of perception toward a model that, once constructed, can be placed back into the system. This is not easily understood but appears useful insofar as a discussion which seems to be largely about semantics manages to arrive at a conception of the cognitive processes involved in the interaction with something, whether tool, instrument or system. The issue of the subject is always lingering on, and the idea of a tool cannot be reduced to merely an instrumental understanding, as if the concept of instrumentalized technology were unambiguous: neither body nor technological instruments are mere tools. There are always ingredients of reciprocal relations (acts and flows), affecting one another. One of the final comments of the session touches on the "visceral feel" the mathematician might have for equations – it begs the question as regards Dan Trueman's adjustments of his hyperinstruments. We assume that this might be based on the development of some intuitive grasp of what changes in the code/scripts might affect the outcome, the behaviors. Can such intuition be considered visceral, embodied thought?

Level two: the artistic

Bebe Miller facilitates: the following is a list of keywords and phrases shaping this second cluster conversation.

> Application of technologies in artistic contexts: application, integration and translation across diverse practices, virtuoso (skilled) performer and 'composed instrument' vs. unskilled audience, and shared kinetic experience of music, audience interaction, user behavior. The sonic costume, sonic masks, characters in the interface. The 'staged' interactive performance vs installation, virtual worlds, mixed realities, dynamic and responsive environments and 'game' structure or navigation-interface design.

Lali Krotoszynksi launches the discussion with further reflections on the nature of a system as a situation with "shared properties" where one element changes space and time (a chaotic system). She also shares an interesting proposition as regards the senses as a "tool," inspiring us to vividly consider the *gestures that carry tool information,* such as a "hammering gesture," or a "turning the wheel gesture," etc. Marlon Barrios-Solano proposes that by getting locked into a discussion about cybernetics/systems/tools, we are cordoned within an "epistemological

FIGURE 2.2 Bebe Miller dancing in *Necessary Beauty*, 2008, Wexner Center for the Arts. Photo: Julieta Cervantes. Courtesy of the artist.

model," caught within certain metaphors. He posits a set of new metaphors for interaction under the concept of "coupling," whereby we can define any relation through (1) proximity, (2) similarity and (3) simultaneity. This is a fruitful proposition as it takes us a bit further away from the techno-functionalist language of HCI toward a deeper acknowledgement of the somatic and the flow of proprioceptive acts (Figure 2.2).

Miller steps in to ask the question, "Are technologies a way of getting us out of the studio, making dance more accessible"? The question is less related to how the internet might disseminate/broadcast more dance, but how technologies, the discourses and practices related to emerging devices, might function in multiple ways to bridge certain gaps between those who make dances and those who do not. Robbie Shaw wonders if perhaps there is a greater interest in process, as evidenced by consumer items such as DVDs making available the behind-the-scenes of movie making, for instance. Curtis Bahn, picking up the discussion about what will remain when those making these hyperinstruments "die" (referencing Coniglio's observation already mentioned), suggests that perhaps this priority on process means that *what remains is the process* – the methodology. It is interesting to consider this as perhaps something like notations or scores, instructions that refer to *process methods*: given the rapid change of digital technologies which seems to resist the sort of "rehearsal" or long-term practice that we are accustomed to, and as the "tool" or "instrument" is constantly changing, *process becomes the tangible outcome*. This is a radical positioning reminiscent of the late Umberto Eco's comparison between the poetics of a "work in motion" and the aesthetics of the fixed and closed form (in his seminal *The Open Work*).

The discussion turns to the experience of being inside the interactive system. Dawn Stoppiello argues that she is "different" in the midi suit, implying that this is a form of awareness that can be practiced/trained – the suit adopted into somatechnic proprioception. Miller and Iris Tenge both pick up on this, with Bebe addressing the potential transformation of the body with regard to habitual movement logics, letting the body take care of what it does well, and tracing the changes affected by the midi suit. Tomie Hahn also mentions feeling "huge" in the PikaPika character/costume where she wears external speakers on her arms and can make the building shake and tremble with sonorous sound. Barrios-Solano points toward the techno-scientific discourse we have gained the rights to use, and elaborates on this notion of the extended body dancing with data in regard to the older history of improvisation. Steve Paxton danced with gravity, he suggests, but we are dancing with data, in a web of causality.[9]

Level Three: the social/ behavioral

Hahn, Trueman and Bahn facilitate: the following is a list of keywords and phrases shaping this third and final cluster conversation.

> How are these art/technology works absorbed, anticipated and transformed across communities, event-spaces and cultures, audience/participant behavior; the 'character' behavior of the dancer in the interactive and responsive environment; the 'player' behavior of the visitor/audience in a responsive environment; who or what controls the 'activities' and inframedia modulations within the system (environment); what happens to 'translations' of movement or sound once we control/modify them with technological systems (MAX/MSP/nato, VNS, Isadora, etc), and can we infuse interfaces and interface design with cultural traditions and expressions?

Trueman surprises us with a fresh schematic connecting along the axis of player to culture and presentations to process. The diagram locates player over character over the concept of training (virtuosity), training runs toward presentations and process (in both directions) and on the other side connects downwards to balance, through preservation, tradition (transmission is added) and culture, with Barrios-Solano requesting that perception is tagged on between tradition and culture. As a thought tool the schema was provocative, and a good starting point for discussion.

The term "virtuosity" caused a few wrinkled noses, on the one hand, and nodding heads, on the other; it is loaded with connotations of exclusivity, on the one hand, and entitlement for hard work and practice, on the other. Birringer ponders the question of "new" aesthetic environments and reminds everyone that youth culture has few problems with new technologies and constantly adapts, thus it remains not very clear what is new to whom, and what "virtuosity" might mean to different generations. Ainger sums up his view by suggesting that what artists are doing is "world building" and "telling stories." Todd Winkler offers his observations on audiences-participants in the context of his new direction into *installation work* where "social interaction" is the material – designing an experience for someone where they "touch" others. With regard to the issue of socio-cultural content and specificity (on a global scale), Birringer wonders if these interactive designs are simply electronic façades, with little or no depth of diverse cultural or community relation. Winkler comments on the earlier observation on process and the DVD, suggesting that in his experience audiences respond well to an instruction, some form of guide beforehand. He provides evidence in the anecdotal account of the guitar player in a rock concert you could only see from the head up, hiding "how" the sound is made. Marc Ainger mentions that so much of what artists now make is seen only once. It is suggested that artists could expose the process more by creating more outputs or ways of entry into the work. Ainger replies that sometimes the worst person to talk to about the process is the maker.

The spectrum between explanation and touch/sensory experience is intriguing. Coniglio later suggests that the concept of "intimacy" becomes freshly compelling for him after the conversations. Bahn offers the provocative comment that when he designs a new instrument to play with Dan in a duet, this instrument/tool or system cannot be played solo – it is designed to be played with someone. deLahunta indicates that his thought matrix/axis, which places research to rehearsal and experimentation to production, is enhanced through Bahn's idea suggesting "rehearsal" might be augmented with the notion of "tradition." The issue of "character" (actor) in an interactive environment is not pursued but remains on our minds after Hahn's account of the very different personae and "sonic masks" she performs in her collaborative work. Her own expressive dance vocabularies are rooted in her Japanese training (tradition), but she considers her performance personae hybrid and reflective of her biracial and polycultural experience (unlabeled). Unlike Xu's insistence on knowing his Chinese dancing body, Hahn thinks of her roots in a more fungal way – as sporulating and drifting, trans-planting into surrounding vegetations. Unlabelable.

Midi-Dancer and group improvisation: telepresence

The second day of the think tank ends with a series of practical sessions that include Kelly Gottesman's documentary presentation of digital slides and video from *L'Entre deux*, his environmental dance installation-concert that features a duet created through telepresence, a live internet link to a remote site. Gottesman explains his sculptural and choreographic process and the learning experiences he had undergone in Birringer's "Environments Lab" and the collaborative work with the ADaPT consortium (Association for Dance and Performance Telematics), which represents one part of OSU's research paths in dance technology, along with research in interactive environments and motion capture. Coniglio and Stoppiello offer a hands-on workshop with the Midi-Dancer suit, and the group watches Tenge, Miller and Barrios-Solano test the coupling. The main recognition for the performers working with the Midi-Dancer is that it is a *wearable interface*. It comprises flex sensors and a transmitter worn on the body (where the small wires run from sensor to transmitter) and takes some tugging and pulling to get it in a functional position relative to the joints of the body. The "flex" data from the sensors is converted to MIDI input data for

Coniglio's Isadora software that maps the input to some form of output (sound, video, text, graphics). We notice these are different bodies in the (same) system, and Bebe has a chance to watch Marlon and Iris improvise. Miller then gets into the Midi-Dancer and triggers a simple sound/spoken text loop and a video clip (prerecorded outdoor footage of one of Troika Ranch's dancers in a wheat field). A camera feed also inputs her movement to the Isadora system. This is a real-time video image of her, but it is delayed because of the firewire speed.

Miller's comments/questions after working for about ten minutes could form the backbone of further research: [1] how quickly a "narrative" appears, [2] how quickly it becomes compositional, [3] how to keep the presence in her own dancing, [4] how to learn what the improvisational "headset" should be with this system, and [5] she felt like she is "listening" to the triggered movement-sound loop – and how does listening change in relation to different sounds or self-generated spoken texts? Her last comment was eye-opening: the dancer listens to the environment she creates interactively, and can thus compose dance movement almost as if she were playing an instrument and sensing the acoustic and visual (video) co-resonances in the space.

These are important findings, especially Bebe's "compositional" comment. During her earlier presentation she had offered an observation on the hard edges/harsher processes of the digital in relation to soft ideas/softer processes of the bodily dance exploration. The Midi-Dancer is constantly "generative" – you move and it delivers data. This "databody" is different from the telematic visual bodies of Kelly's experiments within remote spaces and with which we would next experiment during the webcast.

Web cast experiment

The internet webcast is introduced by Birringer as a warm-up "thinking rehearsal" for the public session on the last day of the workshop, and it is intended to be spontaneous and improvisational. It starts up with a physical "interaction" improv (which is telecast) prepared by Lali Krotoszynski. The guest from Brazil also mentions that she had spent several weeks with the Environments Lab teaching a special Odissi dance workshop that resulted in the members of the Odissi group sampling the percussive footwork sound created in the rehearsals. Krotoszynski offers the sound samples for processing, and then introduces two large red and black fabrics (with holes in them) she had brought with her. Inspired by the cultural precedents in her own history – the participatory performance actions with *parangolés* or non-art objects which Brazilian artists Hélio Oiticia and Lygia Clark had instigated in the 1960s and which can be reviewed in their conceptual relationships to Fluxus, happenings, live art and body art experiments in Europe, Japan and the United States in the 1960s and 1970s – Krotoszynski encourages participants to play with the fabrics and the spatial transformations they enable through interaction. As it turns out, the performers/players not only improvise with the shape- and space-transforming qualities of the fabrics as they touch

FIGURE 2.3 *Handsdance*, Helenna Ren in networked performance interface with telematic partner Ruby Rumiko Bessho in Japan, ADaPT, 2007 © Video
still Johannes Birringer.

upon themselves, but also incorporate and envelope other "audience" members
sitting on the sides of the studio, thus creating physical, sensory and metaphorical connections between everyone present in location. This visceral enveloping,
however, cannot easily translate into the webcast since on the internet it will
only be perceivable as a screen-based medium for viewers – a problematic we
explore with ADaPT collaborations numerous times, as we become interested in
how "touching" feels through a telematic embrace, how its simulation can still
have a visceral affect (Figure 2.3).

In the subsequent "Saturday night live" round robin conversations on camera, we freely roam through a series of questions and concerns that had come up
during the day's cluster talks: deLahunta ponders the group's focus point, "what
is at the center of our collaborative investigation?" He suggests it is the body,
and within this, the gesture is the focus, capturing that gesture in data forms
and mapping this data to output. Coniglio and Stoppiello are building a system
for gesture capture and data mapping that is meant to be useable with a smaller
learning curve. Trueman and Bahn rehearse their instruments and the related
gestures, making adjustments to the quality of play. Hahn not only captures gesture data but also wears the speakers – becoming the *sonic mask*: thus being both
(embodied) input and output.

Rubidge, Winkler, Tonnesen and McConnell are building installations for user-participants, creating the conditions for interactive experience, and exploring the working body as interface, observing and analyzing impact. The question arises how various installation designs concretely include an interactive dimension – in the "performance" or the "play" of the user – that in fact feeds back and affects the system or leads to effective, evolving experiences in the user's cognition and kinaesthetic relationship to given materials, objects, sounds and visuals.

Findings and outcomes

Prior to the public presentation, everyone is asked what they might like to show. A short discussion is launched about the most important issues that might have emerged. Although the group had perhaps not arrived at a clear threshold of "findings," given the short time of the collaboration, several arguments emerged such as: (1) Coniglio noting that we still are looking for a common language; (2) Winkler stating that when everything is new and changing rapidly within a diversity of working practices, it makes everything very "personal"; (3) Rubidge claiming that work should be led by art not the technology; (4) Bahn emphasizing that it is important to conduct research, but maybe not too many people need to see something or have an (aesthetically unsatisfying) experience during stages of interface development. deLahunta stresses his interest in determining how research can be initiated, framed and carried out so that it can also be documented and shown as a work unto itself in specific contexts.

The final public presentation, to which local audience are invited into the studio, is a combination of demonstration and a few new things – work that the participants had not been able to show before. The repeated demonstrations allow for distillation and re-analysis. A few summary comments are offered here, such as Tinapple's description of working all day on a MAX/MSP/JITTER patch for the evening performance, prompting Miller to comment on how that sounds like working "in the body." Roesler remarks that watching the improvisation on Saturday evening was very inspiring to him – simply to see bodies moving in space. Tenge suggests that the general public perception of dance does not tend toward any understanding of where dance and technology experiments might be co-generative/productive. This leads to a heated debate over such perceptions, with Birringer emphasizing the unacknowledged deep history of, at least, a century-old close relationship between dance and other media (photography/motion studies, film, animation, experimental and electronic music, movement simulation, fashion, etc.). Miller proposes that more funding is needed for any future developments, a proposal which dovetails with Birringer's suggestion that such experiments require appropriate studio or lab conditions and curatorial and critical/discursive practices to provide contexts and reference points for the new work with interactive systems. deLahunta ends the round table with the proposition to engage performing artists and dancers with computer scientists and software developers, in particular working in HCI or physical computing contexts, where the body as the interface research could clearly benefit from dance knowledge.

Critical outlook

The collaborative workshop as an opportunity for dialog and experimentation is unanimously considered by the participants to have been a rewarding experience, especially as it drew attention to the practical levels of process in interactive design and to the improvisational and compositional levels of embedding the performer's body in/as the interface, as the composer/composed instrument and choreographic agent, as symbiotic vitality. Furthermore, the overtly interdisciplinary nature of the lab raised awareness of the conceptual, experiential and also political dimensions of such research: ideal conditions for conducting artistic and computational research in interface design are by no means to be taken for granted, and institutions that house separate dance, art, music, computer engineering, design and communications departments have by no means shown the understanding and foresight to encourage younger artists-students to engage in interdisciplinary research projects and programs, which would allow them the kind of cross-fertilization and transcultural cross-media practice required to develop new artistic and poetic methods. Not to speak of the logistical difficulties (access, equipment, expertise) that independent artists or performance companies, visual artists, or musicians might encounter if they were bent on incorporating software-based interactive systems into their art making.

Nevertheless, interactive tools and systems, ranging from multimedia and real-time synthesis software to the construction of proximity sensor (haptic, pressure and flex sensors, etc.), distance-sensor (e.g. ultrasound, laser) or camera-based responsive environments and "seeing spaces" (deLahunta's term), along with telepresence interaction, are in fact already in the hands of practitioners in the independent communities. Some of these artists have developed their own (DIY) custom-built instruments and tool-kits for their work. It is to be expected that the underlying shareware ethos of the independents, together with the facilitation of university programs and residencies, allows for greater dissemination of the practical knowledge that practitioners have gained and that could benefit future generations of digitally born artists who are now growing up with computers, portable and smart media devices and internet access.

For our critical evaluations of the cross-over paths and collaborations, for example, between trained artists from the performing arts (dance, theatre, music), the visual arts (sculpture, painting, holography), fashion (textile design, smart fashion, wearables, etc.), and digital artists and programmers coming from the computing sciences, engineering, design, architecture, robotics, VR and AI, it remains to be seen how we find clearer and more succinct parameters for the understanding and definition of the aesthetic and social construction of the work shown and promoted as "interactive media art." From the recent discussions in the dance and technology community it should be apparent that a broader recognition or critical acceptance of interactive danceworks does not exist to date, and from Curtis Bahn and Tomie Hahn's recent experience of being featured in the "Science" section of the *New York Times*, with Tomie pictured in full gear, blue whig, accelerometers, speakers and all – her PikaPika costume – smiling at us in front of a dance studio mirror ("Making the Music Sway to your Beat," 11/29/2001), one has to assume that it is the same in the contemporary music world, since the research article mentions the "inventors" and builders of new hyperinstruments and the interest provoked by a recent Computer/Human Interaction conference in Seattle ["New Instruments for Musical Expression"], but then ends in a souring downward spiral. Interface design is questioned, then measured against "true" musical instruments. Michael Gurevich is quoted saying that "we will never be able to pack the intrigue and expression afforded by real instruments like saxophones and guitars into a wearable device…"

Dawn Stoppiello expresses her concern that

> most of the information being passed around about the actual art making and the processes of the artists working with new tools/instruments/gear are in PhD thesis papers and so are in university journals and not getting out to the other more main stream arts publications.

She asks "where are the major works of dance that use specific technologies that have been taken seriously by art critics/writers besides *BIPED* (Merce Cunningham/Paul Kaiser)?" Doug Rosenberg, Scott deLahunta, Kent de Spain, Jeffrey Miller, myself and others already had engaged in such discussions on the dance-tech listserv. For example, de Spain suggested that

> as a field, as a movement…we keep getting stuck at the stage of experimenting with the technology instead of honing the images into art. Part of that is inherent in the aesthetic of technology (a driving aesthetic of our culture these days) which is highly linear/developmental (meaning that we drop the old as soon as the latest/greatest is issued). Much of the work I have seen in dance and intermedia (with the exception of some very fine single-media film/video work involving images of dancing humans) seems be continually in a Beta release artistically. Part of that stems from our marginality and lack of access to the tools and funds it takes to make higher level work…At the same time, I know that I would like to see dance and intermedia artists forego new tools sometimes until they have really honed the last ones.
> (dance-tech list, 12/05/2001)

This issue of aesthetic development (with wider appeal and recognition), and the integration of interactive tools and systems into our embodied performance practices, brings me to two critical questions regarding the current use we are making in our interactive design processes, and in our thinking about the often unscrutinized concept of "interactivity."

Questioning interactivity 1

"Interactivity" – with all the conceptual issues this term brings along – cannot be applied equally and uncritically to stage performances, on the one hand, and responsive environments, intermedia installations, telepresence events and immersive VE, on the other. These genres or parameters involve quite different considerations, even if they use the same interactive tools and systems. These considerations are design issues reflecting upon artistic outcome or social, aesthetic or spiritual experience desired by the designers, and thus they are also highly site and context specific, since an installation can happen in an art world context (gallery, museum) but also take place on a side walk, in a disco club, a shopping mall or an abandoned warehouse.

It also became clear during the Think Tank that there is a huge conceptual, practical and aesthetic difference between the use of either "invisible" or "transparent" interactive tools on a stage —– presented in a rehearsed work for an audience to watch – and the use of either "invisible" or "transparent" interactive devices and interface designs in an environment which, as Winkler argues, "offer audience members an opportunity to become actively involved in the creative process by influencing image and sound output from a computer" (2000: 1). This active involvement of the unsuspecting (or suspecting?) audience needs to be examined precisely along the axis of the critical and conceptual criteria we have delineated for the trained performer whose body (as moving and sensing instrument) is or becomes the interface in what artists like Bebe Miller, Robbie Shaw or Iris Tenge would consider an expansion of choreographic/compositional intention and intuition. Tenge spoke adamantly about the choices we make in movement exploration and composition, even in, or especially in, *improvisation* which arguably is always already based on a particular movement system (a rule system/conceptual grid and a culturally informed cognitive process with which the dancer knows how to locate herself in space and in temporal flows in/through such space).

Whereas the dancer experiments with the vocabularies of dancing and their extensions, in a given situation which is knowledge-based, the user of an interactive installation, then, would presumably operate, under the conditions of navigation and intuitive exploration of the responsive environment, with an undefined conceptual grid which is neither based on trained movement nor choreographic knowledge. It could be based on sensory and motor experience, on prior experience of having attended such interactive installations, or on social experience or game skills of playing around, aggressing and defending, engaging and disengaging. These experiences could then be traced also along the cultural vocabularies and narrative or "theatrical" sensibilities that people with different backgrounds may bring to an interactive environment, which invites becoming involved in a creative process (something that not all audiences can be assumed/ expected to enjoy). The creative process, in other words, can also be an aggravation: it can cause distress, alienation, even trauma, as well as joy or exuberance, extroversion and all manner of serendipity. Who knows what a crocodile might imagine seeing inside an immersive van Gogh installation, walking around *Nuit étoilée*.

In light of the presentations by Marc Ainger, Sarah Rubidge, Todd Winkler, Lali Krotoszynksi, David Tinapple and, especially, Liza McConnell, it is apparent that sound installations and visual media installations (involving objects to be touched, handled, explored, read and discovered, etc.) may not operate on the same level of (movement, action) exploration as an interactive dance installation. No one really addresses the issue of what an interactive dance-installation would appear to be like, if it involved more (or less) than the haptic, sensual touch dimension or proximity-relationality implied by interactive video projections (cf. Rubidge's new installations *Phases* and *Hidden Histories*). The interaction with a digital video projection which changes in response to audience behavior raises issues that are neither related to choreography nor to improvisation, but to unsuspecting audience-reactivity to "rules" that they will begin to suspect being built into the system and its reception-structure, allowing for their re-application and (one must assume) subversion or de-articulation. Let us be blunt: some audience reactions (after spotting a camera) may simply involve posing for the camera, waving at it. The reflective reaction these days may be equivalent to a selfie: one poses to the camera to (hope to) see oneself in a good light. Or in a grotesque light, with an odd grimace.

Yet each socio-cultural system, for example, fairground, cinema or shopping mall, will evoke their specific culturally learned reception and action behaviors. Thus we can assume that interactive installations can indeed be experienced along the axis of such rule-bound and flexible (since intermixable/rewritable) reception behaviors which, according to Jacques Attali's cultural theory of the evolving political economy of music, imply a transversal: the "listener" becomes the operator and composer, thus "negating the division of roles [rules] and labor as constructed by the old codes. Therefore, in the final analysis, to listen to music in the network of [digital, interactive] composition is to rewrite it" (1985: 135). In Michel Gaillot's more recent analysis of techno music (*Multiple Meaning: Techno – An Artistic and Political Laboratory of the Present*), he develops this idea of "living the possible" further by examining both the DJ practice of "making music with music" (live re-mixing) and rave dancing as a collective performance in a constructed environment: "techno does not create a work but an environment, a situation to be experienced in common... an experience of the 'with'" (1999: 62).

Regarding the "experience of the 'with,' " Barrios-Solano perhaps comes closest in suggesting that all art experiences involve the creation of relations, of a Gestalt, which is modeled upon models (there is never a "right" or one model only). He understands interactivity to mean a "coupling" (embedded in couplings of systems) that invokes questions about how we know, how we create wholes and fragments, how we might observe a system that is generative and whose environment is unstable, unpredictable. The unpredictability of the boundaries is what is essential in performance not modeled upon the older model of the finished work that presents the finished façade (illusion). The performance we are addressing here is modeled on the "unfinished" or "unfinishable" (process); it is also a paradigm for all internet-based link-communication.

We are able to build on the small improvisational lab experiments we made during the Think Tank, since they have larger ramifications. One experience would be based on the exploration of the "wearable" – since wearable interfaces could easily be built into an interactive environment and would also allow the unsuspecting user a threshold (application of a "costume") and an induction which encourages the assumption of a playful "character" for the experience of the unfinished world of the interface.

Furthermore, as in Krotoszynski's offering of the *parangolés* for audience interaction, the use of fabrics and objects that are "sensitive" wearables, yet understandable as everyday objects, could inspire users to be as creative as they wish in the exploration of how they can move with them, how they can build something with them or use them to interact with others in a *T-Garden* like play-space – a proposition also built into Oiticica's late work, the "Quasi-Cinemas" (1973), where he built multimedia environments (slide projections, music) with hammocks and mattresses to invite visitors to lie down or play around, bringing attention to bodily comfort, exuberance, liberation and trance while subverting the formal logic of the cinema and traditional notions of aesthetic contemplation and artistic value.[10] Interactive installations follow the same logic and privilege audience activity, and thus an activation of playful behavior and subjective kinaesthetic or synaesthetic "processing" of experience with multimedia-communicational phenomena. This breaks down any notions, in dance, that are based on formal considerations of choreography. The correlation of design and choreography is thus completely undermined.

We have to assume that the vast majority of trained dancers will not want to design themselves out of business, abandoning presentational performance on stage for audience interactivity. We also have to assume that musicians will continue to compose and play with their instruments, and visual artists and sculptors will prefer work in presentational exhibitions that allow a certain amount of artistic and creative control for the implementation of their ideas, not to mention now the basic economics of needing to sell work. Thus the actual playing field for audience-participatory interactive environments will be limited to those media and performance artists and designers who are explicitly interested in investigating the path that Hélio Oiticica and others have traveled (toward "quasi-cinema," and not the crocodiles) or who comply with the "impact museum" world's evolving interest in engaging audiences, providing interactive and VR experiences as new modes of outreach and pedagogy (or revenue).

The implications of wearables will be investigated further, also drawing from deLahunta's reports on the London "Software for Dancers" [http://www.sdela.dds.nl/sfd/scott.html] and the TRANSDANCE lab in Athens [http://www.sdela.dds.nl/transdance/report/print.html). In the context of the latter, deLahunta broached the subject of constraints in the wearable interface, which was not investigated in the OSU think tank. This same idea will be explored in some of the later chapters when I address the work of my own ensemble (DAP-Lab).

Questioning interactivity 2

My second proposal concerns the level of mapping procedures and processing of video and audio material generated in interactive performances or installations. In other words, the issue of "scratching" and real-time video/audio creation, live processing (post-processing) and manipulation via Isadora, Max/MSP, Jitter or other interactive systems and hyperinstruments – demonstrated in our Think Tank – needs to be pursued in greater detail in a manner that allows practical understanding. In the case of Isadora, Coniglio and Stoppiello offer a software system for gesture capture and data mapping useable with a relatively small learning curve. Such systems could then be incorporated into our training practices with performers, allowing them to get a hands-on understanding of the concepts involved.

The main conceptual shift implied by our Think Tank observations concerns the manipulations on the material level of "inframedia." First, the practice of data manipulation and data mapping is of course polymorphous and not easily defined. One would have to examine the various forms of "scratching" in different contexts of computer music, interactive media, sound art, video art, graphic design, motion capture-derived animation, telematics, etc. What makes inframedia manipulation a common media arts practice today is the capability, afforded by digital technology, to manipulate digital objects on the level of data, "interfering," as deLahunta has suggested, with the "numeric properties of digital media image or sound." It means that in artistic terms, the basic materials of the digital artist are not necessarily the *image or sound itself* which is essentially a representation or manifestation of the underlying numeric representations or mathematical formulae (although this view does not take into account the needs of an audience). Essentially, these underlying numeric representations can be broken down further and used to represent a variety of "surface" media (Figure 2.4).

Surface media refers here to the image or sound, text or graphics that are the commonly accepted new media means for communicating and producing meaning for the viewers/users. Generally speaking, today's average computer user/consumer does not grasp the underlying numerical systems that lie at the heart of computation. However, for an experimental (non-traditional) artist working with new media, it is normally not sufficient to simply manipulate the surface media as this does not allow for an interrogation of the basic materials or principles of the digital media. What is beneath the surface is the underlying numeric system or code; what we hear or see is the signal. In audio terms, we could say that what we will hear is a signal that is constructed, assembled, processed, manipulated and layered, and in real-time interactive processing there will be

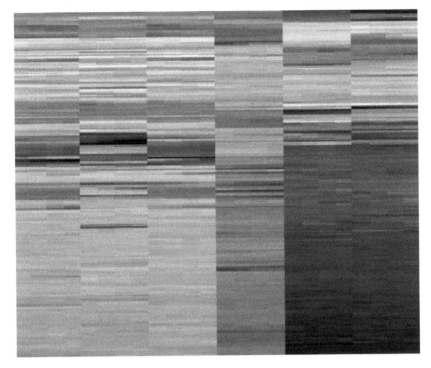

FIGURE 2.4 David Tinapple, installation *Pacific Slip,* 2002 © Video still courtesy of David Tinapple.

a continuity/non-continuity of such constructions. This is neither sound as a transparent substrate for organized expression, as in Western classical music, nor sound being "itself" in a Cagean sense, but sound being mediated, synthesized, generated and collaged (cf. Whitelaw 2001: 49–52.).

I became interested in this inframedia operation when I saw Jarek Kapuscinski's interactive piece *Yours* (created with Nik Haffner, Antony Rizzi) at CrossFair 2000 (Choreographic Center, Essen, Germany). His scratching of the digital video of the dance/the dancer eliminates the physical dance and places the construction of the interactive experience entirely on the side of the "player" playing the touch-sensing keys of a piano. The piano keys manipulate the digital video image on the screen by intervening in the order, continuity and speed of the dancer's movements. The interactive set up allows control of the surface media down to the finest molecule (frame-rate, pixel-rate), and the generation of what the audience sees (the live videodance) is done as a manipulation of the image-frame as instrument. In other words, this is no longer a dance or an interactive dance, in conventional understanding. Kapuscinski plays the media as an *image instrument.*

If we apply this lesson to our Think Tank experiment with Bebe Miller generating the Midi data for the processing of sound data in a software patch of

Isadora (or MAX/MSP), it follows that a dancer could be the interactive generator and techno-actor of a signal process which will not at all cohere into a logical surface media representation of dance movement. The kinetics here are atmospheric and more subliminal, they float and vibrate, underneath visibility. Once the signal processing is interfered with, the connection between movement and sound becomes unreadable (in dance terms). And yet it becomes an audible artifact of sound media or, rather, an artifact that resembles noise. A different ecology emerges.

Sound artists have noted that there is widespread interest today in the abstract, tweakable flows of audio-data. This is exploration of *audio infrastructure*. Often the artists enjoy precisely the contingent and unexpected occurrence, glitches, clicks, pops and skips, similar to the unpredictable and messy stream-disintegrations in telepresence or the visual scan-data slicing David Tinapple showed in his *Pacific Slip*. I am not sure how this would compare to infrastructure and surface of digital dance, but in sound art the glitch has become very nearly a fetish. As we have seen in live concerts by Oval, Aphex Twin, LTJ Bukem, Alec Empire, Funkstörung and other underground musicians, picked up and taken to other levels by Drake or the Atlanta hip hop artists, it is only one of a larger repertoire of media noises and beats generated from infrastructure, along with tape hiss, digital aliasing, and the sharp clicks caused by discontinuities in a digital waveform. Although these sounds arrive at the surface of a recording/processing medium, many artists draw on the entropic internal workings of audio processing systems – often carefully entangled networks of hard and/or software. A kind of sonic residue emerges, refracted from its source into subtle decay-patterns, which allows us to think of digital media also in regard to ecological ideas (cf. Anechoic Media, the underground label founded by Kim Cascone; or Amanda Steggell and Per Platou's Glitch Festival held in Norway, January 11–13, 2002).

The inframedia approach I sketched here would seem to work more on a formal, abstract level of data-reflexivity, thus evoking perhaps the specter of modernism or even the kind of self-reflexive scratching we see in video art (Nam June Paik, Douglas Gordon, Paul Pfeiffer, Pipilotti Rist). Or, rather, it will to appeal to a more maverick understanding of the *informe* (formless) – the persistent/undercurrent performative and operational force within the history of modern art which Bataille saw as slippage, ink spots, or quacks, sliding outside the opposition of form and content (cf. Bois and Krauss 1997). Butoh artist Tatsumi Hijikata had become attracted to the surrealism of Bataille's *acéphale,* i.e. the idea of headlessness and expenditure, or what the choreographer began to explore as sideways, horizontal movement, corporeal extremity, breakage, abject postures. We have not yet seen it very often, if at all, in interactive dance or digital performance, but it inspired me for the research on *Mourning for a dead moon*, my last large-scale dance composition. One of the questions that remains open, then, is the way in which we can delineate this inframedia process (hypermediated digital dance-images) for dance and make it understandable as a dancemedia.

What equivalencies in dance might there be to the experimentations in digital audio, for example, this manipulation of hypermediated audio material, with errors, slips, accidents, and entropy as deliberately cultivated material that can be worked into textures? In audio art, some of the results of this texturing can be surprisingly beautiful, in spite of the disintegrated or broken materials, but we have noticed the same in our ADaPT telepresence performances (https://www.youtube.com/watch?v=ucNM0ax3Sik), when movement-images break up and become surreal architectural blobs and blotches. The sound artists tend to pay close attention to sonic textures and timbres, to layering and recompositions of elements, to repetitions (loops) and various elements cycling against and through each other, as well as evolving as spirals, small cycles nested into larger cycles, etc. One effect of accumulation lies in the duration of these cycles: macroscopic change is a product of accumulated changes in the microelements. At other times there is collapse, breakdown, interruption and disintegration of the material. The code or machine language is shredded or decompiled; deliberate mistakes in the programming produce unexpected but fascinating qualities. Thus there are gentle cycles of sonic build and decay, but also of violent gaps, cuts, bursts of noise and silence.

What I have just discussed are impressions I have taken from the constructed landscapes and environments of real-time sound inframedia. They are surface features and structural features that reflect a practice which has an aesthetic/compositional/improvisational or architectural vocabulary just as we saw it in Tinapple, Roesler, and Ainger's demonstrations. Before I end with a description of this work's affective power, I want to ask whether we have a vocabulary of inframedia practice for digital dance. I think not, and this is perhaps the decisive disadvantage of dance as a partner in interactive system design: dancers and choreographers are generally working in real-time space, which is its own self-referential system, and not on the material level of digital datastreams derived from dance. In order to make a transition to another system, the idea of mapping dance data (cross-platform/cross-systems data or notation transfer and synthesis) would have to gain wider acceptance into the practice of choreographers/dancers or into their collaborations with digital artists.

Constructed, processed dance emerging in interactive environments – what does it look like? How does inframedia work in real time and space, how does it translate into sensory experience? Whitelaw's provocative description of inframedia audio persuasively refers to a synaesthetic experience that connects it to movement. In performance contexts, he suggests, sound systems and volume levels are tuned for a balance between aurality and corporeality; so too are the sound palettes. Low frequency drones, rumbles and pulses are prominent, and vary in their effects from hollow abdominal shocks to gentle seismic flows; crackles, clicks and hisses go straight for the ear and the head, and in between the warm, overdriven midrange sound (as found in the work of Minit, David Haines, Vicky Browne and others) works like a phantom voice in the thorax and sinuses.

This is an exquisite somatechnical exploration of auditory phenomena. If we follow Whitelaw's "noise" ontology (see also Hainge 2013), we delve more deeply into the immanent materiality of sound as we must imagine it in performance environments (and in all environments). So the body vibrates, is pushed, pierced, bathed, comforted, swayed, occasionally assaulted. The body swims in contingencies. It is pebble skipping. These are varying textures and flows of sound but also of shared sensation, moving through a specific space and duration. Extended repetition creates plateaus of acclimatization, a soaking-in process; attention moves to the audio's interior structures, and to psycho- and physio-acoustic processes. Temporal perception loosens; subjective durations alternately compress and expand. The performance acts as a mesmeric sound-field, an involving process which opens lateral spaces for thought and affect.[11]

This beautiful short text from Australia comes back to my mind to haunt me, repeatedly, after the experience of our Think Tank and the questions it raised for many of us, including the doubts we have concerning the interactive future of performance. Some members of the public audience asked about the aesthetic merit of interactive systems, and how the systems would become integrated and productive for new (or old?) choreographic processes. For digital artists, this would be a surface question only. Dance, to my knowledge, has not yet been explored on the level of the subsurface, even though we might argue that it is inside of our bodies, in our nervous system itself, and thus beneath representation.

Sustaining presences in interactive futures

The future of sustainable acting/performing will affect the choices and desires of performance artists who partake of the same energies that we experienced, in all our questioning. One apparently banal question that also comes up occasionally is in regard to interactivity's visibility. If we do not recognize the data processing in the subsurface, does it matter that it exists? Could an analog performance generate the same façade? What difference do façades make?

The lively discourse on interactivity has slowed down in the past few years, and one reason for this could be the growing naturalization of the digital. Many performance artists are freely mixing analog and digital techniques, conceptually re-framing physical theatre and digital theatre, sampling widely, for example, adopting acrobatic vocabularies and burlesque humor from circus and cabaret, as I observed in Martin Zimmermann/Dimitri de Perrot's exuberant *Öper Öpis* and in Galactic Ensemble's acrobatic *Optraken,* or pushing the facticity of the real to its very limits (Rimini Protokoll's re-staging of broadcast TV documentary news coverage in *Breaking News*). Or transposing physical performance into a cubist montage of webcam conversations between present and absent performers, as demonstrated in several recent stage productions by The Builders Association (e.g. between callers and the workers of a call center in *Alladeen*; or between a girl and her peripatetic father in *Continuous City*; Figure 2.5).

FIGURE 2.5 *Alladeen,* The Builders Association with motiroti, dir. Marianne Weems, 2003. Photo courtesy of The Builders Association.

Taking a subjective notion of space and body, the inner workings of some of these performances include the creation of sensorial and perceptual mechanisms in *immersive* and *augmented* environments (sometimes they are telematic and involve remote networked sites or Skype/face-time conversations), drawing attention to role played by viscera, bones, muscles, nerves, chemistry and various sensory regimes, thus also raising questions about prosthetic supplementation and sensory abstraction as well as the spectators' (or experiencers') positions and involvement vis-à-vis mediating apparatuses and relational techniques. We detect in contemporary theatre a considerable fascination with compounding living beings and technical objects; the emphasis on enframing recalls philosopher Henri Bergson's theory of bodily perception and its consequence, namely that the affective body, as a center of indetermination, selects its images from the universal flux of images according to its own embodied capacities.

Multimedia theatre places performer and experiencer into active temporal processes of (re)configuring living images. The performers in *Alladeen*, playing Asian operators in a fictive call center in Bangalore, deal with US customer requests and project their own wish-fulfillments onto avatar-characters chosen from a popular US-American TV series (*Friends*). On the screen suspended in front and above them we see projections of graphics, text, recorded video and live webcam images of their phone interactions blended with morphing shots of the "Friends," their fantasized US-American alter-egos. The mixed staging of live actors and avatars, of acting and image projection, generates a complicated façade upon which emotions and confused cultural imaginations are laminated under the sign of global capitalism functioning as a perverted Arabian Nights story. All this sarcastically spiced up with extravagant scenes from Bollywood movies and karaoke clubs.

The borderlines between characters, dialog/calling/skyping and multiple images are fluid. Thus it makes little sense to distinguish between actors and images in this amalgamated digital scenography. The actors call the shots, and the shots call the actors. In this circuitry, it is also difficult to distinguish the cognitive and physiological processing of multiple simultaneous images, or ask how the viewer experiences the physical impact of the projections, and yet it is apparent that such multimodality is a particular effect of contemporary digital stagings pushing beyond cinema and video. From the point of view of the "shots" and the software that drives the intra-activities, if one were to raise questions the new materialisms bring to the assemblage of *animate matter*, it would be intriguing to speculate on the cybernetics involved, how machines break down barriers between mind and world, real and imaginary.[12]

Whereas the physical theatre traditionally defends the unremitting immediacy of theatrical experience as lived and present-living enactment, digital performance can engage different simultaneous temporalities, fusions or juxtapositions of the real and the virtual, thus altering and reorganizing our sensations as well as our understanding of such layered temporalities, refrains, loops and reappearances. This is a sustainable practice, with living "presence" redefined. Thinking of the sonic history of the loop – and embracing a media-archeological approach to the evolution of sound technologies from early electro-magnetic experiments to the current era of "cracked media"[13] – one could devote a whole chapter to just interrogating the various functions and psychoacoustic effects of reverb devices, echo, delay and other mediating apparatuses, all experienced in present presence. The augmented space of intermedial theatre affects not only the experience of presence (material and virtual) but also of cause and effect, action and reaction. If experience is (always) mediated and translated across both material and virtual space, embodied perception is challenged to reconsider its sense of self, its images of a system that was the self. Intermedial performances explore new subjectivities; they examine how patterns of consciousness, perception and identity, absorption and distance, emerge in such settings and engage our sensorial experience.

> We will assume for the moment that we know nothing of theories of matter and theories of spirit, nothing of the discussions as to the reality or ideality of the external world. Here I am in the presence of images, in the vaguest sense of the word, images perceived when my senses are opened to them, unperceived when they are closed. All these images act and react upon one another in all their elementary parts according to constant laws

which I call laws of nature, and, as a perfect knowledge of these laws would probably allow us to calculate and to foresee what will happen in each of these images, the future of the images must be contained in their present and will add to them nothing new. Yet there is 'one' of them which is distinct from all the others, in that I do not know it only from without by perceptions, but from within by affections: it is my body.

(Bergson 1991: 17)

New spaces of the affective body and the perceptional and affective image are opening up, especially in the experience of non-linear, virtual time and the virtual ecology of images here implied by Bergson's beautiful meditation on the recognition of images. Over the last years I have been specifically interested in interaction design and the collaboration between performance, design, film, music and a digital aesthetics, which concerns the body as affective interface. The results of recent productions, with my own ensembles as well as with other collaborators, are multimedia plays and installation-environments created by way of *digital animes* – moving projections which interact with material objects, costumes, fabrics and architectures or respond to the range of performers' body signals in real-time. The incursion of digital image modulations into the theatre of course challenges the role of the actors in the physical space, and some chapters in this book are particularly dedicated to the performance demands arising for the actor or dancer in such kinetic multimedia architectures.

The role of costume agency has moved to the foreground as well, due to the many years of collaboration with fashion designer Michèle Danjoux, with whom I founded the Design and Performance Lab in 2004. In the development of interactive performance and the continuing research of the DAP-Lab, a main focus has been directed at functions, affordances and impediments of *wearables* for the actor/dancer – custom-built sculptural or intelligent garments that augment and measure the tactile sensory experience of moving in/with them. Wearers engage with sensing-processing systems which respond to intimate gestures and expressions, transmitted in proximity or distance, with visible and audible objects (outputs). Theorists of moving images and animation have spoken of "technical ensembles" when they refer to organisms co-evolving with their environment; in this context we can also think of the technical objects as co-evolving with the technical ensembles, which include the human performers.[14] In such kinetic atmospheres, we create dynamic models to detect changes in behavior patterns and the equilibrium within digital and physical space.

One concern is to show that the boundaries of the self extend beyond our skin and can connect wearable clothes to wearable space and projection environments (which are both material and virtual) – thus carrying up into atmospheric dynamics. Specifically, in the work of the DAP-Lab we are interested in sensing and sensorial behavior, in the synesthetic processes and consciousness of sensing that connect performers to projected (visualized and sonified) environment, and in how presence, proximity, touch, motion and buoyancy can redirect the way we understand ourselves and others, bodies and images of bodies, other bodies, fabrics and material objects. Fabrics, after all, are closest to our skin, and our malleable openness or protectedness toward the outside elements. One dimension of such performance and design works could be called *contact improvisation*, revealing a sense of impermanence and openness, and exposure to the climate – *sensing improvisation*. But there are other dimensions that go further, involving dramaturgical and narrative structures, the psychic dimensions of atmosphere, altered states and extended operationalities of the biological body.

Stelarc's performances of extended operational anatomies have been path-breaking in this respect, and his body of work also relates to technogenetic performances of artists who pioneered the recombinant poetics of intermedial theatre and installation (Woody Vasulka, Laurie Anderson, Robert Lepage, Paul Sermon, Bill Seaman, Lynn Hershman, Yacov Sharir, Diane Gromola, Char Davies, Guillermo Gómez-Peña, Teiji Furuhashi, to name just a few). Laurie Anderson's extended voice was heard in *Night Picnic*, broadcast to outdoor visitors gathered on the Landiwiese at Zurich Theaterspektakel during the "Summer of COVID" 2020 ("one infection community per picnic blanket allowed," as the protective protocol announced). Seated on the blanket, one listens to a radio transmission of Anderson's songs and stories "from the middle of a disaster zone, notes from a country that is in a deep dream state, unable to separate real and unreal."[15] Do countries dream? How do I listen to a country dreaming? At that moment I am reminded of an anatomical discovery I made with Anderson's *Handphone Table*, an early sculptural work incorporating a concealed soundsystem, which emits low-range vocal tones through one end of the wooden table and instrumental music at the other. Just like the sound-system that produces them, the sounds are hidden, inaudible in the absence of a listener who needs to position their elbows onto two depressions on the tabletop, using their hands to cover their ears. Sharing the wood's porous properties, the bones of the listener become conductors allowing the sound to travel through the arms to the ears. To encourage my dreaming, *Night Picnic* includes strategies for survival, hypnotic experiments and stories delivered in Anderson's typically laconic oral combination of rumor, history, sci-fi speculation, fantasy.

Hypertextual and computational interactivity is not always used in my own work, although stories often form fertile subtextual mycelia. Nor is it a common feature in the performance repertoires of most theatre and dance companies. But it is a more frequent characteristic of installation art. When talking to designers and design companies, I am told that the future of generative software is now – an understatement, since interactivity in the arts has been around for four decades and could be traced back to Jasia Reichardt's seminal "Cybernetic Serendipity" exhibition at the London ICA in 1969 as well as to many rule-based algorithmic practices in conceptual art or Fluxus event scores. Synaesthetic and multimedia playwriting/directing is bound to explore the possibility of performing in networked, shared (streaming) spaces to extend the local site and reduce the distinction between performers and spectators by regarding the spectator as a potential player or co-author, in the sense in which interactivity frames the inclusion of a proactive experiencer.

The success of multiplayer online games and shareable virtual universes such as Second Life challenges the theatre to rethink its system of presentation, its involvement of audiences as participants. As I suggested earlier, new notions of relationality have been used increasingly often in the visual arts and today's "participatory museum" where immersion in installations accentuates sensorial experience or provokes interactional play. Some performance companies have adopted such relational strategies to resituate their work within the commons, utilizing public space and information networks or creating theatre involving the audience in imaginative ways, as Yan Duyvendak's provocative *Virus* demonstrated at the 2020 Zurich Theaterspektakel. For the theatre to widen its ground and build more diverse (and younger) audience communities, such dramaturgies need to be explored more vigorously.

Collaboration and resources

At the end of this workshop chapter, it is crucial to recognize a collaborative ethos, to redefine the way we understand collaborative process and interaction in processual artworks, comparing how aesthetic perception of environments can influence our social interaction within that very environment – interaction in terms of the "intimate," the "personal" and the "collective." This is one preliminary definition I can offer for the *kinetic atmospheres* examined here. Practically speaking, the book ought to be written in the plural voice, as it was tested in the laboratory discussion above, and grew throughout my work with AlienNation Co. (Houston), Environments Lab (Columbus), Interaktionslabor (Göttelborn) and DAP-Lab (London). It grew through adaptation processes. The testing was a matter of investigating how performance environments actually *perform* – how they move and are moveable, how they engage visitors and the actors that have rehearsed with the environment and are ready to wear it, play its instruments. In an historical or archeological sense, kinetic atmospheres – in a manner in which we think of how species are descended – evolve as ecosystems; they often depend on variable elements that conjoin forces, tangled webs that lead from ecosystem to ecosystem, from assemblage to assemblage. Over the years, the collaborative work I have been engaged in has fused many elements into interactive multimedia installations that incorporate a variety of combinations of real-time process. Initially thinking of them as performance-environments, I now want to call them *kinetic atmospheres* – dynamic spaces where sound, architecture, gestures, visuals, material moving objects and wearables play an important part, alongside language or dialog (which I tend to think of as audibles), software and code. When using the term *atmosphere*, I refer to the creation and the exigencies of concrete sculptural-scenographic as well as virtual/artificial densities, volumes and energies, each aspect containing potential realities, some mirroring natural phenomena or being affected by them, some not.

The core of future-oriented explorations would be the creation of real-time environments in order to instigate interactive processes and – this will not come as a surprise – to learn about the restrictions or downfalls of interactivity/participation. Interactive learning is a slow process, depending on the resources available, and thus on the contexts in which such performance research can flourish. It took time for the media arts to develop their support systems, and although numerous media labs, residencies, commissions and collaborations were established over the past decades, the same cannot be said for the theatre. In fact, the idea of virtual reality labs or digital performance studios in a theatre (or conservatoire) may not be sustainable in a time of cut backs and strained budgets, when commercial theatres are under market pressures and smaller non-profit theatres or independent artists may find it hard to survive at all. On the other hand, the tools are out there. Cellular phone cameras, recorders, iPads and laptops are ever more widely accessible, allowing any studio to be transformed into a lab. Professional training for digital performance requires a setting and resources, and a growing number of universities, art schools and art centers provide contexts for advanced R&D, but functional support for productions would need to be created on a much larger scale to allow practitioners to benefit from the sharing of knowledge as we have seen it in independent media arts initiatives founded in many parts of the world. Workshops and festivals offer frameworks for training and exhibition, but stability and continuity – and thus organizational bodies/institutions – are vital for the long-term development of digital vocabularies and techniques as well as critical discourses.

While it is obvious that financial resources are strained, and market expectations in the creative industries may favor conservative mainstream aesthetics or gimmick VR impact shows, the impact of new media technologies on creativity is unquestionable. Digital performance, therefore, is driven and sustained by the same collaborative networks and open source environments that nurtured the new media arts. Much of the exploratory, even wild re-mixing and remashing, which is happening all the time, reflects the various fusions (most obvious in music) and alliances shaped within the Net community – DJ and VJ practices, hacker communities, machinima, amateur manga production and cosplay are obvious examples of a large spectrum of cross-overs and digital hyperimprovisation with data that can be translated across cultural codes and technological platforms.

On the side of training and education, digital creativity today affects young practitioners and audiences alike. We know that our environment has been changing, and that it is populated by new communications systems and new modes of creative operation and distribution. Overlapping interests in related fields – film, electronic music, digital art, science and technology, game design, engineering, robotics – advance our understanding of the complementary thinking processes that drive interdisciplinary research and teaching, opening up training opportunities for young artists. Like music before it, theatrical performance has begun to incorporate "instruments" (cameras, video-projectors, monitors, microphones, sensors, microcontrollers) and software tools which allow it to structure and control the various components of any technical ensemble. The Brasilian Cena 11 company, directed by Alejandro Ahmed, is a typical case of a company working directly with software programmers and developing their own custom-built systems. They are currently exploring the combination of dance and robotics, similar to the productions of Åsa Unander-Scharin (*The Lamentations of Orpheus*), Margie Medlin (*Quartet*), Pablo Ventura (*kubic's cube*) or Garry Stewart's Australian Dance Theatre collaborating with robotics engineer Louis-Philippe Demers on the perplexing dance work *Devolution*. Their works have toured and been enjoyed by audiences in different contexts (festivals, science museums, universities, etc.). When I took part in the kedja Festival in Oslo (2009), where I also met Unander-Scharin, Norway-based choreographer Amanda Steggell taught a workshop for a large group of participants. "Invisible Twin – an exploration of electromagnetic Oslo" proposed the idea that every city has its own invisible twin-city – a parallel architecture in flux made up of electromagnetic waves emitted by its numerous electrical facilities, transmitters and receivers. It is a hotly contested private, commercial and political territory that people pass through every day – at home, work and play – yet it is difficult to perceive without some kind of technological intervention. During this DIY workshop, participants used video camera, headphones and an electromagnetic detector that converts electromagnetic activity into audible signals to explore and record their experiences of moving around in Oslo and its uncanny twin. The recordings, which simultaneously capture the image of the visible city with the sound of the invisible one, were presented at Oslo School of Architecture and Design.

I mention this further example of a digital workshop to inspire reflection about such imaginative dramaturgies for shared creativity, especially as "Invisible Twin" appears to have little to do anymore with Steggell's earlier stage works. But she insists that her background as a choreographer helped her to become interested in various types of cross-over between disciplines, having practiced and taught choreography, media, video art and more, working in many parts of the world. Drawn to the complexities of working with people from diverse cultures and disciplines (artistic and otherwise), she found that learning from each of them offers each discipline new avenues for exploration and understanding of their own fields. As she demonstrated in her workshop and the much admired interactive public artwork (*Electromagnetic Fountain*) she created in Stavanger in 2008, Steggell's choreographic interests now focus on social behavior and perception filters. Communications technologies, their media and related devices that invite tinkering and DIY methods, have become instruments for her imagination through which she envisages creating playful and thoughtful experiences.

During the Oslo workshop, Steggell told us that she believes in depths that can be achieved by multiple types of collaboration; she felt they were obtained during the process of visualizing, designing, constructing and exhibiting *Electromagnetic Fountain*. Before the process got underway, she had herself attended an art and engineering workshop at Atelier Nord, where she first gained the opportunity to try out an electromagnetic detector. I can only confirm the underlying story of practical tool sharing, which I have experienced over many years. Today they call it "knowledge transfer" in the academy; funding is funneled to so-called clusters of excellence. But the independent performance and tech community arguably grew over time because flexible programming and electronic tools (such as Max/MSP/Jitter, Arduino) had become hugely popular, and beta tester and user communities freely shared information and new plug ins. Initial custom-built systems such as LifeForms and VNS (Very Nervous System) were soon shared as well. Troika Ranch's composer and software writer Mark Coniglio, who created the Isadora software, demonstrated its application in innumerable workshops. International platforms such as IDAT, Future Physical and bodydataspace, CYNETart, Boston Cyberarts, Digital Cultures, and Monaco Dance Forum, along with important studios such as STEIM (Amsterdam), V2 (Rotterdam), Harvestworks (New York City), Raqs Media Collective (New Delhi), and other media arts festivals or conferences (ars electronica, ISEA), helped to bring practitioners together and provide occasions for the sharing of tinkering.

In the history of such tinkering, live artists have often mixed old and new media, especially as live art practices tended to emerge from the visual arts and later commingled with video and installation art. The poetics of improvised live creation with video, webcams and network connection characterizes the work of the Indian collective Raqs Media, the *Distant Feelings* (bram.org/distantF/) and *Distant Movement* projects led by Annie Abrahams, and the Brazilian group Corpos Informáticos (www.corpos.org).[16] Corpos' installation events often take place in galleries or museums but always involve online participants joining the composite action from afar. This is a participatory philosophy also promoted by Ghislaine Boddington's London-based *bodydataspace* programs (www.bodydata-space.net) which evolve creative and interactive experiences for the public. Boddington considers *bodydataspace* as a design unit that delivers artist-led projects into architectural applications and uses interactive technologies to enable people to learn, develop and extend within intelligent data spaces, ultimately enabling the public to have a direct impact on the content and atmosphere of the space they inhabit. In a recent collaboration, occurring frequently now in Europe, *bodydataspace* fueled a networked coproduction (*PostMe_NewID*) with CIANT (International Center for Art and New Technologies, Prague), Trans-Media Akademie Hellerau (Dresden), and Multimedia Centre KIBLA (Maribor) to investigate the complexity of 21st-century European human identity with an exploration of the evolution of cyborg culture through technologies of the body. Two "research engines" were held in Lisbon and Prague to generate learning exchanges between diversely skilled practitioners, followed by a Public Forum at the international media arts festival CYNETart_08 in Dresden, and creation processes culminating in public performances in London and Prague (2008–09; Figure 2.6).

FIGURE 2.6 Screenshot of *Distant Movements #6*, jointly created online by Annie Abrahams, Muriel Piqué and Daniel Pinheiro, 2018. Video still courtesy of the artists.

Similarly, Sher Doruff's projects for the Waag Society for Old and New Media (Amsterdam) have included multiplayer online collaborations and a performance ecology, which bridges remote sites (via webcams and interactive software), playing with the tension between determinate structures and indeterminate potentials. In an early work, *Cassis Caput* (2003), she linked Amsterdam with Berlin, London and New York utilizing public webcams as found objects to create "conditions of possibility" from which events could have occurred once dancers moved into the camera-sites and improvised relationships with observers (seen and unseen), thus exploring performance concepts now associated with scientific notions of emergence or *autopoiesis*.[17] The ropes and pulleys, if we use the image again, here comprise an increasingly widely spread out net of CCTV cameras. In China there are an estimated 300 million CCTV surveillance cameras in operation, and the number is expected to be doubled soon. Visual artist Xu Bing's *Dragonfly Eyes* (2017) is a provocative response to the government's tactical digital infrastructure of surveillance, creating a filmic assemblage stitching together images from CCTV, social media, live streaming sites, residential and commercial surveillance systems, and facial and object identification technologies to create a strangely angled fictional tale about a young woman traversing life in a contemporary hall of mirrors China, amidst internet and cloud based data of the daily documented mayhem. The fruits of surveillance, Xu Bing suggests, make actors unnecessary: "Without actor or actress, could the best performance be done by you?" he asks in one interview given during the Locarno Film Festival (Halligan 2017).

Well, as experiencers we want to be at our best, surely. "Move: Choreographing You" was the title of an exhibition at the Hayward Gallery, Southbank

Centre, London (2010), inviting audiences to do the best performance they could, and I will return to this stipulation repeatedly in the following chapters, further pursuing the comment just quoted from Xu. Rafael Lozano-Hemmer's "relational architectures," for example, his large-scale outdoor *Body Movies* installation, which caused much attention at ars electronica in Linz (2002), uses a similar but different approach, asking for your best shadow. The "shadow casting" is done in combination with a large data bank of prerecorded portraits and a real-time camera-tracking system. The digital images only become visible when passersby block the lights that wash out the projections, while the participants at the same time superpose their shadows onto the digital ghosts. The participants enact some wild somatic ghosting, they are puppeteers, they are us. Doruff's and Lozano-Hemmer's practices echo the Situationist concern with psychogeographic experience and political and affective situations in everyday urban life.[18] Without the need of actor or actress, we shall have been there, and the rest you know from the movies.

Notes

1 In microscopy, the science of investigating very small objects and structures using an optical light instrument to enlarge the object under scrutiny, the device enables the human eye by means of a lens or combination of lenses to observe enlarged images of tiny objects: the microscopic view magnifies, the viewed only becomes visible to the eye unless aided by a normal microscope (there are several types, e.g. the compound light, stereo, electron, scanning probe and acoustic microscopes). See: https://eco-globe.com/types-of-microscope/. In comparison, X-ray capture could be said to penetrate; X-ray microscopes use electromagnetic radiation in the soft X-ray band to produce magnified images of objects: unlike visible light, X-rays do not reflect or refract easily – they are invisible to the human eye.

2 Cf. Zarrilli 2010: 37–38. Haein Song, a contemporary digital choreographer who has practiced for many years in *kut,* the traditional Korean shamanic ritual performance, recently completed a series of works that intermesh the traditional and the digital, and in her writings she describes the ritual techniques (*mugu*) deployed to achieve the desired collective healing and well-being effect of the practice: *Ecstatic Space: NEO-KUT and Shamanic Technologies*, PhD thesis, Brunel University London, 2018. Song joined DAP-Lab as an associate artist in 2016; DAP-Lab is affiliated with the same research program that Zhi Xu has joined as well, but both artists also direct their own companies. Xu's last creation, *Unexpected Bodies,* premiered in December 2020, and I was invited to be one of the camera operators for the filming of the work (no audience were allowed to the Artaud Performance Center during the COVID pandemic).

3 The term was first used by choreographer William Forsythe (artistic director of the Frankfurt Ballet from 1984 to 2004) who over the past two decades began to create installations proposing movement possibilities of interaction to participant audiences; he explains the concept of a "choreographic object" in the catalog for the exhibition *Suspense* (Forsythe 2008). See also Birringer 2012a. Together with researchers at Ohio State University's Advanced Computing Center for the Arts and Design, Forsythe has also published *Synchronous Objects* (http://synchronousobjects.osu.edu/), a web-based research archive detailing numerous provocative recombinations of visual, descriptive and sonic analyses of his dance work, *One Flat Thing, reproduced,* transformed into a creative resource for exploring space making, movement, spatial composition, and the complex, multi-layered, 4-dimensional construction of kinetic events. Forsythe's

work, obviously, has been a wonderful inspiration (and my colleague Scott deLahunta was one of the research coordinators for *Synchronous Objects* and the subsequent *Motion Bank*, first initiated in Frankfurt in 2013 and now moved to Mainz (http://motion-bank.org/).

4 See Zhi Xu's writing on his practice-based research, *Techno-Choreography and the Embodiment of Chineseness*, PhD Thesis, Brunel University London, 2021.

5 Jalylah Burrell, a writer, oral historian, deejay, and audio editor, cited Marfield's wonderful and also haunting phrase in a talk on "Black Women's Comedic Performance in the 1970s and 1980s" (Anthropology Department, Rice University, February 27, 2019). Here the notion of playing into roles (or expectations) gained deeper and much more complicated historical layers of resonance as Burrell also commented on the legacy of minstrelsy in the United States, and the misogynoir prejudices black female performers face when standing up or taking center stage. The citation is taken from the 1984 video documentary *I Be Done Was Is* by Debra J. Robinson.

6 For a persuasive study of black radical performance traditions, see Gaines 2017. See also Moten 2003.

7 With crocodile paradox I mean to pun on a great moment in Salazar Sutil's book when I felt that after his splendid deconstruction of Werner Herzog's film, *The Cave of Forgotten Dreams* – having demonstrated how *unlike* cinema the painted cave is and how biased Herzog's assumption is to regard cave paintings as proto-cinema and animated motion pictures – he leaves Herzog with the best line in the coda of the film. There Herzog shows some crocodiles he filmed in a biosphere near Chauvet Cave; they are floating in glass tanks and one is staring at its own reflection. Herzog then asks:

> Nothing is real, nothing is certain. It is hard to decide whether or not these creatures here are dividing into their own doppergängers. And do they really meet or is it just their imaginary mirror reflection? Are we today possibly the crocodiles who look back into the abyss of time when we see the paintings of Chauvet cave?
>
> (Salazar Sutil 2018: 188)

8 The lab rehearsals and experimentations included Marc Ainger (OSU, Columbus), Curtis Bahn (Interface, Rensselaer Polytechnic Institute, New York), Mark Coniglio (Troika Ranch, New York City), Kelly Gottesman (OSU, Columbus), Tomie Hahn (Interface, Boston/Tufts Univ.), Lali Krotoszynski (São Paulo, Brazil), Liza McConnell (Columbus), Bebe Miller (New York), Axel Roesler (OSU, Columbus), Sarah Rubidge (Chichester, England), Robbie Shaw (OSU, Columbus), Dawn Stoppiello (Troika Ranch, New York City), Iris Tenge (Frankfurt Ballet, Germany), David Tinapple (OSU, Columbus), David Tonnesen (FoAM-Starlab Belgium/California), Dan Trueman (Interface, Colgate Univ.), Todd Winkler (Brown Univ., Providence, RI), Marlon Barrios-Solano (OSU, Columbus), Oles Protsidym (McGill Univ., Canada), Mitchell Tsai (UCLA, California), Gary Lee Nelson (Oberlin, Ohio), Robbie Shaw (OSU, Columbus) and Anthony Bowne (Laban Center, London). Scott deLahunta (Coventry University), a preeminent international researcher and curator in the dance technology field, had initiated a workshop series – "Software for Dancers: [phase one]" – in the fall of 2001 in England (in collaboration with Arts Council England, Sadler's Wells Theatre and Random Dance). Our lab was conceived as a follow up and Scott helped me to organize it at Ohio State University's Dance Department where I was teaching at the time. It was an unforgettable experience, and we published a joint report on the *Environments* website (which no longer exists), titled "New Performance Tools: Technologies/Interactive Systems." I am very grateful to Scott for giving me permission to quote from our findings and paraphrase as well as extend our workshop ideas. I am grateful to all the artists who participated in the lab and granted permission to quote them or use images of their artwork.

9 At this point of writing and rethinking the collaborative workshop, the reference to Paxton makes me pause as I have had numerous discussions with members of the

movement and somatics community about different dis/abilities and what Viennese choreographer Michael Turinsky calls "productive deviations" (e.g. in his curation of the 2019 *No Limits* festival for Disability & Performing Arts at HAU Hebbel am Ufer in Berlin); my oldest brother's movement capacities collapsed this year and he has been for months in a rehab program at a clinic for paraplegic patients, and I remember vividly that it in fact cheered him up when I sent him Paxton's DVD *Material for the Spine* (2008) and his book *Gravité* (2018), which like his performance project *Swimming in Gravity* traces a lifetime of exploration of the conditions of moving and dancing, and the physical forces that affect each of us – and each of our differing potentialities.

10 For helpful literature on Oiticia and participatory performance, see the exhibition catalogs *Out of Actions: Between Performance and the Object, 1949–1979*, with essays organized by Paul Schimmel, The Museum of Contemporary Arts, Los Angeles. New York: Thames and Hudson, 1998; and *Hélio Oiticica*, Galerie Nationale du Jeu de Pomme, Paris, with Projeto Hélio Oticica, Rio de Janeiro and Witte de With, Rotterdam, 1992. The exhibition *Hélio Oiticica: Quasi Cinemas*, organized by Carlos Basualdo, was shown at the Wexner Center for the Arts, September 18–December 30, 2001. For an analysis of the relations between architecture and digital performance, see my "Suite Fantastique," available online: https://www.bstjournal.com/articles/10.16995/bst.254/. On the subject of "wearables," see *New Nomads: An Exploration of Wearable Electronics* (Rotterdam: o10 Publishers, 2000), and Birringer and Danjoux 2009.

11 Whitelaw concludes by suggesting that such audio art, although it shows up "the opacity of the medium, flawed technology, failed and false representation," is fundamentally reconstructive:

> It spins artefactual audio out into rich and complex streams, which are rarely composed or formally determined 'works,' more often durations felt out, traversed through improvisation, real-time processing and sensory/kinaesthetic feedback cycles. In that process the transgressive quality of the glitch, the click, or the exotically artefacted sample, fades. There is nothing for it to rupture, no clean surface to crack, just a buzzing cloud of other artefacts. When it's no longer an error or a failure, the artefact is simply the self-identifying sound of the media substrate, something approaching a raw apprehension of the signal, the microactivity of fluctuations in data and/or voltage which subtend digital and electronic media. So what gets reconstructed is an aesthetic whole, certainly, but more interestingly it's one which works with the sensory and affective textures of a media substrate, rather than media 'content.' This aversion to 'content' is evident here in the prominence of process and improvisation…. There is often a reluctance to make a mark or a gesture, soundfields are carefully balanced 'grounds' of microfiguration, immanent activity, and (near) stasis. Once again this formal tendency has an experiential side; this rich absence opens an undetermined temporal hollow…This process draws out the media substrate, the subsurface infrastructure, out into the material world, into temporal, kinaesthetic and affective experience.
>
> (Whitelaw 2001: 50)

In my research on the "image instrument," conducted in the context of the telepresence experiments at OSU, I am also indebted to Manovich 2001: 64–75.

12 With new materialism I mean the emerging trend in 21st-century thinking on the primacy of matter and its properties, actions and assemblages, which can be encountered in philosophy, sciences studies, cultural theory, feminism, political science and the visual and performing arts. A much referenced book is Jean Bennett's *Vibrant Matter* (2010); see also Barad 2007; Coole and Frost 2010; Salazar Sutil 2018.

13 For a media archeology of sound machines, see Gethmann 2010. For contemporary sound art experiments and tools of media playback expanded beyond their original

function as simple playback device for prerecorded sound or image, see Kelly's book on *Cracked Media: The Sound of Malfunction* (2009) and his more recent *Gallery Sound* (2019). A fascinating exhibition, *3D: Double Vision,* excavating the history of image projection machines (from 2D to 3D) was staged at the Los Angeles County Museum of Art (July 2018 – March 2019). The catalog of the show is available as: *3D: Double Vision,* Munich: Prestel (2018).

14 Lamarre 2009: xxxiii; 197–206.

15 See https://www.theaterspektakel.ch/programm20/produktion/laurie-anderson/. Anderson's *Handphone Table* (1978) was exhibited as part of *Laurie Anderson, Trisha Brown, Gordon Matta-Clark: Pioneers of the Downtown Scene, New York 1970s* shown at the Barbican Art Gallery, London, in 2011. For reports on Yan Duyvendak's *Virus,* see https://www.hellerau.org/de/virus/ and https://www.nzz.ch/feuilleton/yan-duyvendak-wenn-der-virus-zum-gesellschaftsspiel-wird-ld.1571713.

16 For Corpos Informáticos' work, see Medeiros 2006. In 2018, Annie Abrahams collaborated with other network artists (Antye Greie, Helen Varley Jamieson, Soyung Lee, Huong Ngô, Daniel Pinheiro, Igor Stromajer) on an "entanglement training" workshop for Online Ensemble, within the framework of the online symposium Art of the Networked Practice – Social Broadcasting: an Unfinished Communications Revolution symposium, School of Art, Design & Media, Nanyang Technological University, Singapore (March 29–31, 2018). I participated in this event and also performed in several of Abrahams' *Distant Feelings* encounters.

17 I have written about my own early involvement in telematic performance (the ADaPT network initiated in 2000) elsewhere, and presented our findings at numerous conferences and symposia, starting with "Connected Dance: Distributed Performance across Time Zones," a panel discussion at the 2001 CORD conference in New York. Here is an extensive dance-technology bibliography that includes references to the literature on telematics: https://sensorwiki.org/isidm/dance_technology/bibliography. For my collaboration with Sher Doruff on the autopoietic installation, *East by West,* at DEAF 2003, see: https://v2.nl/archive/works/east-by-west.

18 For a critical documentation of his interactive art concepts, see Lozano-Hemmer 2000. For Doruff's workshop, see Doruff 2003: 70–99. For the theoretical debate on interactivity, see Hünnekens 1997; Quinz 2004; Camurri and Volpe 2004; Corin 2004; Couchot and Hillaire 2006; Hansen 2006; Birringer 2008a; Chatzichristodoulou, Jeffries, and Zerihan 2009; Kwastek 2013; Salazar Sutil and Popat 2015; Paulsen 2017, and Wiberg 2017.

3
REALLY ACTUALLY WINDY

The space is cramped, a small chapel-like venue constructed out of white wooden planks. We sit huddled together. Yet the ceilings are high, giving us breathing room, and a Victorian arm chair mysteriously rests up there on one of the cross timbers, a seat for angels to look down upon the crowd of 50 or 60 that have gathered for this evening of improvised dance and music. Our venue, I'Klectik Art-Lab, lies hidden under trees to one side of Archbishop's Park, on the south bank of the London Thames across from Westminster Parliament. In the courtyard, we notice that artists-in-residence at this Art-Lab also tend to vegetables and flowers that grow in the yard. When the dance and music begin, we are beckoned inside and for the next three hours become enveloped in this green social ecology – an environment of very diverse practitioners and international visitors drawn to experimental contemporary art that ranges across all genres and takes place in a working enclave, where members can rent space to develop projects. During intermissions, we are asked to go outside and linger in the garden (Figure 3.1).

Could we think of *technologies* in a different way altogether again? Do we hear too much about terror and violence, causing dizziness, vestibular disorders, tinnitus, and hyperacusis? Does this accelerated political sensationism (not quite foreseen by Fernando Pessoa's claim that "ideas are sensations") make us sick?[1] And what is it about listening (the aural) that has obsessed me lately, turning me toward different somatic places of investment in acoustic embodiment, other site contingencies, and away from that recent paradigm of immersive theatre and participatory social-works? Are we not listening to other forces of things now – climates, atmospheres and heterotopias? And how does renting space connect to growing vegetables?

The combination of herbs, planting, and performing hints at new hybrid materialities and interrelations. Could the title of this evening of eclectic work, *Really Actually Windy*, in the oddly named site I'Klectik, point outside the common

DOI: 10.4324/9781003114710-3

FIGURE 3.1 *Really Actually Windy/R.A.W. Vol. 1*, I'Klectik Art-Lab, Old Paradise Yard, London. May 7, 2016. Poster announcing the event.

parameters of theatre and performance to other assemblages or "confederations" as Jane Bennett calls them in her book on *Vibrant Matter?* At one point Bennett mentions the strange concatenation of stuff she discovers in a storm drain – a glove, a bottle cap, a dead rat, a smooth stick of wood... (Bennett 2010: 4). In this chapter, I want to talk about such confederations.

The works I hear at I'Klectik are introduced by (two of the ten) performers themselves, Anita Konarska and Mirei Yazawa. The performers are also curators – a familiar trend in many alternative venues. When Macarena Ortúzar enters to the fine, almost inaudible sounds of Bruno Guastalla's cello, we are instantly mesmerized by a quality of strength and fragility that she conveys through her slow moving, contorted postures. We inhale them as sensations. The tones of fragility also come from Guastalla's strings – the sinewy mezzo, low frequency overtones, unleashed by bowing at the bridge. Two wooden sticks help Ortúzar to stand upright; they are her crutches and yet they become so many other things – branches of the wind, bones, walking sticks, lightning rods, spines, arrows. They are thin and smooth, one of them later seems attached to her chin, her face resting on it. A Chilean dancer who had worked on Min Tanaka's Body Weather Farm while training in butoh, Ortúzar here wears black blouse and leggings, a white apron wrapped around her hips (is she a maid?), her dark hair framing a face that is intensely focused, serious, and sorrowful. We see her movement reflected in every inch of her strongly muscled body, the way she can bend, twist her balance and shift her center of gravity, lean without falling, fall without breaking, hovering in horizontal a few inches off the floor, as if weightless. Ortúzar's dance, performed to the highly sensitive improvised music of Guastallo, who touches the wooden body of the cello with hands more than he plays the strings, falls into place with the later solo by Anita Konarska. Beginning in the garden, leaning her weight against a pine tree, Konarska slowly, slowly slides down against the trunk of the tree, then sinks her arms into a flower bed. Later she performs in the chapel, but we are blindfolded as we enter to witness her actions. So I can only listen to what I cannot see, imagining what is nearly inaudible, growing what I collected from the outside, the nearly dark, unlit permutations of contingency that are also relations to the surrounding architecture, relations to that night, that urban context. Eventually someone invites me to remove the folds and Konarska is revealed: she stands *en pointe*, balancing a huge tree branch on her head that stretches almost across the entire width of the space. Tree woman, agent of near-silent sounds that we imagine hearing while blind, her performing conducts sensory power and a strangely shamanic vibe – I am not sure what to call it (Figures 3.2 and 3.3).

I recall a peculiar announcement released by composer Richard Povall out West in Cornwall in early 2016:

> Are you an artist, writer or performer looking to take your practice in a new direction? …We will explore the shift in perception that comes from tying yourself to a tree … the discombobulation of acoustically penetrating a tree's internal workings (Tree Listening), and the mind-opening excitement of embodying tree-being (Other Spaces), among other innovative

FIGURE 3.2 Macarena Ortúzar, *Really Actually Windy/R.A.W. Vol 1.*, I'Klectik Art-Lab, Old Paradise Yard, London. May 7, 2016. Photo: Johannes Birringer.

FIGURE 3.3 Anita Konarska, *Really Actually Windy/R.A.W. Vol 1.* I'Klectik Art-Lab, Old Paradise Yard, London. May 7, 2016. Photo courtesy Anita Konarska.

tree-led strategies designed to remake your sense of human-tree relations. *Branching Out* fosters a radical reconception of the ways we inhabit the world in relation to other organisms.

(Povall 2016)

Why not shift our attention to art and performance that makes visible what is, by nature or by design, often unseen or undervalued when working in a range of performance media, physical theatre processes and animated materialities such as the ones, for example, that Min Tanaka's students learn when they train in *body weather* techniques (in the landscape) or work on the farm planting rice? Why not branch out into shamanism and pataphysics, into discombobulated soundings, and what Konarska calls the "raw pieces" that can be felt, heard, touched, but not necessarily seen?

Shamanic rituals are quite common in the South Korean *kut* tradition, where dance and music blend but where the voice of the (usually female) shaman, or *mudang,* performs a high wire act between earth and the air. She intones repeated words, special phrases, and movements to mediate between the everyday world and another realm (of spirits and demons) with the aim of healing, protecting the community from evil, comforting the troubled. When I watch the *kut* dances and listen to contemporary musicians, like the group Jeong Ga Ak Hoe, who play classical instruments to accompany the dance, I am reminded how Ortúzar's performance alongside Guastallo's strings was an atmospheric occurrence. She walked the earth, with her wooden bones, and sounded out a vibrational sensation that connected us to energy landscapes, forces of molecular configurations.[2] In a shamanic sense, she connected us to spirits, and I felt the rods, the bones, the spines. She used her body weight and somatic sensations to develop, as a sustained improvisation, a certain "technology" of movement-design. I can interpret such design in an immanent, material sense, looking at the architectures of her labor and the social spatiality she created through her use of space, but also in a psychogeographical sense, through the imaginary narrative she intimated and made me listen to.

Here I am interested in how space is occupied, how atmospheres are elemental and induce specific affective responses, how atmospheres are engineered (performed) to direct or trick our hearing, seeing and sensing. I stumble trying to define kinetic atmospheres more precisely, whether to stay with the term as mood, ambience, weather and environment that conveys an aesthetic of emotional states yet lives and breathes and has its own forces and material presences, or whether I see them also at the same time as engineered, controlled, architected and designed, as in VR and games and AI worlds that tend to have their own algorithmic intelligence. And could exist independently from us, not caring so much about what we feel or do not feel.

Exploring this seeming paradox further, I move through four transborder productions I witnessed, two of which struck me as employing an aggressive loudness that descended upon me, squatting on me, whereas the other two seemed to hover more elusively, working through an almost sacred calm and an air that expands to another plane. I wondered what coastlines one might see in this pairing, and where the lines withdraw and so that site-contingent projects, like *Really Actually Windy*, come to clearly remind us of the precarious, much less accommodated side of contemporary performance. But first, let me return to the larger implications of this kind of listening to the atmospheric, which tunes me into different architectures of performance, and thus distinct accommodations of our becoming drawn, like the wind that moves and becomes trees and grass and spreads itself. This is an interest of mine (partly embedded in my own choreographic practice) that I explore here across a range of other manifestations: the "technologies" in question are not necessarily technical, digital, or software-based but methodical techniques able to conjoin human and non-human, organic and non-organic matter.

The choreographic is a vibrational and tactile occurrence – reminding us historically of significant endeavors to connect movement and architecture as practiced, for example, in Anna and Lawrence Halprin's "Experiments in

Environment" (the workshops they directed in California, 1966–71). Speaking of traces and source code, then, and the kind of *PASTForward* reworkings of original Judson Church pieces that White Oak performed in 2000,[3] I sense that contemporary postdramatic and immersive performances tend to reconceptualize kinetic environments and happenings which the Halprins had tried to investigate carefully in order to heighten perceptual experience through the intersection of component media. As became apparent in the exhibitions *Experiments in Environment* (Graham Foundation, Chicago, September 19 – December 13, 2014) and *Mapping Dance: The Scores of Anna Halprin* (Museum of Performance + Design, San Francisco, March 17–June 4, 2016), Halprin was aware of Kaprow's Happenings but sought a much more rigorous, interdisciplinary engagement with the media and materials incorporated (lights, slide projections, transparencies, found objects) into what she and her architectural partner called "collective creativity." Half a century after the Woodstock generation, and Situationism's critique of capitalism, the psychogeographical and political resonances of work that follows the Halprins' call for collective creativity may have fresh significance.

Halprin's choreographic scores are visual and instructional mappings of the temporal, spatial and participatory dimensions of the performances she imagined and created. It may not be easy to recognize her Situationist counternarrative to the politics of the era, or her critique of Warholesque psychedelic environments at the time (e.g. *Exploding Plastic Inevitable* at the Factory), but from her records of the workshop activities we learn of contact-based exercises with the environment, blindfolded walks or "departure rituals" instructing participants to isolate and then reassemble different parts of their bodies – an exercise I enacted a little while ago at the "E/motion frequency deceleration Choreolab" in Krems (Austria). The deceleration workshop links up to pertinent concerns debated today in the Netbehavior and Rhizome online communities.[4] I see such practice/research-driven debates as a very fruitful corollary to the performances, exhibitions or workshops I encounter; they provide a more dynamic narrative context than reviews or scholarly studies in print media. Deceleration and data mapping stand in an intriguing relationship with verbatim theatre or documentary practices used by Rimini Protokoll, Walid Raad's Atlas Group, Rabih Mroué and others (in the wake of earlier performance artists like Anna Deavere Smith) which seek to "channel" words, testimonies, interviews, confessions, remembered speech patterns and movements, etc. By slowing down the constant overproduction of data within the technological, political and socio-economic infrastructures that bind us, we shift our attention to time and touch, listening more carefully to such words and testimonies – and to their necessary translations between different media, bodies, voices and contexts. Given much of the transborder work we see today, could not translation also necessarily slow down intensities and saturations, and the curious frenzied rush to participatory performance? To what extent can a ritualized slowing down protect us from the fetish of speed and compression of time?

I experienced such *motion frequency deceleration* in Austria during a blindfolded walk across a hillside trusting a partner on whose shoulder my outstretched arm rested. Instructed to do so by choreographer Gill Clarke during the Choreolab, this walk was a departure ritual of sorts that brought greater attunement to breath, hearing, smelling, the careful touch of my feet on the ground – proprioceptions of being-together with the other person, of being "dividual." Deleuze distinguished between a disciplinary system that sees the individual maintaining a distinct position within the mass, where the dividual is interrelated, always formulated anew in relation to a network.[5] This jives well with Deleuze and Guattari's image of the rhizomatic (in *A Thousand Plateaus*), the spreading out of horizontal, heterogeneous growth. What I try to evoke here are examples of contemporary performances that grapple furiously with the transfusion of environment through the sonic, the complexities of transcultural tones, volumes, cadences, and textures (instrumental and bodily gestural), *the atmospheric* where one can discover within relational saturation (or subtraction) that "individual" elements in fact always traverse. There are no individual elements. They are, rather, "dividual."[6] And here we are, imagining a few performances that I need to visualize for you, even though you may not have heard of them or will not encounter them, physically and behaviorally. And what shoulder can I offer your hand?

Staying for a moment with artistic projects that perform data visualization, consider Catherine D'Ignazio and Andi Sutton's *Coastline: Future Past* project in Boston Harbor in June 2015, where 30 participants walked through the core of Boston tracing a route from the predicted coastline of the city after climate change to its history, as a way of physically understanding the future and past at scales that are difficult to see and comprehend (https://vimeo.com/160370905#at=0). Sutton encouraged the activists and performers – she calls them "poetic protesters" – to walk holding onto a rope and carrying stencilled messages, engaging in conversation with passers-by, and at key points climbing a ladder marked with the depth of the flooding scenarios projected for the year 2100. This left listeners, according to the story Sutton and D'Ignazio tell on the videos documenting the work, *under the water* at most locations.

This is an image to be savored – an audience under water, embodying calamitous climate change having traced the future (of a past) of their urban environment. When I earlier mentioned *PASTForward*, I was not concerned with recreation or Baryshnikov's homage to minimalist conceptual and instrumental work, but with a contemporary interest in interrogating the theatre's ill-equipped negative potential to resist an enduring reality, and a disastrous climate of performance within the automatism of the marketplace. In these weeks, as I write, the terror repeatedly comes to cities and villages in European countries – vividly imaged in the media and reframed in speculative social media as if it were new and radical – while wars continue to generate multitudes of refugees and migrants. The latter move across the waters, they crowd forward having risked their lives, while the locals ponder and suspect the (suddenly dangerous?) participatory nature of public space, their relations with the emancipated communities. "Wie sicher ist der öffentliche Raum?" asks a feature article on the "epidemic of suspicion" (Assheuer 2016). The paranoia in question is a defensive reference to Islamic fundamentalist militant terrorism (symbolized in the ISIS). More than that, it is a pernicious use of language that invokes biopolitics and shame about what Alan Read has called the "immunisatory" logic of the West (and the theatre).[7] Migration now is suspicious too. The notion of immunity, of being immune against an other (or an outside), of course conflicts with the phenomenon of the atmospheric, the swirl of movement, the excess of passage. In the following, I evoke transborder performances – in the context of the international theatre festival – that inevitably deal with cross-cultural translation and, at the same time, with the confederations of the incommunicable.

The ethereal beauty of the late Kazuo Ohno's hovering presence, or the consanguinity of Eiko and Koma's performances with watery or scorched landscapes is still alive in my memory. Their slowed-down metaphysical extrusions amaze when contrasted with the frenetic contemporaneity of Toco Nikaido's *Miss Revolutionary Idol Berserker* drastic pop culture send up – a crazy mashup of TV celebrity reality shows and high school musicals that wants to immerse its audience while facetiously protecting them from an onslaught of noise, lights, water, foodstuff and glitter.[8] We are not quite under water, but are given water-proof ponchos and earplugs, our seats covered with cling film. A cacophonous tribute to fan culture – *otagei* are geeky dance routines performed by superfans for their Japanese pop idols – this company of 25 performers practically envelops us in a projectile theatricality, with Akimi Miyamoto's video designs washing over the stage backdrops in a discombobulating, deliberately shapeless mass of incoherent graphics, and the dancers prancing around in an absurdist hyper-*kawaii* (cute) style.

The concept of the cute or small, however, is here turned anarchic and pummels us blind with these wilder flashes of troubled youth. For a moment in the beginning they remember tradition, sporting Kabuki masks, lanterns and parasols of Japan's old theatre, but soon they turn hyperactive and silly, a riot of noise, color, and nightmarish animation. As 21st-century Japanese disco drill the atmosphere makes a kind of pataphysical sense, perhaps inadvertently riffing on *Exploding Plastic Inevitable* and our own Euro-American idol-celebrity culture. Yet even more interestingly the environment self-consciously comments on trends in promenade and immersive theatre, surrounding us from all sides, with dancers shooting water pistols at our necks as we squirm in our ponchos, sarcastically celebrating a frantically choreographed study into relationships forged between performers and audiences. The elements encroach. Apparently the performance uses around 2000 props: cardboard signs, samurai swords, paper masks, Matrushka dolls, buckets of seaweed and water, leeks, ticker-tape, etc. Creator and director Toco Nikaido embodies the *Idol Berserker* intensely, on the edge of a darker chaos and excess that may not translate as easily as the banners proclaim when held up by the *otagei:* "NOMAL THEATRE SUCKS" [sic].

During the 2016 edition of the Wiener Festwochen Festival, Oliver Frljić's *Naše nasilje I vaše nasilje* [Our violence is your violence][9] offers a similarly spectacular work of performance art, mixing dance, visual choreography and electronic sound collaged into drastic physical theatre scenes that are meant to shock. Yet the images here, including religious symbols and references to rape, torture, terror, fascism, and Islamophobia, evoke an almost old-fashioned sense of a bygone political theatre aesthetic. Frljić, who was born in Bosnia and now works as artistic director of the Croatian National Theatre of Rijeka, was commissioned by the Berlin HAU Hebbel Theatre/Wiener Festwochen to devise the performance as a critical homage to Peter Weiss' novel on radical resistance, *Ästhetik des Widerstands* (1975–81). He had previously provoked attention with his performances of *Aleksandra Zec* and *Balkan macht frei [The Balkans set you free]* (both 2015) – the former dramatizing Croatian war crimes against a young Serbian girl and her family, the latter a more personal and intentionally stereotypical depiction of discriminatory policies present in every society. The main character in *Balkan macht frei* is Frljić's alter-ego, performing his struggle to meet/overcome the expectations placed on him as a director coming from the Balkans.

Watching the relentless stereotypical violence of *Naše nasilje I vaše nasilje*, one cannot but sense Frljić's overcompensating furor, trying to "explain" Islamic terror in the wake of a long history of Western colonial and religious terror, fascism and capitalist exploitation. He delights in attacking the hand that also feeds him now. The production floods the stage (in front of a back wall comprised of dozens of oil barrels) with refugees and prisoners, who at one point perform a hallucinatory trance dance in orange Guantánamo detainee uniforms, and in the next moment appear naked, with calligraphic Arabic inscriptions on the skin as if they had walked out of a Shirin Neshat video. Jesus descends from the cross to rape hijab-wearing Muslim women; the dancing Guantánamo prisoners now sit in a circle and torture the new "Syrian" refugee just arrived, while voiceover announcements request us to observe a minute of silence for the victims of terrorist attacks in Paris and Brussels. Then there is the attack against the audience (of NOMAL THEATRE) itself: *Am meisten schäme ich mich für Sie, das Theaterpublikum. Denn für Sie ist der Tod ein ästhetisches Ereignis* (I am most ashamed for you, the theatre audience. For you death is an aesthetic event).

The religious and political symbols function as a kind of ritualized semiotic – the iconography of signs of terror are meant to provoke shock on both right and left ideological spectrums, attack the violence of terror and show the radical illusions of consensus, complacency or "feel-good humanitarianism."[10] The West has no moral superiority at all in the current political context. A young director from the former East plays havoc with the left liberal mindset in the former West – a mindset that can be easily debunked now in light of the state's necropolitical violence which sustains contemporary racism as a primal ideology of global capitalism. With its blatant, fetishized violence, Frljić's heavy metal theatre can be called *plakativ* (in German), i.e. trotting out shrill political signs, shoving them into our faces and casually intermixing them with the archive of performance gestures that once resonated (e.g. mimicking Carolee Schneemann's iconic *Interior Scroll*, a hijab wearing actress in *Naše nasilje I vaše nasilje* pulls an Austrian flag from her vagina).

An aesthetics of resistance, if one were to follow Weiss' study of historical fascism and the worker's movement, would have to grapple with material phenomena of resistance (e.g. strike, protest, activist organization), not with terror as aesthetic choreography. The propulsive in-yer-face theatre tends to privilege its political content through spectacular gestures that heighten theatrical affect. It is the loudness of the affect that turns me off. I wonder whether current dance theatre productions pursuing a more abstract spiritual technique of ritual and more subtle tonalities make us listen differently. And, whether their withdrawal from political sensationism can shape other awarenesses, or mobilize other creative collectivities that are not whole or united and do not share the same cynical despair or political disappointment.

Formed by Marcos Morau in 2005 with artists from dance, film, photography and literature, the Barcelona-based company La Veronal perform a fascinating example of such an abstract ritual with *Voronia,* named after a geographical location (a deep cave in Georgia, Caucasus) which must have inspired the dark vacuum of the stage space.[11] The group conducted their research there, descending into a kind of bottomless pit that Morau compares to an empty center of gravity and to Dante's Inferno. "¿Qué o bien dondé está el Mal?" (What or where is Evil?), he asks in the program notes. As we enter the theatre, a young boy is seen standing alone in a vast gray horizontal stretch; then we note the cleaners that hover on the edges, slowly scrubbing the floor. Figures in black and white emerge in front of the long gray curtain: they twitch and contort in slow and fast cycles of mutated body-popping. Short volleys of clapping hands or slapped hips evoke chittering insects, sounds which recur often during the performance, and we hear them moving back and forth across the stage as words are projected onto the wall. Those words come from nowhere and disappear; only later did I realize that the sinister biblical refrain of gnostic sentences belong to the prophet Ezekiel and St Augustine. Their origin and destination are unclear, but they evoke an atmosphere, as the rear curtains part to reveal various scenes set behind glass: a dream surgery in an operating theatre where surgeons bend over a human body, or a boy trapped like a fly in a glass box, his hands bloodied. Dark clad monks pass across the stage in a slow procession. Animal puppets and real animals appear now and then in a hallucinatory landscape that draws on Christian iconography and prophetic allusions to the Valley of Dry Bones (Ezekiel; Figure 3.4).

Midway through the performance, the glass transforms into metallic elevator doors through which a table emerges, set for a supper at which the dancers gather. We are taken through repeated changes of "location" in a dance that is highly cinematic, allusive and allegorical yet without a guiding narrative (thus following the tanztheater collage technique but avoiding all literalizing or epic tendencies that once marked the theatre of Pina Bausch). The choreographic work is extraordinary, fast-paced and often disaligned, distorted and fragmented.

FIGURE 3.4 *Voronia,* La Veronal, Sadler's Wells London, 2015. Photo courtesy of Sadlers Wells and La Verona. Photo: Josep Aznar.

The twisting, cavorting bodies now move or sit rigid around the table at the absurdist supper, and we hear the sound of electronic and human babble, interwoven with fragments of classical music, sacred chorals and the percussive sound of the performers clapping hands. A woman (Sau-Ching Wong) shouts what appears to be a long tirade of abuse (in Mandarin) while pushing people away from the table. She tries to escape, but finds increasingly strange things lurking in the elevator every time she calls it: a military figure in riot gear, naked people flailing in a dynamic Rodin frieze; the young messiah boy being measured for a suit, an old man, etc. One dinner guest has quietly turned into a polar bear. Near the end, the boy pulls out a casket, as if he were in a morgue, discovering the corpse of the polar bear.

Choreographer Morau and his group tend to trust the associative visual imagery even if it risks being oblique. I find it gripping and also hypnotically strange because it is offered in such a detached, ritualistic calm, allowing me to think about the underground and religious (apocalyptic) violence, revelation and rapture in many new ways, complicating the weird surrealism with claustrophobic references and the blatant poetic beauty in biblical phrases that I had not expected (or long forgotten). The dancers are mesmerizing, pushing, pulling and stretching themselves into exquisitely grotesque positions, and El Greco and Goya come to mind when imagining the allusive structure of this performance that still resonates in my ears through the quick, tiny clapping sounds made by the dancers. Then again, during his last years when he painted the Black Paintings hiding locked up in his house, Goya, the painter of dark phantasms, was deaf – he could not hear. There are moments when I thought the choral passages, of dancers huddled together or piling on top of each other (as in the table scene), constitute quiet musical ruptures, brief instants of the eye of the storm, when all is quiet: in Japanese anime one calls these moments *udokanai* animation (stilled animation).

Akram Khan Company's *Until the Lions* is staged as a prophecy as well, not in a biblical sense but in the narrative mythological contexture this strong dance work evokes.[12] Khan's stature as a choreographer has grown consistently over the past decade, and his deconstructions and transformations of the codified languages of *kathak* are much talked about as he now clearly inspires a younger generation of artists who do not so much politicize their ethnic or racial bodies but push the creative potentials of their diversely trained corporeal instruments, blurring all boundaries between codes and abstractions, and classical, modern and contemporary performance idioms. As in the case of Morau's work, which relies on specific improvisation technologies he terms *Kova*, Khan has refined his aesthetic of collaboration through formal experimentation with multiple movement vocabularies that allow shifting (or queering) gender roles and masculine and feminine energies.[13] *Until the Lions* offers a beautiful, haunting example of how a performed gender identity can become self-divided or dividual.

Ostensibly a trio, with Khan partnering Ching-Ying Chien and Christine Joy Ritter, the work actually features seven performers. The choral presence of four instrumental and vocal musicians, who are placed in four corners of the circular stage (inside the massive circular Roundhouse, a famed rock arena in Camden Town) and move around the circle as well, defines the overall choreographic, kinetic and aural atmospheres of the work. The sensual atmospherics were very noticeable even before the dance began, for the space appeared misty, as if a fine sawmill dust hung in an air suffused with a strange scent. In front of us, the round stage designed by Tim Yip (known for his art direction in *Crouching Tiger, Hidden Dragon*) resembled the stub of a 30-foot-wide tree-trunk, sawn through just above the ground. Cracks later opened upward to create an uneven *mesa*, and through them smoke seeped up insistently, while Michael Hulls' lighting meticulously framed luminous enclosures and clearings (Figure 3.5).

The clearing, I take it, is for the gods that populate this dance drama, for the ancestors on the other side of the ritual curtain. As Morau reaches back to the Hebrew and Christian mythographies, this production is an adaptation of *Until the Lions: Echoes from the Mahabharata*, a retelling in verse of the *Mahabharata* by Karthika Naïr. Danced in an elliptical manner, it would be practically impossible to follow if one were not apprised of the tale. Khan had turned to the Hindu epic before (e.g. in his 2009 *Gnosis*) after having performed the role of the Boy in Peter Brook's controversial *Mahabharata* in the mid-1980s. But now he chooses to focus on the story of Amba, a princess abducted from her wedding ceremony by the powerful and obdurately celibate Prince Bheeshma, who then takes revenge on him by killing herself and assuming the form of a male warrior.[14] Taiwanese dancer Chien portrays the fierce Amba; Khan takes on the role of Bheeshma;

FIGURE 3.5 Ching-Ying Chien, Christine Joy Ritter and Akram Khan in *Until the Lions* at The Roundhouse, 2016. Photo: Jean Louis Fernandez.

Ritter (who trained at the Palucca School in Dresden) is a kind of animal presence, skittering and slithering around the clearing with intensity, a possessed figure of destiny who becomes the spirit driving Amba's revenge.

Bathed in a shimmering, sand-colored light on a giant slice of tree trunk, the performance envisions the world as a living organism and a continuum. My eyes travel with an inner and outer wind, as if rustlings and movements of plants, trees, things, landscapes, living beings, kangaroos galloping on all fours and supernatural actors combined into a collective whole. The trunk, with its rings and bark, becomes a platform for a strangely erotic mating ritual during which Chien and Khan embody Amba's attempt to persuade Bheeshma to marry her; she reaches to touch him and grasp him, yet he alternates between pushing her off and reciprocating, increasingly confused by transactions that we can also imagine as internal transformations. Later, the trunk becomes the battleground on which Amba, Bheeshma and their invisible armies rage against each other. They are watched over by the blackened, severed head of an old warrior that is mounted on a stick, and in the final scene the musicians join to throw innumerable long wooden arrows onto the scene as if preparing a funerary pyre.

In conclusion, after this bracingly physical, multisensorial dance, it is the sounding that lingers prominently. A score by Vincenzo Lamagna underlines the action: a low electronic drone with whirlwind percussion from Yaron Engler and impressive vocals from Sohini Alam and David Azurza who prowl the perimeter of the stage environment, joining the action from time to time. Most stunningly, they use the (amplified) tree trunk itself as percussion instrument, making it as ritually threatening and earthly as the pounding rhythms in Stravinsky's *Sacre de Printemps*, or shifting into lyrical, melancholy registers with Gaelic love songs (accompanied by guitarist Lamagna). Azurza surprises us near the end with his remarkable countertenor voice, enriching the piece's gender fluidity.

I cannot describe the sound of this dancework any closer, but it touches me on levels of experience that exceed the semantic or syntactical dimensions of the epic narrative or the movement enunciations. It is no longer solo *kathak*, but dividualized and disjointed, diversified through the collaboration with dancers and musicians working from other, sharable vocabularies. If we take *Until the Lions* as a post-authentic work that messes up its (inter)cultural sounds to the point where listening to performance is precisely challenged (and East and West are interwoven and hybridized to a point where the mythic text is perceived as invented and the ritual force only a pataphysical prank), then the juxtaposition of the ridiculous *Miss Revolutionary Idol Berserker* and the grave, overly sincere *Until the Lions* becomes stranger. The juxtapositions between abstract, expressive, political and pop certainly dissolve slowly, as well as the differences between the tree (with its roots) and the torqued grass, cunningly captured in the invitation photograph for *Really Actually Windy*. The relation of the post-authentic to sound (an acoustic transculturalism?) suggests that any notion of an essential identity or individuality is at odds with the rhizomatics of atmospheric exchange.

What then happens to accepted ontologies when movement techniques are no longer recognizable, words and translations fail, voice and music no longer demarcate cultural histories and spaces, and acoustic and visual relations drift apart? What kind of synaesthetic listening do we perform in the face of the incommunicable? One could argue that among the collectives of audiences, connotations of sounds, of languages, of gestures and sensorial phenomena will always be discerned, mobilizing a potential expansion of multiplicity. We are quite capable of somatic identification; the calm and slowness of the shamanic, ritual performance dimensions, which I addressed may qualitatively contribute to an enhanced perception of condividuality. Theatre always communicates even if it cannot grant immunity, then, and what I called, in the beginning, the less accommodated, site-contingent performance, whose dividual dispersion in space is more difficult to fathom, may in fact challenge the very denomination of site and identity position itself. For example, how do you resolve the paradox of an occupied theatre, a theatre under occupation, unable to move or unwilling to move, compared to the commonplace dispersed production and diasporic actors (and privileged migrants) within globalized economy?

Let us listen to the wind, one more time, in this case to a company called Iraqi Bodies – their incongruous or ironic name evoking a kind of "national theatre." They came to my attention when I was doing research on Middle Eastern performance (Noura Murad's Leish Troupe, one of the few independent and experimental theatre groups in Syria; the Khashabi Independent Theatre, a group that has just relocated to an old building in the historic Wadi Salib area of downtown Haifa from which the majority of Palestinian residents were forcibly expelled in 1948). All three companies were afflicted by war and occupation, to the point where Leish Troupe had to cease their creative activities for a while; Khashabi were homeless for several years after they founded the group in the occupied territories in 2011. Iraqi Bodies, founded in 2005 in Baghdad, had to re-form in exile in 2009 (Gothenburg, Sweden) after director Anmar Taha was forced to flee the increasing violence of sectarian conflict in Iraq. They kept the name Iraqi Bodies even as the company now included Swedish, Greek, Dutch, and other international performers and musicians.

When I saw Iraqi Bodies' *Possessed,* I was struck by its dark intensity, which connects it to other works in the field of contemporary dance theatre (e.g. Hofesh Shechter, Sidi Larbi Cherkaoui, Rachid Ouramdane), but it surprised me on an aural and kinetic level because of the ironic sensibilities that suffused the choral pattern of the work, giving it a ritual quality similar to the one observed in *Voronia*. *Possessed* is a world of sparse lighting, silence and near immobility disrupted by seemingly erratic, repetitive movements. The opening sequences are nearly invisible (under a dim red and a blue light spot one can only discern two flailing bodies). When the light grows, we seem to be in a smoke-filled landscape, as if after a fire, jutting white lines mark the space as if delineating an architecture to be built, or the contours of buildings that were once there. Then the chorus of nine dancers huddles together tightly (separated from a figure that lies motionless on a different spot in the empty environment), moves together, halts, moves, halts again, the only sound is that coming from the bodies themselves, their tiny steps, shuffles, molecular conjoined movements, later the whisperings and mutterings of incomprehensible words.

In a similar piece, *Vowels,* Iraqi Bodies uses a grouping of four actors (two older, two younger, perhaps a family) in an infinitely long motionless opening scene. The two men and the younger woman are wearing traditional dress from the Middle East; the older woman (mother) stands still, untouchable, in the background, Western-styled and beautifully dressed. Working in butoh vocabulary, the actors' motions, when they occur, are minimal, reduced, the "vowels" remain unspoken. There are so many steps not addressed that the audience constantly have to fill in the gaps: we cover up the silence (in the relationship between the former West and the Middle East). I imagine the interrelated shapes of bodies in motion – like the slow erosion of coastal lines – cast outside of the lawscape, papers and applications processed, welcomed as refugees in detention centers or make-shift camps, broken buildings, waiting quietly with ardor to escape a violent and proxy war, dreaming of a better life for their families ("is this Europe?"). I want to end on this note. Perhaps this is what it is going to be like: we are under water, and everything we hear is strangely muffled, the movements we perceive slowed down, in a thicker medium than air that offers more resistance, yet the echo waves travel. We, on the other side of the ritual curtain,[15] open our ears wider, to listen to the ancestors behind the glass: "I, body spittle, laughter dribbling from a face/In wild denial or in anger, vermilions" (Iraqi Bodies, *Vowels*).[16] The ancestors hope we understand, or everything will be lost.

Notes

1 Cf. Fernando Pessoa, "Sensationism" (1916) on http://arquivopessoa.net/textos/4103.
2 For a more scientific exploration of visualizing/sonifying the atomic dynamics of energy fields in intra-active audio-visual dance installations, see Mitchell, Hyde, Tew and Glowacki (2016: 138–47).
3 Cf. Ramsay Burt on such reworkings and what he calls "political disappointment" in *Judson Dance Theater: Performative Traces* (2006:186–201).
4 For example, Furtherfield's forum on accelerationist performance (April 2016), and also the empyre list investigation (July 2016) into feminist data visualization. See http://www.furtherfield.org/features/articles/accelerationist-art and http://lists.artdesign.unsw.edu.au/pipermail/empyre/2016-July/thread.html.
5 See Deleuze 1992.
6 I first learned of the notion of the "dividual" while attending a workshop with Yoko Ando, where she tested her *Reacting Space for Dividual Behavior,* an interactive dance created at YCAM (Yamaguchi Center for Arts and Media, Japan) in 2011. I reflected on this experience of multiple fluidities in "Gesture and Politics" (2012: 380–88). My own current work with DAP-Lab explores multisensorial environments that enable dividual proprioception, for example, we recently invited an audience of blind persons to touch, listen to and play with our dancers and their costumes in *metakimosphere no. 3* (London, April 2016); each of the visitors was led through the environment by a performer and invited to play duets with them.
7 See Alan Read's provocative writings on community and immunity in *Theatre in the Expanded Field: Seven Approaches to Performance*, London: Bloomsbury, esp. pp. 192–99.
8 Staged at, among other sites, the Festival Theaterformen, Braunschweig, June 2016; and LIFT Festival, The Pit, Barbican Centre, London, June-July 2016.
9 Text adaptation from *Die Ästhetik des Widerstands* (Peter Weiss) by Oliver Frljić and Marin Blažević; directed by Oliver Frljić; May-June 2016, Festwochen, Vienna.
10 In December 2015 Latvian theatre director Alvis Hermanis canceled his production *Russia.Endgames,* projected to open at Thalia Theatre Hamburg in the spring, as he felt uncomfortable with the theatre's political engagement for refugees, dismissively calling it a "Refugees-Welcome-Center."
11 Staged at Sadler's Wells, London, October 2015.
12 Staged at The Roundhouse, London, January 9–24, 2016.
13 In a video where company members present aspects of the technique, Kova is referred to as "geographic tools": https://www.youtube.com/watch?v=NA3ACE927qs; see also: http://www.laveronal.com/work/kova-%C2%AC-geographic-tools/. For Khan's transcultural body of dance work, see Mitra 2015.
14 The dramaturgy is more convoluted, as Naïr's book wants to foreground the voices or viewpoints of the epic's hitherto silent female characters. Princess Amba is abducted by a warrior, Bheeshma, but released when he discovers that she has a lover, Shalva. Shalva rejects Amba because she now "belongs" to another man; yet Bheeshma refuses to right the situation by marrying her, since he has taken a vow of celibacy. Amba wows to revenge herself on him, and after years of penance, Lord Shiva prompts her to self-immolation so that in her next life she is reborn as Shikhandi, a woman-turned-man who trains as a warrior and kills Bheeshma – who recognizes and accepts his doom in battle. Naïr's poem takes its title from an African proverb: "Until the lions have their own historians, the history of the hunt will always glorify the hunter."
15 See Raunig 2016.
16 Text quoted from the website introduction to *Vowels*: www.iraqibodies.com/body--identities-vowels.

4

THE OPENNESS OF THE ATMOSPHERE

Internal and external imagination

The digital and the not digital

Let us rehearse our common senses and uncommon senses, bringing attention to the sensorial and embodied perception approaches that underlie everything here. Looking at thresholds of perception – here in the context of performance environments – takes us across not only various materialities but also divergent modes of *physical thinking*, to use a choreographic expression. Such physical thinking strays across sensing actions – listening, touching, smelling, swaying, reaching, flailing, rising, tumbling, disaligning, falling, blurring – not necessarily based on visuality. This includes subliminal and peripheral sensing, rhythms of sensation, vibrations, proprioceptive and imaginary kinaesthetic relations, dancing as a kind of morphing, detailing the imperceptible (Figure 4.1).

FIGURE 4.1 Yoko Ishiguro, rehearsing with large stage dress and Kepler object, *kimosphere no. 3*, 2016 © DAP-Lab.

DOI: 10.4324/9781003114710-4

The *un-common senses*, neuroscience philosopher Barry Smith[1] proposed on a BBC Radio 4 program, are the ones we are less conscious or clear about – thermo or mechanoreceptor nerves in fingers, arms, or the spine, giving us tingling sensations; skin and hair sensing temperature and wetness or feeling textures, though not reliably; muscles and ligaments that "hear" how our anatomies, the bones, minerals and water in bodies, move along and stumble about; how organism and metabolism are comfortable or tensed, affected and afflicted by the environment as well as internal biophysical processes.

If we leave vision aside, for a moment, and think of the digital context of art and other popular cultural media of spectacular consumption, then we need to invoke numbers and the computational (*les numériques*) – counting, measuring, moving forward, progressing, recounting, retracting – and thus the rather large underside abstractions of what is considered our ubiquitously dispersed networked data world: our *digital ground* (cf. McCullough 2004). I like to think of it as under-ground, less visible on the whole, and yet more vital, like the underground motion of water we know but cannot *behold*. Criticism of the naturalized digital ground is also vital (cf. Kluitenberg's *Delusive Spaces*), and necessarily so, if the electronic networks of predatory capitalism's transnational governing agencies are not to remain ungraspable and incontestable (Kluitenberg 2008: 368). The delusional also forms an undercurrent of my references to the non-visual and non-privileged, as you lean a little forward and a little backward, cocooned inside VR wearing your head-mounted device, but missing your visible body, trying to remember where it is now. This chapter slowly opens out toward virtual reality art, the graspable and tangible kind.

The be-holding is a matter I actually want to connect to the hallucinatory, heightened sensuous pleasure of immersive aesthetics, as its inspiration has to do with the trans-sensory fluidity I wish to describe.[2] The trans-sensory experience of dance that most excites me is when I experience movement as ecstatic, as most physically concrete and yet hallucinatory, perhaps much in the way Antonin Artaud envisioned the poetic reforging of what does and does not exist. Or my Nigerian-British dance collaborator Olu Taiwo envisions the 5th, 6th and 7th dimensions in our practice. When I first met Olu, he taught me the "return beat" and encouraged me to believe in my capabilities of moving in complex rhythms. Olu is also passionate about basketball, just as I loved goalkeeping in soccer all my life, and we can intermingle the sports and the dance rhythms. Tap, tap, tap, tap, the sound of the ball hitting the floor. The digital ground is also the ground many of us walk on, even hold on to when developing control over corporeal rhythms, and if networked and social media are ubiquitous, so is our motion, our data traffic that is harvested. We are everywhere. We are tracked. It is now commonly agreed that in advanced post-industrial metropolitan sites our refrains of living and communicating are deeply infused with technical dispositions, as well as continuous capture. But I cannot quite see this as a *natural* condition, and thus it makes less sense to think of our era as post-digital.

I grew up in a nature environment – river valley, forests and hills in agricultural seasonal rhythms – and I return to it every summer. I am then offline over many weeks. During that time I am neither pre- nor post-digital. In his book on *Motion and Representation,* Nicolás Salazar Sutil speaks of the "ecstatic position of digital technology," as if it were outside of us to challenge or lure us, so to speak, to forget ourselves and the ground we walk on, to "move beyond a present state of kinetic being…moving in 'ex-stasis'" breaking with our established kinetic spheres (2015: 75). These spheres, I gather (Salazar borrowing the "kinesphere" from Laban), are topological movement ideas – physical movement understood in relation to basic properties of expenditure and recovery, like the breath that animates us.

The era of the digital has only just begun, slowly, and for many who live in rural areas and still follow this rhythm of the seasons feeling the moist shadows of the dawn and the light breaking through the branches of trees, being alive and knowing the ever-changing skies is not a digital experience at all. Being alive, as anthropologist Tim Ingold calls it, is not a matter of technology but of immersion in a continuously unfolding meshwork of relationships and wayfaring (moving through). The kinespheric idea, as Laban suggested, is dynamic and complex, and it helps us to imagine a movement of inner and outer dimensions or spatial articulation. And when I look up into the air, or listen to sound waves, the bird song and crickets of my native valley, I also improvise, I feel vibrations of after-images, dreams, memories – expanded sensorium of the hyperreal and of imagined exuberances.

Immersion, therefore, takes on a certain significance as a category of experience if the term is now often used in conjunction with Virtual Reality, with games and with engineered atmospheres that range from the architectural, built environment, the urban spectacles of light and consumerism, to the various intimate aesthetic experiences designed by performance and sound makers, fashion and interaction designers, or biotechnological and bioscientific experimenters. Prostheses create affordances that point to their *Umwelt*, enabling new qualities of existence, relationships, inhabitations. My own sense of immersion as a technique, however, is derived from movement and from dance, and to a large extent also from sports (soccer, swimming, running). When I was small, I dreamed of becoming an acrobat.

Sensory environments

First, I wish to argue that in the increasingly anxious times we now face in the post-Anthropocene, and a potentially catastrophic climate crisis, the aural and the kinetic senses connect us in a fundamentally elemental manner to the apprehension of physical space. We clearly feel the storms and the fires, we smell ashes and see dead trees crumbling. I suspect ritual is on the increase, as many people come to realize the depredations and deformations of planet earth. In such a time frame, performative ecoscenographies move closer to a ritualized exploration of atmospheres, climates and constellations. They confront bewildering and erotic, disturbing yet incommensurate, alluring atmospheres, inciting us to bathe and surf in them, feel their relational force, touch the wind, making us dream that we are enveloped or scattered or that perhaps flying on those same aerial currents that animate the kites of the Palawan Highlanders. Here is Tim Ingold's delicious retelling of the tale of their becoming like birds:

> The Palawan Highlanders of the Philippines have a very special relationship with birds, considering them to be their close yet ephemeral companions. Their understanding of this relationship is epitomized in the practice of flying kites. Constructed of leaves or paper with split bamboo struts, kites are regarded as the copies of birds. Flying a kite is as close as terrestrial humans can get to sharing in the experience of their avian companions. Playing the wind, flyers can feel with their hands, holding the connecting strings, what birds might feel with their wings. 'Anchored to the earth,' as Revel puts it, Palawan kite flyers 'dream in the air, their thrill equal to the splendour of the whirling of their ephemeral creations.' Becoming like birds, their consciousness is launched on the same aerial currents that animate their kites, and is subject to the same turbulence. Armed with their kites, the Palawans have achieved the precise reverse of what modern art historians have achieved with the concept of landscape. Where the latter have confined the world within the ambit of its surfaces, the former, reaching out from these surfaces, have regained the openness of the atmosphere.
>
> (Ingold 2011: 135)

I now sketch some constellations referring to creative research engaged with kinetic architectures for moving bodies in augmented-reality environments that approximate such "openness of the atmosphere." Open, affective environments must have existed since ancient times when sacred dramatic festivals took place in amphitheatres, sanctuaries, temples (e.g. Javanese *Wayang kulit* shadow puppet plays were staged in village cemeteries). Theoretical discourse on *atmospheres* is fairly recent, derived from philosophy (Sloterdijk, Böhme), cultural geography, spatial studies, architecture and legal studies (Thibaud, Pallasmaa, Zumthor, Philippopoulos-Mihalopoulos).[3] Even if the notion of atmosphere may be originally indebted to perspectives of geography, meteorology, physics and chemistry, it now often relates to architecture/design, to aesthetics and social psychology, thus to questions of how designed space surrounding our bodies affects our emotions and moods, or channels our perceptual embodiment (Figure 4.2).

When speaking of augmented reality in performance, and more commonly now in installations and aesthetic settings (e.g. museum exhibitions), we imply that the physically affective is amplified through technical means (sound diffusion, digital projection, lighting, VR, interactivity). Thus an expanded multi-channel sense of the choreographic is evoked: the experiential in motion, with various alchemical dimensions, corporeal and perceptual irruptions. With

FIGURE 4.2 Yoko Ishiguro, Azzie McCutcheon and Elizabeth Sutherland entangled with "large dress" in *metakimosphere no. 3*, DAP-Lab 2016 © DAP-Lab.

alchemical I here mean to suggest the *tekhne* of design strategies with its often quite wonderful sensory messiness. Augmented space enters us and our receptors receive many (often ambiguous) clues. The fullness of the real returns in such densely sensorial atmospheres of *performance rituals* we long to rediscover – immersion is craved, the sublime, the erotic and rapturous desired.[4] In the current experience economy, *immersion* and collective experience radiate. This could be delusive, of course, since we live in an age of hyperindividuation. The ritual side, even if it has little to do with the sacred or the spiritual, is always close by, like the collective jumping at a rock concert when crowds screech with wild abandon and joy. It is incipient in the flow of expansion and contraction, the breathing of a charged environment. Exploring such contagious material conditions, our DAP ensemble members have become builders.

In a recent rehearsal, after having suspended large amounts of delicate, sensual gauze and white fabric from the ceiling of a warehouse, attached to aerial wires that allow the fabrics to fly, I watched how dancer Yoko Ishiguro slowly emerged, like an amphibian creature, from under the fabrics, still shrouded by them, then extending the large dress so that it stretched out almost the whole length of the building (Figure 4.1). As we walked around, trying to disentangle with eyes, ears and sensory touch what was un-folding, a tiny whirring sound was heard coming from the cone-shaped origami object she held in her hand: a sound instrument, reflecting dimly the blue light that shone on it. We call it "Kepler" – named after the recently discovered 452b exoplanet and constructed by my design collaborator Michèle Danjoux out of the same polypropylene material as the costume for one of the dancers.

The costume, in turn, was inspired by an interactive architectural origami structure we had been asked to perform with by a group of architects.[5] Materials, in other words, transitioned and became transformed, from architectural animation to wearable, from conductive costume to sound-object-choreography – re-contextualized morphed kinetic characters and accessories. The sensory atmosphere thus impregnates aural and tactile experience while it implies movement, a "trans"-motion across.

Performing (with) architecture, then, is one of the sensory challenges I propose here for embodied scenography. How does scenography and movement choreography enjoin with spatialities both material and virtual? Do we even need to call such scenography immersive? A haptic feedback relation seems inevitable, when we speak of these different, animating materials that can be felt. Sensing and feeling spatial visuality constitutes one cycle of perception. However, multisensory processes, especially in regard to the proprioceptive stimuli (bodily position and inclination sensed in space through receptors in muscles, ligaments, joints, etc.) and environmental signals – and thus intra- and extra-corporeal awareness – go beyond visual immersion in performance. They are far more tactile, auditory and somatosensory (Figure 4.3).

The visual relation (as a feedback loop to oneself, to one's "own" body) in such immersion is unstable and not necessarily crucial, since performing (with) architecture or the environment is interfacial and implies an *outside* of our own bodies – other bodies and objects. Furthermore, when the dancer moves in the stiff polypropylene dress, two aural events happen. First, the synthetic dress itself

FIGURE 4.3 *Metakimosphere no. 2*: Vanessa Michielon performing with "Origami-iDress" designed by Michèle Danjoux, in front of {/S}*caring-ami* architectural structure by Hyperbody. Azzie McCutcheon moves inside foreground gauze [right]. Madrid 2015 © DAP-Lab.

creates sound: each time the dancer moves and tilts the dress, one hears a crackling or popping sound. Second, on her left arm she wears a sensor band: her interaction is meant to explore conductivity and sonic feedback (an aspect of our alchemical *tekhne*). A small wired metal sheet sits in the corner of the space, and as dancer Vanessa Michielon moves closer to it and eventually completes the conductive circuit, she elicits a sonic reaction. She is drawn into an inside-outside cycle. How, then, do we become ensounded in such orbiting?

Engineering atmospheres, flying kites

Becoming ensounded is a fundamental consequence of movement, of the "trans." If we invite our audiences to listen, and to touch, they will instinctively follow where the sound comes from and where it goes. They will follow the dancers, listening to them and the sound their costumes make, similar to the ways in which the dancer is drawn to an outside, the conductive exchange (the dancer is flying the kite, so to speak, dreaming in the air; Michielon also dreamt of giving birth to the little Kepler object – the birth-giving scene happening as a complete surprise to me and the others in our team). They also will be orienting themselves through the lighting, the changes in lit areas, the color moods. If they are handed an object, as Michielon did when she passed around her Kepler child, the audience will want to touch it, handle it, listen to it. If I give you a conch shell, you will hold it against your ear.

The theatre's relation to engineering of atmospheres is commonplace, having been adopted as a paradigm for such operations by philosophers or architectural theorists from Vitruvius to Gernot Böhme. At a recent conference – "Staging Atmospheres: Theatre and the Atmospheric Turn"[6] – the organizers claimed that within the current interdisciplinary *atmospheric turn*, theatre has presented itself as an heuristic paradigm in which the social, material and political elements of "atmosphere" are thought to resonate, albeit in an idealized manner. Yet if the theatre had been adopted as a paradigm for augmenting or engineering atmospheres for a long time, why then, the organizers asked, does it present such an acute example of the "affective tonality" of aesthetic experience in today's cultural obsession with audience participation (Thibaud 2011: 2014)? And why did Böhme's "The Art of the Stage Set" – suggesting that atmospheres can be engineered rather than just being contingent like the weather or the diffuse mood sensed in natural or urban environments – become a key text concerning the production and reception of atmosphere?

Speaking to the participants of the conference, I mentioned that my company had been creating installations for years, without having heard of Böhme and an "atmospheric turn" – scenography of course always having implied a basic concern for environmental composition. And did not Kandinsky and Schlemmer, or the Russian constructivists and cosmists, take us deep into the technical/spiritual dimensions of architectural tonalities? Did not Eisenstein, from his drawings to the stage designs and films, explore the challenge how to reconcile elemental sensuality with forms of logic and artistic abstraction to produce "ex-stasis," that nearly mystical foundation for aesthetic appearance? Did not Alejandro Otero, in 1950s Venezuela, pioneer a visual language of exhilarating rhythms in line and space, in urban architectures and through the colors of his public sculptures (*coloritmos*), just as Yayoi Kusama's color dots vibrated in the voids, their seemingly infinite horizons? Did not Hijikata Tatsumi delve into forbidden erotics and a meta-physical accretion of movement that made him ask dancers to imagine themselves melting or being made entirely from nerves, stretching everywhere, from back of the head to the ceiling?

Spatial performance, music in particular, powerfully generates affective tonalities and perceptual resonances that can link to metaphysical concepts such as the ecstatic. Böhme suggests that *ecstatic materialities* adhere to properties of *things,* and vibrant matter emerges along with what actors do or designers fill the stage with (1995: 33). There are many examples (in the European context at least since Appia, Craig, and Piscator; in the Latin American context one immediately thinks of Oiticica, Meireles and Neto) of theatrical spaces filled with tensions, their intensity contours tuned with uncanny, unnerving or soothing affect, with compelling rhythms, timbres, shadows and presences. Appia's staircases for Dalcroze's staging of *Orpheus and Eurydice* in the Hellerau Festspielhaus (1912) spring to mind, since Appia's concepts for "rhythmic spaces" and his radical ideas for environmental and indirect lighting (what he considered "creative light") were path-breaking for 20th-century design, influencing directors, choreographers and composers (from Bob Wilson to Kirsten Dehlholm, William Forsythe, Manos Tsangaris and Ragnar Kjartansson).[7]

Scenographic exhibitions at the Prague Quadrennial have revealed such affective tonalities in works of designers who do not just build sets. Installations such as Tomás Saraceno's *Biospheres* or Olafur Eliasson's *Mediated Motion* or *The Weather Project* have drawn special attention to lighting, color, air and liquidity of materials, whereas sound artists, for example, the Finnish group that created WEATHER STATION[8] for PQ '15, have been equally drawn to changing auralities subject to environmental conditions, harking back to John Cage's aleatory concept of music as weather, inspired by his study of Zen and nature processes. Observers have pointed to German stage designer Katrin Brack's recent productions, her minimalist yet excessive use of single materials – fog, foam, snow, confetti, balloons.[9] The use of stage fog as a special effect is common, but Brack's filling the space continuum with dense and uncontrollable fog throughout the production of *Ivanov* (Berlin Volksbühne, 2005) alters conditions, making the fog a performer, so to speak, thus requiring the actors to improvise with the aerostatic atmosphere, "weather" conditions as they evolve and change, hovering, drifting (Figure 4.4).

Such hovering presences, where atmosphere also *appears* uncontrollable, emergent and not engineered, evoke complex ontological and spiritual questions; the wildness of nature is perhaps harbored deep inside our skin and bones, our muscle and genetic memory, internal perception and emotional conditioning. Wildness is inside. It may be imaginary – the forests, hills, and the fog a scenography of ghost stories, myths and fairy-tales with which I also grew up and which is

FIGURE 4.4 Olafur Eliasson and Günther Vogt, *The mediated motion*, 2001. Water, wood, compressed soil, fog machine, metal, plastic sheet, duckweed (Lemna minor), and shiitake mushrooms (Lentinula edodes) Installation view: Kunsthaus Bregenz, Austria, 2001. Photo: Markus Tretter © 2001 Olafur Eliasson.

refreshed when I am exposed to the smell of moss, the touch of mist on my skin, or the aura of diffused light when sun beams flicker through tree branches, and hundreds of flies somersault. How are we to think, then, of trans-sensory hallucination as other than an effect of elemental materiality in contagious synaesthetic constellations? Is it a kind of delirium, an effect of sensations of the animate? The external becomes internal or always already has been internal, animistic like the fungal, the bacterial and other cosmic sides of the unconscious?

The production of such atmospheric-auratic conditioning through design, with the phenomenological impact on sensory perception, gestures and also ethical perspective (how to react to affective presences and interact with lurking environments), thus points to an assemblage of becomings already explicitly at work in Cage's *Lecture on the Weather* (1976), a concert which, on one hand, seemed un-engineered (with chance operations performed on Thoreau's *Walden* and *On Civil Disobedience*), while on the other gathered a storm of text fragments, images, music, voices and lighting. There was a *score* in Cage's aspirational lecture, and so we can also think of atmospherically orchestrated scenographies as audible-visceral environments that are not seen from the outside but are shared, taken in – they are meant to overtake us, perhaps in the sense of shamanic rituals where spirits are invoked to inhabit and possess us, emerge from within us and heal us with their powers. Or make us dwell in a shared circle of continuous community (along with the ancestral spirits). A more recent example of such aural scenographies is *Der Klang der Offenbarung des Göttlichen* (Volksbühne Berlin, 2014), directed by Icelandic artist Ragnar Kjartansson and scored for orchestra by Kjartan Sveinsson: a four-part opera without performers, set on an empty stage filled only with slowly moving hand-painted *tableaux vivants*.

Bodies of color and intercourses with weather

Possession rituals involve trances that are often induced by music, drumming, chanting and dance. In the visual anthropology of such trances we recall Maya Deren's powerful films she recorded in Haiti (*Ritual in Transfigured Time*, 1946; *Divine Horsemen: The Living Gods of Haiti*, 1977).[10] In her writings Deren speaks of the intercourse with Vodou possession ceremonies as transformational rituals that allow a de-centering of self, ego and personality (Nichols 2001: 8). Such de-centering I associate with Fayen d'Evie's suggestions about blundering and be-holding, when she notes that her installations for vision-impaired audiences shift sensory attentiveness to tactile and movement perceptions, encouraging "vibrational strategies" (d'Evie 2017: 48) for audiences that "lean" into the work differently through tactile and kinaesthetic entanglement. This leaning could also be considered central for *wearing* and incorporating (in terms of embodiment) exhibitions which choreograph audio-visual experience differently.

Here I am reminded of my experience of Hélio Oiticica's work in two drastically different environments. One was the major retrospective of his works at Houston's Museum of Fine Arts (*The Body of Color*, 2007), where I was invited to wander through an amazing consecutive arrangement of installations that encompassed Oiticica's paintings, reliefs, suspended three-dimensional sculptures, *Nuclei*, and *Bólides*, on to his *Penetraveis* (the architectural environments), and then the *Parangolés* (wearable color fabrics). The *Parangolés*, in particular, which we were invited to put on and wear, created energetic encounters with Oiticica's habitable cloth-objects (at the original opening in Rio de Janeiro worn by samba dancers from the *Mangueira* favela where Oiticia had studied the dance). I vividly remember the color-in-motion stimuli that also made these objects both actionable – in the sense that they had to be worn, manipulated and felt – as well as psychedelic, enabling me to drift off into the smell and sensation of the fabrics (their motion anticipating Oiticica's later sound-projection works of his quasi-cinema *Cosmococa*, 1973).

On the other hand, the Tate Modern (London) includes one of Oiticica's *Penetráveis* in a permanent collection exhibit seductively titled *Performer and Participant*. But the problem here is that I cannot participate but am shown a non-immersive history told through artifacts, display cases, photographs and documents. When I pass by an installation of Oiticica's *Tropicália* (Penetrable PN 2 "Purity is a Myth" and PN 3 "Imagetical," 1966–67), I do not even realize that

I might be allowed to walk in or lie down on the sand among the tropical plants, listen to the live parrots in the corner (recordings?), feel the textures, light and ambience being formed if I were alone with them. No one else realized it either on the day I was there. *Tropicália* seemed untouchable, an artifact on display.

Let us briefly meditate on being immersed alone, in a museum or gallery exhibit. I walk, stand, move around; sometimes I go to a corner, to look at the room from an angle, to see the whole, or I feel tired a little, moving to sit down on a bench if there is one provided. I sit and peruse. I try to look, then I look away. In installations that persuade me to be very active, I often comply, I like playing around and enjoying active engagement with the choreographic objects or interactive devices. I am demonstrative, then again I retreat and observe how others behave or let go of themselves to dream in the air. Then I step up, I let the installation "choreograph" me (see next chapter and my comments on Forsythe's *The Fact of Matter*) and I am thrilled. If the challenge is lesser, I am disappointed.

This sense of being "alone," at the same time, is perhaps a contradiction in terms, since immersion seems to imply a social and collectively experienced ambience, and also a permeable perceptional experience, beyond the confines of one's skin, a kind of intermingling, dreaming or co-meditating – an experience I have, for example, when sitting on cushions with others inside Houston's Rothko Chapel, surrounded by the painter's 14 monumental black canvases. After a while, I always close my eyes, as there is nothing more to see in this vast monochrome stillness, in midst of these immense floating anti-mimetic and anti-naturalistic dark appearances. But I am still aware of the others who had joined me, or were there before, meditating, contorted in some strange and beautiful torque (Figure 4.5).

FIGURE 4.5 Rothko Chapel, Houston, Texas. Visitors on benches inside the chapel architecture. 2007 © Getty Images.

Australian choreographer Fayen d'Evie speaks of stories "told through blindness, with a vibrational narrative that will blunder amongst macro propositions, with intermittent be-holding of sensory recollections" (2017: 42). Her notion of *be-holding* is fascinating as she explains grappling with fugitive, partial, hallucinatory, and tenuous "threads" in an aural environment. Or an environment of sculptural objects that are touched and grasped (not seen), through the word's etymological root (before the ocular), namely, holding, handling, guarding and preserving. Such be-holding, in the curatorial practices d'Evie describes, means haptic engagement and interaction in the way in which she admits learning from choreographic practices, and in particular William Forsythe's "choreographic objects," which she analyzes and then proposes to reorient through blindness (2017: 50–54). In her own work, for example, during a residency in Moscow, she developed *Tactile Dialogues* (2016) with choreographer Shelley Lasica, inviting participants to share actions handling objects or architectural structures, giving attention to materials, temperatures, textures and tactile surfaces – and also the kinaesthetic angles of navigations around the materials. The orientations and phenomenological instigations addressed in d'Evie's provocative work with "handovers" and intersensory translations point to an important new understanding of kinetic atmospheres – and what d'Evie calls an "epistemology of hallucination" (2017: 58) – in their unfolding through affective connections that may not depend on, or exceed, conscious apprehension. The repertoires in dance which I have become interested in over the past years (tanztheater and butoh inspired and yet also firmly invested in digital practices of augmented reality) have helped me to become more attuned to complex transitions between ephemeral performance, medi(t)ation and design (Figure 4.6).

FIGURE 4.6 *Tactile Dialogues.* Artist: Fayen d'Evie and Shelley Lasica, performed by Fayen d'Evie and Irina Povolotskaya. Medium: Performance, 2016. Exhibition: Human Commonalities at V.A.C. and Vadim Sidur Museum (Museum and Exhibition Association Manezh). Curators: Anna Ilchenko and Yaroslav Alyoshin. Photo: Evgeniya Chapaykina.

Changing repertoires/The suprasensorial scenographic

In the conclusion of this chapter, and in the next, I will link the series of kinetic atmospheres created by DAP-Lab over the past years to political questions raised by the METABODY project, wondering whether the aural and hypersensory dimensions I suggested are applicable in the sense in which Olafur Eliasson imagines the control of affective movement:

> Like the weather, atmospheres change all the time and that's what makes the concept so important. An atmosphere cannot be an autonomous state; it cannot be in standstill, frozen. Atmospheres are productive, they are active agents. When you introduce atmosphere into a space, it becomes a reality machine.
>
> (Qtd in Borch 2015: 93)

How productive is the concept? Eliasson's fascinating idea of the active agency of a "reality machine" is ambiguous, as he is aware of the contingency of materials, their psychosocial content, yet admits that productions of atmosphere are manipulative. They always have been, as we know from the history of movie sound tracks, for example, or the illumination tricks and saccharine muzak design for commercial shopping centers, or the erotic fetish couture of Jean Paul Gaultier. It is also suggested that atmospheres can be made explicit (say, if they are normative) by being ruptured, implying a Brechtian approach to becoming-atmosphere, pointing up its machining, its product-ness and not-inescapable social choreographic. The rupturing may be far more difficult.

METABODY took as its premise that bodily motion and non-verbal communication, understood as *changing repertoires of emotional expression and cognition*, constitute a fluid matrix of embodied knowledge in permanent formation. The in-forming diversity (and this of course also implies differently abled bodies, queer, gender fluid and trans bodies, inscribed with a diverse racial and class-based body schema), however, is being undermined by the impact of digital information technologies, which tend to induce an unprecedented standardization of non-verbal, bodily and kinaesthetic communication processes. The METABODY project claims that a sustainable diversity is also undermined by the ways in which *design*, in an expanded sense (incl. Robotics, Biometrics, Virtual Reality, Human Computer Interaction, Ergonomics and Artificial Intelligence), reveals a problematic attempt to simulate and repeat reduced repertoires of human emotions.

This suggests that the reality machine tenders repetitious and homogenizing scenographies (the aestheticized spectacular), whereas I am arguing here on behalf of the sensual, the poetic and the subliminal. All of DAP-Lab's kinetic architectures, from a first prototype produced in an intimate studio at Brunel University's Artaud Performance Center to large scale versions exhibited in Madrid, Paris, London, Durban, Trondheim and Montréal, have been composed to flow, widen repertoires, percolate and resonate, and also saturate, strain and create a sense of the vertiginous. The architectures are unspectacular and messy, to the extent that fabrics stretched across a void tend to create irregular shapes and folds, and participants hovering inside also tend to get entangled. Such *Metakimospheres*, as we initially called them, are atmospheric habitats staged for unrehearsed visitors who pass through them, listen to them and feel them, attentively, distractedly, blindly, trustingly or skeptically, invited to dwell for a time and do whatever they desire, including noticing the apparent absence of actors. Performers are in fact present, embedded in the kimospheres (cocooned inside the gauze and draperies) – perhaps one could call it co-present and co-extensive with the materials, exploring the tactile and sonic interfaces, as well as the visual moisture that animates the growth or stillness, scale and direction, the breath of their movement, their gauzeous entanglement. They may be nearly invisible but their incubating presence is felt. Perhaps they do not invite looking, as their role is not necessarily one to be looked at. Yet their bodily presence, and what I imagine to be this particular de-spectacularized *expanded choreographic*, is affecting the body of the architecture in-between or beyond the thereness (*meta* referring to such "between" and "beyond" notions of presence/atmospheric space) – in the duration and circulation of space-time. The habitat itself is the main object of contemplation: an immersive, immanent medium. The architecture's thereness can also be a wave, a flutter, touching bodies, thus its atmosphere might be that of the wind, electric energy, seismic force. Perhaps also of imagined water, oceanic medium, cetacean habitat, the non-airy atmosphere where marine counterparts of humans exert their fine-tuned sonar senses. Presences, their motion or stillness, animate the elastic veil-like gauze draperies that are suspended from

FIGURE 4.7 *Metakimosphere no. 1,* intra-active graphic projections tracking motion from performers inside white fabric architecture, DAP-Lab, Artaud Performance Center, 2015. © DAP-Lab.

the ceiling and slouch down on the floor. Presences breathe and their breath (as it animates their bodies) animates the architecture (Figure 4.7).

There are degrees to which I have asked our performers to be "there" in the raw, so to speak, to be unself-conscious specters or uncharacteristic characters in the landscape – fruiting bodies, if one were to use plant or fungal language, using their "chemical sensitivity" (cf. Tsing 2015: 45). Anna Tsing, discussing how smell is elusive but effective in attracting or repelling, posits that smell, unlike air, is a sign of the presence of another, to which one is already responding. She also, unsurprisingly, associates such understanding of relations among living things with "smell memories," and with nostalgia. I believe fabrics (silk, wool, gauze) also evoke such smell memories. They mobilize fantasy, in the same sense in which apparent stillness incites the sensorium and its threshold experiences, drawing attention to small tremors and activities that may go on underneath the skin. In the expanded choreographic, I would add, there is naturally also indeterminacy (as John Cage noted when he studied mushrooms and fungal growth), something always different in the here and now of relations. There is no real stillness as the breath not only moves, inhaling/exhaling, expanding/contracting, but also is audible. And in our *Metakimospheres* the biophysical, etheric sound is amplified. Microphones and special speakers (Fils Sound Film, suspended from the ceiling) live inside the metakimosphere. The elemental thereness of the environmental atmosphere includes the audience as experiencers who are "inside" the atmosphere, and the atmosphere is in them. *Meta*: through them. Both, so to speak, reciprocally make up the materiality of the interaction merger.

There is black porous gauze on the perimeter, and soft white veil net inside, and these insides-outsides – or "interskins" as Hae-in Song, one of our dancers, called them – are housed inside a darkened gallery-studio space (circa 12 by 12 m wide). This was the first envelope, for a test performance in London in March 2015. Later that summer, the second envelope was a huge auditorium in the Medialab Prado, Madrid. The envelope is to be developed further, envisioned as an architectural skin with its own properties and behaviors. The studio-envelopes were prototypes, in future years meant to grow into an architectural pavilion and scaled up. The first kimosphere had tighter, narrower skins. These skins are also a kind of costume that stretches close and far between, an entangling fabric that can be touched, grasped, stretched, squeezed, pressed, unfolded, pulled over. Here is how our collaborator Nimish Biloria describes the larger "HyperLoop" structure he is developing with his architecture team for the METABODY project:[11]

> The HyperLoop is an attempt to develop the world's first large-scale real-time intra-active pavilion structure, which pro-actively augments its physical state via real-time information exchange with its environmental, social and technical context. The structure geometrically takes the form of a loop, which can fully re-configure its skeleton in real-time. The entire loop is a fully dynamic structure, which harnesses generative movement, sound and light as an active mode of interaction with its visitors.

The HyperLoop is the very first iteration of the proposed large-scale pavilion structure and in its current format is a scaled version, outlining basic tactile properties of the proposed structure.

(Biloria 2015)

The Loop structure embodies material agency and performative dynamics that will reveal behavioral tendencies and exchanges with the flux and flow of the physical and technical (analog/digital) feedback context – the *RSVP cycle* as Lawrence Halprin once called it – namely the environment that surrounds body or "enters" body as much as bodies enter into and move through it. The term "hyperloop" is an apt title for such a concept. I am thinking of the visitor/experiencer as the embodied subject, but the architecture is here also understood as a hyperobject having physical states that are looming, precipitating, changing, reacting, as it became very clear to us when we collaborated with one of Biloria's teams that had fabricated a moveable origami wall structure.

The Loop scale model Biloria introduces only shows the bones and joints of the skeleton, as there is no agreement yet about the skin textures. The physical states of the skins may be subject to mechanical motor enactment of the legs and joints, the embodied artificial intelligence of robotics. Or they may respond to surrounding temperature and touch (if they are made out of thermochromic textiles), and manifest color changing abilities, say, based on levels of carbon dioxide in the environment and transformative light or sound transmission patterns. They may also be inflected by human, physical animation. The small-scale prototype of the Loop had motors on the knee-links, and some of the other modules of the future pavilion skin also are operational through motors that actuate the motion of the skin through small pulley systems, for example, in the origami wall with folded polypropylene sheeting which we worked with during the second *metakimosphere* installation in Madrid. Thus, engineering and a physical force dimension enter the environmental conditions, while the dancers who are present in the space wear costumes that may be connected into the spatial structures, exoskeletons and materials, thereby also affecting the thereness of the material architecture. The architecture, in this sense, can be likened to a choreographic object or sculpture.

The metakimosphere as hyperarchitecture needs to be discussed further, in greater depth. In this chapter, I tried to cross over between performance matters, choreographic and design processes, sensorial and experiential perceptions, to new forms of spatial and kinaesthetic composition stretching from the real to the virtual (VR). Choreographing atmospheric conditions, in AR or VR, is not something I had pondered before the METABODY project, although physical arrangements and sitings of skins (also considered as wearable costumes) are design challenges that thrive on the various contingencies implied by different materials, or situations where human and non-human agencies might be enfolded, as Bruno Latour explains in an interesting reference to the elastic connectors or filaments spanning across a large space to produce the shape of networks and spheres in Tomás Saraceno's *Galaxies Forming along Filaments, Like Droplets along the Strands of a Spider's Web* (2009 Venice Biennale). This is an amazing nodal work that also implies vibrancy, reverberations along the links and points of the network paths if a visitor shook the elastic tensors. That was strictly forbidden by the guards, Latour remarks (2011: 2), even as it is hard to imagine how the envelope (the institution), in this case, could prevent visitors from touching or bumping against the lines if they climbed through the threaded environment. It seems obvious to me that the spatial composer envisioned such encounters – the pickers, as Tsing calls foragers "dancing" in the forest, scanning for "lines of life" (2015: 243; Figure 4.8).

The untouchability (or encapsulation) of such an environment is of course a paradox, and we can see through it in Saraceno's spider web. But the question of the continuum is critical, and whether metakimospheres can also be enacted outside an inside I shall again address further below, while I hope to contribute to current thinking on performance scenography, wearable space, and immersive participation. Participation in an immersive choreographic has socio-political and ethical and not only aesthetic implications, and I see our practical work in performance as part of a larger investigation of "virtuosity and precarity" within

FIGURE 4.8 Tomás Saraceno, *Galaxies Forming along Filaments, like Droplets along the Strands of a Spider's Web*, 2009. Installation view at the 53rd Biennale di Venezia, Italian Pavilion, Venice, Italy, 2009. Curated by Daniel Birnbaum. Courtesy of the artist. Copyright Tomás Saraceno. Photography by Alessandro Coco.

the political realm of art: performative and critical empowerment after democracy (the theme of a research/workshop series I curated at Brunel University, 2015–18).[12] The "after" is meant to be a Menetekel, disrupting Belsazar's and neoliberal capitalism's feast. Immersion is a biopolitical technique, and can also refer to various perceptual illusions or phantom phenomena tested or generated by biomedical, neuroscientific and industrial (military, aeronautic, gaming, interaction design) VR applications. There are risks, getting caught in the spider's web.

The openness of the atmosphere is my primary concern here, teasing out some of the filaments or lines of life that stretch between real co-presences (also between humans and other, non-human materials) and virtual ones, and between roles or inter-changes of roles and mixed abilities. Even if one were only to begin to sketch an outline of performers as guardians and guides, for example, or to interpret some of the modes of entry delegated to immersive audiences (as players, test persons, pickers or voyeurs), with immersive acts having recourse to ethical implications and affecting the spatial/temporal circumstances of an installation, such an outline of immersant behavioral experience may not yield detailed methods of defining immersive technology or VR theory. I am not aiming at a comprehensive critical framework for the study of immersive theatre or performance, others have already begun to do so (including Gareth White, Adam Alston, Rosi Braidotti, Lyn Gardner, Rose Eveleth, Alan Read, Liam Jarvis). And Mark Hansen's *Bodies in Code* still has one of the most incisive chapters on "Embodying Virtual Reality" (2006: 107–37) – thrilling in fact as it examines in great detail Char Davies' path-breaking *Osmose* (1995), an early virtual reality artwork that many of us admired.

The scholarly rush on immersive performance, with Punchdrunk as its flame thrower, will surely subside. I am content to watch from the sidelines, not wanting to queue for tickets to see Anne Imhof's latest *Sex* installation "taking over" the Tate Modern Tanks (2019), as its promotion warned. Hearing about her award winning *Faust* at the 2017 Venice Biennial made we wonder how I would have behaved in the glassy German pavilion, perhaps desperately trying to avoid all the audience members rushing around with their cell phones filming the cool, beautiful models in Imhof's company. Yet, I hope at least to add in a small way to the literature on immersion design by evoking my basic ecological commitment to the poetics of distinct enactments: performances of technological assemblages which relate bodies to landscapes without exploiting or instrumentalizing either.

Whereas *Metakimosphere no. 2* constituted one of the stations within a much larger METABODY parcours constructed by the 11 partners of the project at MediaLab Prado (Madrid), enabling the DAP-Lab to forge a stimulating cooperation with the Hyperbody architects and with sound engineer Jonathan Reus (who designed the conductive circuits), *Metakimosphere no. 3* moved forward to dis-articulate the intelligent architecture and displace some of the material agents and leitmotifs of the Madrid prototype, inside a very large-scale environment staged in London during the following year (2016). The environment combined a small number of mobile "metakinespheres" built by Jaime del Val (which he had prepared to use outdoors during the *Metatopia 1.0.- Occupy 2.0* event in Madrid 2015) with a massive "stage costume" of suspended fabrics and cloth.[13] This main concept of a wearable space, underlying the third prototype, was owed to our questioning of the architectural utopia of a self-organizing or quasi-autonomous building-creature or artificial intelligence that exhibits binary behaviors (withdrawal or aggression). The latter, even if it appealed to the idea of a shapeshifting inanimate/animate character, made less and less sense the more we pondered the anthropomorphic or animal subtexts, if the {/S}*caring-ami* architects in their form-finding process indeed subscribed to the narrative of a defensive Minotaur-like "monster" whose heart needed to be unlocked (a narrative mesh up of *Edward Scissorhead* and *Alien* with JJ Gibson's proximal perceptual psychology and the notion of architecture as actant, in the sense of Bruno Latour's Actor-Network Theory). The inanimate animate in our design for installing the very large costume implied largely a shift from synthetic (hard) to soft materials and textiles. Or rather, a shift in-between, thus altering the manner of tangibility and also the ambient light qualities obtained by projecting film animations and ecological images through the cloth. One element in the suspensions of fabrics that interested us in particular was the manual operation of pulley systems that enabled gauze to fly up and down, or be lowered by visitors in whichever way they liked to change the overall shapes. The other main element of exploration was the tactile-auditory dimension of this kinetic atmosphere, put immanently to the test by a performance arranged one afternoon for a group of blind and vision-impaired audience members (from London Ealing Association of the

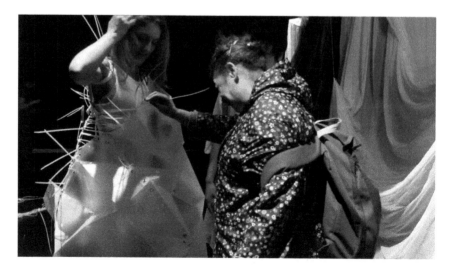

FIGURE 4.9 Vision-impaired audience member touches dancer's costume in *Metakimosphere no.3*, OrigamiDress design by Michèle Danjoux. London 2016 © DAP-Lab.

Blind), who were led through the environment by facilitator Karen Staartjes and by our dancers, or who moved through it quite on their own, following the sensory stimuli and engaging very concretely with all the performers and materials co-present. The visitors conversed with our performers, asked questions or gave feedback, touching costumes, pulling strings, listening to sounds emanating from the atmospherics. The visitors spent nearly two hours with us. It was one of the most rewarding audience interactions I have ever had, as I was invited to listen to our blind visitors' stories. The stories they wrote into the kimosphere (Figure 4.9).

Kimosphere no. 4, staged in the following year, introduces numerous stations in a large theatrical architecture where, for the first time in my scenographic work for the DAP-Lab, the real and the virtual (VR) merge and transition, the virtual complementing the real in a tangible way as these realities are layered on top of and within each other. Having introduced augmented reality before, in the sense in which theatrical design is expanded through technological media (digital projection, sound diffusion), I now want to introduce techniques of immersion that complement telepresence and VR technologies (computational/networked virtual 3D environments) through a paradoxical mode of reverse engineering. These are very simple techniques, one might even see them as humorous, comic: namely adding organic physical stimuli to the ocular VR experience of the visitor wearing their head-mounted device (HMD). The immersant disappears, so to speak, under their headset, having entered into a visually simulated world. The wearer can no longer see their own body. The goggles, handed over to the audience, one person at a time, are tethered to the computer (with a 2 or 3 m long cable which will have to be carefully held up by the guide). The wearer enters into the abstracted graphical representations in the VE, the HMD obstructing any view or orientation in the real-world space. The virtuality machine, in other words, is largely ocular, generating a first-person/gaming point of view (POV).

We ventured to create some irruptions, using very basic and organic materials. Initially though, the visitor encounters *augmented virtuality* first through an innocent poetry game, "Red Ghosts," playable at a small console in a corner, by each visitor so inclined, their feet stepping on real leaves and twigs. They may not initially realize it, but the leaves and twigs remain in the virtual-real space, and if the visitor moves on to the HMD and the VR station, they will feel them again, alongside other environmental effects that we create manually (waving towels to create wind, touching the body of the immersant with a tree branch, whispering into their ears, touching their elbows or feet, making their hands touch something wet and slimy, spraying mist, etc.). The layering invites different experiences for each audience member, creating a sense of their own emerging views as they construct a narrative that flows through the collective body of the audience. Many immersants are of course quite surprised to feel something touching them, when they imagined being in an unreal world.

"Red Ghosts" is also heard. An amplified voice, intoned live by sound artist Sara Belle, speaks about lemurs while typing the keys of an old typewriter. Belle literally types out the prose poem of the moonlit acrobats and accidental pilgrims of *Shadows in the Dawn* (a field report by primatologist Alison Jolly in Madagascar) – evoking an allegory of evolutionary migration, over millions of years.

> The word lemur is derived from the Latin for ghost, perhaps so named because of the lemur's nocturnal habits. What makes a lemur a lemur rather than a loris or a tarsier?
>
> Some scientists believe the difference is based on certain details of the lemur's inner ear. Others think the matter is much more simple, believing that a lemur is a lemur because it lives on the Red Island.
>
> And of all prosimians, only the lemurs dared to step onto the floating logs and branches and raft across the Mozambique Channel. Today there is an immense dent in the coastline of Kenya where Madagascar once was – for it was not always an island, but once was part of the continent of Africa.
>
> And then a very long time ago, 175 million years ago the continent of Africa broke. One hundred thirty five million years later, the lemurs floated across the water on logs, branches, branches, perhaps even thick mats of seaweed, to reach the island. Earlier the journey may have been just a hop, for it takes a very long time for a piece of land to unhitch itself from another.
>
> Madagascar moved away from the east coast of Africa at the rate of about an inch a year. When the lemurs began their raft trips, Madagascar had drifted 231 miles from the African shore. It was a big trip, and many other mammals did not go. Elephants would have been too heavy to raft across, but what about wild dogs or lions? Most mysterious of all why, among all the primates, were the lemurs the only ones to dare the journey?
>
> It was important for the lemurs to be the only ones. On the continent of Africa and elsewhere in the Northern Hemisphere where fossils of ancestral lemurs have been found, they were constantly outwitted by larger-brained primates such as monkeys and apes. All the prosimians – the lemurs, tarsiers, lorises, pottos, bush babies – were forced to stay in an nocturnal existence, giving the day to the brash upstart primates. Without competition from their more quick-witted cousins, the lemurs had the run of the island. Only on Madagascar were they free to move out of the night and explore the dawn, and the day, to live in groups, to build nests, to find new habitats. As the island moved farther away from the coast of Africa, fewer pilgrims attempted the crossing. Staying behind with the monkeys and apes were salamanders, newts, toads, poisonous snakes.
>
> They all missed their chance and never made the journey to freedom.[14]

Shadows in the Dawn reflects on migration, on leaving a continent, with Jolly having worked for decades in the forests of Madagascar, describing lemurs leaping through tamarind trees (the only place on earth where the animals, all 30 different species of them, live freely having managed to survive after Malagasy broke off the Eastern African continent and drifted away, with the lemurs following onto the island). Slow time/slow space was pertinent for the extenuated experience we had devised for the environment of *kimosphere no. 4*. The atmosphere and the audience are the reality machine; the audience produces a sense of immersion for themselves, tuning into (or out of) a forest of sensorial stimuli they instantiate into their immersive experience of the installation. Sound and tactile materials move this kinetic poetry, disseminate it around the architecture of the whole, with voices, electronic sounds, echoes, processed natural sounds, distorted crackles and hisses, lights, mists, colors and moving textures. The eight-channel installation, with each speaker shrouded in a mosquito net suspended from the ceiling grid, maps a metaphorical forest of ghostly presences in deep red light (three dancers, wearing masks, are hidden inside, still or barely moving), with dense layers of a sound-in-motion that is experienced by visitors while moving around the forest of speakers – the micro-polyphanies in fact only audible if they move across and between the nets, listening and absorbing. There are also stations on the perimeter: a VR-headset (goggles) and five lighter cardboard 3D headsets (with inserted iPhone); an igloo-like Soundsphere where visitors crawl inside to explore a GSR biosignal interface (listening to galvanic skin response turned into sound); and a coral reef sculpture where they can lie down and float inside a deep sea digital projection that percolates over a synthetic origami architecture. The coral reef sculpture is the old repurposed origami wall, built by the Hyperbody architects. In *metakimosphere no. 3* it was hoisted up like a sail, floating under the ceiling as if hovering in the wind above a huge ship (Figures 4.10 and 4.11).

FIGURE 4.10 *kimosphere no. 4:* dancer Yoko Ishiguro, standing still inside one of the eight ghost speakers; the coral reef is on the left, and sound artist Sara S. Belle performs in the background right. The skeleton of the Soundsphere is visible in the far back. London 2017 © DAP-Lab.

FIGURE 4.11   Visitor floating inside coral reef and watery projection, *kimosphere no.4*, 2017 © DAP-Lab.

Now it lies on the floor, stranded, with underwater ocean film projected onto to it to give it a liquid appearance. Visitors are invited to lie down inside the coral reef and dream about floating in the ocean. Many of the audience visitors do lie down, some staying for many minutes. The ritual-communal aspect of immersion and participation is an important concern, otherwise there would be no reason to experiment with these forms of interaction. Atmospheres of suprasensorial design suggest a scenographic strategy involving the audiences to step inside and come closer, touch, listen and act in greater intimacy with unfolding actions. These are also unfolding imaginations. The idea of a coral reef may or may not be convincing, but we suggested it. What our visitors dream, we do not know.

The *kimospheres* are living, breathing spaces; they are currents felt through sonorous, tactile connections. One is corporeally present in them, moving through their *Stimmungen* (the German word *Stimmung*, similar to *Atmosphäre*, implies in its etymological origin also *Stimme*, i.e. voice, an acoustic experience, a tuning), perceiving-listening to the relational, dynamic flows, such as Sara Belle's clicking type writer keys, and her words from the poem of the red ghosts. The surfaces and media require a creative investment from the audience, particularly obvious in the case of the VR "accessories" that need to be worn.

From projection to immersion – it is not a big shift as digital projections are a part of the installation architecture and also part of its lighting. 3D film or VR remains a cinematic projection medium, yet it has enhanced its plasticity and the illusion of absorption (of the viewer feeling being inside rather than looking from the outside in). 3D interaction designers argue that such absorption – and what our collaborator Doros Polydorou refers to as the perception of being physically present in a non-physical world – relies on the *plausibility illusion,* namely that you are not only using your body to perceive in the way you normally do, but that the environment believably responds to your actions to make you think it is real. DAP-Lab's research on formative and wearable space,[15] on mediated and yet highly visceral environments that are not constructed in a stable form but evolve through movement, provides the basis on which I propose to look at current ideas about immersion-dance, perhaps also questioning those very notions of *plausibility* (since they are to some extent ocularcentric). What exactly needs to be plausible in a virtual environment?

The kinetic, I suggest, includes motion of light, pixels and graphic projection, diffusion of sound waves, energy fields, color fields, implausible edgespaces and anomalies, objects that can be touched, handled, prodded and dragged, thus many different forms of embedded *motion sensing* which result in environmental reactions. Augmenting the virtual with analog technics (such a touching a visitor's cheek with a leaf of a tree branch) may not sound particularly exciting, but from our observations we gathered that immersants are surprised, sometimes electrified, sometimes shocked, not having expected such a sensation in a synthetic world that strives to be objectively "realistic" but is obviously not. It tries to have a compelling sensorimotor correlation in the participant, but even a couple of steps beyond the 2.5 by 2.5 m parameter of the VR sensor camera set-up (i.e. beyond the boundaries) will bring up a blue or green grid, a border – and the illusion is gone.

The idea of choreographic wearables, carefully augmenting the virtual and asking performers to be guardians and guides, helping the visiting immersant to feel comfortable and at ease inside the virtual reality that seeks to confer reality to the immersant's perception system, is here conjoined with the exploration of aural scenography. This is quite important – as in all filmic and sonic installations, namely implying a material-sensory audition filtered through a particular fashion design. The fashion design aspect arrives through the accoutrements and the illusory/fictional environment that is created in the kimosphere: making costumes, architectures, analog and digital accessories immersive (not just what is seen inside the HMD) and thus wearable. That I am even daring to use the term fashion here is owed to my collaborator, Michèle Danjoux, who has come into performance design from her fashion and garment design background as well as her experience in art direction. For a while, we both used the concept of "wearables" rather than costumes, yet Michèle's audiophonic garment constructions in particular have larger ramifications than either term, wearable or costume, can suggest.

This *expanded choreographic* materially reproduces itself even when there is only breath (internal movement). Breath not only moves space – inhaling/exhaling, expanding/contracting – but also is audible. In all *kimosphere* installations, as I mentioned earlier, the biophysical, etheric sound is amplified and thus the breath is animated environment. It is the forest polyphonics. The critical exploration of this breath would have to focus on the choreography and the virtual forest, how the installation includes intimate personal (meditative) resonances derived from the floating "coral reef" and the "Red Ghost" poetry game. Each immersant dreams their own floating (Figure 4.12).

FIGURE 4.12 Visitor enacting/embodying what she perceives inside "Lemurs" forest interface with VIVE goggles, conducted by Doros Polydorou, *kimosphere no. 4.* London 2017 © DAP-Lab.

Once the visitors enter the simulated forest via HMD goggles, *kimosphere no. 4* straddles two kinds of atmospheres, real architectural space and virtual computational space, both actuated through the same tactile narrative, neither perhaps all too plausible. We cannot know whether our visitors pick up the evolutionary tale of the lemurs migrating from Africa to Madagascar, or whether they at all ponder the darker ironies of Brexit. The critical aspect for us is the immersant's sensory participation, letting the resonances of real and virtual spaces become rhythmically entwined. Madagascar or England, floating islands. The occurrent gestures become reciprocal: pushing the kinaesthetic into a perceptual virtuality (VR) that so far is largely contained in the visual. The ergonomic challenges with VR headsets are well known; such accessories are tethered with thick cables to computers, thus a visitor putting them on has to be helped by the guardian conductor, and this role is very beautiful, I believe, as it has a deeper ethical resonance. I have observed guardian behavior in a number of workshops, and never fail to be amazed at the delicate, gentle care with which the guardians tend toward the well-being of the immersant, who may have laid down on the floor, pointing, searching, trying to get up and so on.

This inter-change also feeds the virtual "play" back to the corporeal, pouring it back into the player's gestural action (Figure 4.12). The kinematic is the challenge for a VR scenography which does not insulate/isolate the immersant or focus on visuality but allows for an expanded synaesthetic entanglement where imagined full-body perceptual virtuality feeds back into the kinaesthetic.

The momentary insulation from other visitors or friends, during *kimosphere no. 4*, turned out not to be a problem: everyone seemed patient, waited for their turn, observed, chatted and commented upon one another's be-holding, the "choreography" of walking into the lemurs' forest, flying up trying to catch a glimpse of the moonlit acrobats.

Notes

1 Barry Smith, with sound artist Nick Ryan, "The Uncommon Senses Radio Series," BBC Radio 4, March 2017: www.bbc.co.uk/programmes/b08km812/episodes/ player.
2 My writing is based on a lecture I gave on the subject of sensorial techniques of public engagement (engaging audiences or publics through immersive methods), and although I reflected mostly on my choreographies, public workshops and installations, I also drew inspiration from d'Evie 2017 and her manner of be-holding. Some of the ideas raised here will be followed up in the next chapter of the book as well.
3 The German philosopher Peter Sloterdijk 2004 has devised a philosophy of *spheres* and *envelopes,* contributing to the current interest in atmospheres and Gernot Böhme's aesthetics, much as Andreas Philippopoulos-Mihalopoulos' study of "lawscapes" as atmospheres draws attention to embodied social and political norms in the conflict between bodies "moved by a desire to occupy the same space at the same time" (2015: 3).
4 This is immersive experience associated with ritual and the sublime. Examples from the recent history of music and sonic art, relating to this ritualistic dimension, are too numerous to cite, ranging from Xenakis, Paik, Lucier, and Stockhausen to drone and noise artists (Niblock, Fusinato, Ikeda) and sound installation artists such as Cardiff, Heimbacher, López, Wollscheid, Leitner, Suzuki, etc. Brandon LaBelle has written on *sonic agency* as a mode of resistance, commenting also on its phantasmic elements and its insurrections (LaBelle 2018). On occasion, the dramatic reach for the sublime crumbles into the solemn pathos of the quasi-religious (e.g. Rirkrit Tiravanija's *mise en scène* for Stockhausen's *Oktophonie,* staged at New York's Park Avenue Armory in 2013).
5 I am referring to DAP-Lab's cooperation on the METABODY project with architects from the Hyperbody Research Group (TU Delft) who had devised {/S} *caring-ami,* a computationally generated origami pattern based surface with integrated lighting, motion capture and robotic actuation. The {/S}*caring-ami* team (Anisa Nachett, Alessandro Giacomelli, Giulio Mariano, Yizhe Guo, Xiangting Meng) gave us the polypropylene materials to create new wearables (costumes and sound objects such as Kepler and Accordion). Michèle Danjoux's ideas for conductive wearables and proximity-sensing performance had evolved from her work with Jonathan Reus during the e-textile lab at STEIM (October 2014), and my scenographic sketches for "kinetic atmospheres" evolved in March 2015 during the first public presentation of *metakimosphere no.1* (with Azzie McCutcheon, Yoko Ishiguro, Helenna Ren performing). Jonathan joined us in Madrid, and the dancers for *metakimosphere no.2* were Vanessa Michielon, Azzie McCutcheon and Miri Lee. For the 2016 presentations of *metakimosphere no.3,* DAP-Lab has invited additional dance artists to join, including Helenna Ren, Yoko Ishiguro, Aggeliki Margeti, Waka Arai, Elisabeth Sutherland, and Sasha Pitale. Visual interface design was created by Chris Bishop and Cameron McKirdy, with documentary filming and production support by Martina Reynolds. Initiated in Madrid (2013) by a collaborative network of arts organizations and research labs (www.metabody.eu), METABODY posited the rethinking of perception and movement away from the mechanistic and rationalistic tradition, and thus also the dominant Western tradition of visuality or ocularcentrism combined with formal and systemic "built" environments and protocols that take certain embodiments for

granted. METABODY was coordinated by Jaime del Val (Asociación Transdisciplinar Reverso), with 11 primary partners including DAP-Lab, Hyperbody, STEIM, InfoMus Lab, Stocos, Palindrome, K-Danse, and Trans-Media-Akademie Hellerau (See Birringer 2017).

6 Queen Mary University of London, December 8–9, 2017, coordinated by Martin Welton and Penelope Woods, in collaboration with Ambiances (Réseau International) and the *Journal of Sensory Environment, Architecture, and Urban Space* linked to Ambiances (ambiances.revues.org).

7 See Brandstetter and Wiens 2010, especially Birgit Wiens' chapter on rhythmic movement and "Kreatives Licht" (223–54). Appia's spatial experiments at Hellerau were conducted alongside Émile Jacques-Dalcroze's eurythmics; the ideas for "gestaltendes Licht" (creative light) were implemented by Russian designer Alexander von Salzmann. See also Beacham 1993: 53. I am grateful to Wiens for a lively exchange of letters regarding a Lacanian phrase about eyes/gaze and images as dissolving meshwork (*Rieseln einer Fläche*) which she used in regard to "light looking at us."

8 The Finnish Weather Lab, curated by Maiju Loukola, with Heidi Soidinsalo, Antii Mäkelä, Kristian Ekholm, Elina Lifländer, Nanni Vapaavuori, Antti Nykyri, sought to highlight the role of sound as scenographic material – sensual, spatial, performative and unexpected. PQ curator Simon Benham spoke of the *Weather* section as giving room to *wild spaces* of superimaginary processes.

9 Cf. Joslin McKinney's lecture "Scenographic Atmosphere and Spectatorship" at Staging Atmospheres: Theatre and the Atmospheric Turn, Queen Mary University of London, December 8–9, 2017.

10 I already mentioned Böhme's notion of the "ecstatic" above. A recent issue of *Performance Research* (24:6, 2019) investigated the subject of animism and especially non-Western metaphysical traditions and practices that enter forces and energies otherwise, traveling in ways that do not relate to conventional movement. Another approach to the ecstatic is offered by Haein Song, a contemporary digital choreographer who has practiced for many years in *kut,* the traditional Korean shamanic ritual performance. She recently completed a series of works that intermesh the traditional and the digital, and in her writings she describes the ritual techniques (*mugu*) deployed to achieve the desired collective healing and well-being effect of the practice: *Ecstatic Space: NEO-KUT and Shamanic Technologies*, PhD thesis, Brunel University London, 2018.

11 Nimish Biloria, email to all co-organizers, preparing work-in-progress presentations at the International Metabody Forum in Madrid (July 2015), June 17, 2015. Biloria works for Hyperbody, Digitally-Driven Architecture, Department of Architectural Engineering & Technology, Faculty of Architecture, TU Delft, the Netherlands.

12 The concluding laboratory "Shadows of the Dawn: Migration and the Indeterminacy of Community and Immunity" (April 18 and June 13, 2018) was inspired by the political writings of some of the activists and artists I invited to the Series (http://people.brunel.ac.uk/dap/ResearchSeminarSeries.html), but it was also motivated by the alarming refugee crisis we experienced in Europe at the time, to a great extent caused by ongoing wars and conflagrations in the Middle East and Afghanistan but also severe economic disparity in many parts of the world, especially in Africa. The workshop was set up to probe troubling interpretations of the increasing unrestrainment of capital, and its impact on all social-economic, cultural, creative, and educational sectors in the developed world. The sustainability of democracy in the developed world (where I live and work) is an urgent theme for those of us in the performing arts/creative field becoming intensely aware of the multiplication of realities (virtualization; networked infrastructures, militarized societies, migrant communities and diasporas) and growing depoliticization of culture and art. Our main objective is to articulate various perspectives against the emptying of politics (in favor of the aestheticization of pure ideology within the context of precarization and the

operations of unknowable information technologies). In regard to my introduction to my own practice, I wish to question whether *sensorial-participatory performance and installation* have a relevance to questions of political movement and survival, and to research into *bodily practice* formed through and forming larger social frameworks organized by the *movement of bodies*, i.e. social choreography as recognized in migrating labor, social practices, cultural representation or interaction with technology. See also Birringer 2019.

13 See the image on the book cover, with Ghanaian dancer Elisabeth Sutherland wearing NailFeathers Dress, designed by Michèle Danjoux. For *metakimosphere no. 2*, see Birringer 2016, 2018, and videodocumentation at: https://www.youtube.com/watch?v=b0t4jKH3SBI. For *Metatopia 1.0- Occupy 2.0*, see https://www.youtube.com/watch?v=KKo5I9B0VWA#t=750. For an overview of the METABODY project, see the documentary produced by STEIM: https://vimeo.com/117743950; for *metakimosphere no. 4*, see https://www.youtube.com/watch?v=0aIW6Klfm1g.

14 Quoted from the program for *kimosphere no. 4* (2017) and derived from the fieldnotes by Alison Jolly (transcribed in Kathryn Lasky's *Shadows in the Dawn: The Lemurs of Madagascar* (with photographs by Christopher G. Knight), San Diego: Harcourt Brace & Company, 1998).

15 Cf. Danjoux 2014 and Birringer 2010. For the *kimosphere no. 3* staging, arranged for vision-impaired visitors from the Ealing Association of the Blind, see https://www.youtube.com/watch?v=5DdAcv37jmc. For Polidorou's reflections on his VR design and participation in *kimospheres no. 4 and 5*, see Polidorou 2021.

5

TRANS-SENSORY HALLUCINATION

The extended choreographic

In Anna Lowenhaupt Tsing's chapter on "The Life of the Forest" she takes us on a walk through precarious, debilitated landscapes, adventurous nevertheless in all their unstable assemblages and disturbed ecologies, filled with their multispecies entanglements, plant networks, nematode stories and multispecies polyphonies (2015: 155–63). The nematode story of these small wormlike creatures, who had been a minor pest for pines in the United States but when traveling to Asia became very lethal (there the pines were unprepared and vulnerable), is quite peculiar and reminds me, in reverse, of the lemur story of peaceful migrants we told in *kimosphere no. 4,* migrants fleeing a habitat perceived as too threatening and restrictive. The lemurs settled peacefully on the Red Island.

Pine wilt nematodes, Tsing reveals, are unable to move from tree to tree without a carrier, and the help they seek are pine sawyer beetles they can hop on, even though this transfer voyage is far from easy. The nematodes must "approach" beetles in a particular stage of their life cycle, namely just as they are about to emerge from their piney cavities to move on to a new tree.

> The nematodes ride in the beetles' tracheae. When the beetles move to a new tree to lay their eggs, the nematodes slip into the new tree's wound. This is an extraordinary feat of coordination, in which nematodes tap into beetles' life rhythms.
>
> (156)

The way Tsing tells the story of how the effect of nematodes' attacks on pine trees impacts on the tracking of matsutake (and how the mushroom dwells in dilapidated pine forests in Japan, Northern Finland, China, Canada, Oregon)

DOI: 10.4324/9781003114710-5

makes one remember Jakob von Uexküll's *Streifzüge durch die Umwelten von Tieren und Menschen* (*A foray into the world of animals and humans,* 1934) and his stupendous investigations into the POV, the world-view so to speak, of a tick, suspended in the sunlit bush and waiting to smell its prey, onto which it lets itself fall (Giorgio Agamben, in *The Open*, reimagines this fall and the infinite variety and conjunctions of perceptual worlds). There is no objectively and concretely fixed environment, I understand Uexküll to be positing, and for each perceptual system there are carriers of significance (*Bedeutungsträger*) that are specific and consequential for particular organisms in the particular *Umwelt*. The *Umwelt*, Uexküll proposes, is not the same as the objective space or environment; *Umwelt* is constituted by those elements or carriers of significance that are vital to the discrete perceiving organism, be that of an animal or insect or jellyfish or human or plant. In an assemblage of perceptual stimuli, this of course also then implies that different perceiving organisms, say in a forest, constitute different forests, and we should rename our initial passages above into "lives in forests."

This will resonate before long, I believe, once I have walked into the atmospheres of contemporary digital art installations and explored with you the notion of an extended choreographic. This extension, and the lives of "being alive" (cf. Ingold 2011), reaches far beyond the theatre and leaves behind older distinctions between actor, media, things, settings, words. It touches upon current materialist and ecological concerns with *contaminated entanglements*, as Tsing considers them evoking or provoking collaboration, gathering, happening, crossing over – something she also calls "interwoven rhythms" (Tsing 2015: 34). In her research, such affective rhythms and entanglements embrace human and natural histories. The particular actor or life form in the assemblages and trajectories she pursues is matsutake, a sought-after mushroom that was iconic in Japan but became very rare and thus had to be harvested and found in other human-disturbed forests across the northern hemisphere, leading to the development of complex commodity chains, rhythms of foraging happening in strange fungal ecologies. Tsing explores contaminations and collaborations that constitute transpacific trade. In our case of *kimosphere no. 4,* the focus is on lemurs and tamarind trees (on a subtextual level of allusions to migration), and on elements of sensory choreography through performative constructions of reality and virtuality.

In analogy to my introduction above, we can now turn to such constructions and foraging/tracking constellations in the performance arts and in Virtual Reality (VR); I am particularly interested in asking how performative atmospheres move us and affect us, how material and immaterial variations are expressed and experienced, and how amplification and intensification work as transformational effects through somatechnics – through an integration or entanglement of body and environment. Again in analogy, we might even think of such performative sensorial habitats as precarious environments, and a concern with the climate crisis was never far from our minds when we started to work on the *kimosphere* series. The *things* in the installations I evoke here have their lives. They could be said to have their various rhythms and thus belong to or contribute to constituting

FIGURE 5.1 Visitor wandering through augmented virtuality environment, with her right hand feeling imaginary resistance. CNDP Bucharest 2018 © DAP-Lab.

complex relations between rhythm figures (or refrains, as McCormack explores them in *Refrains for Moving Bodies*), light, sound, spatial milieus, and elemental infrastructures. Rhythms, and trancing, you will have guessed, are an important dimension of trans-sensory stimulation explored here (Figure 5.1).

The *extended choreographic,* if we were tracking Rosalind Krauss' notion of the "expanded field" (Krauss 1978), implies not only the sculptural-architectural, but the unhinging of clear sites of artistic forms/practices. It connotes a spiritual kinship with Oskar Schlemmer's sculpturized figurines (or Hélio Oiticica's *Parangolés* with their flying colors) – things that move – and the wild growth of the Schlemmer costumes that would spatially and atmospherically embed the figurine/human being (as *Kunstfigur* or artificial figure) into a technically designed environment that is also a sphere of lives, of animate matter. The technically designed environment could be an organic one at the same time, "site-specific" in the sense of being else-where, outside of conventional theatre architectures. It could be contaminated, a kind of artificial plantation. It could be a dilapidated former industrial site, such as coal-mine or port. Or an underpass, a public square, park or an actual *Urwald* (wild forest) as I encountered it in Darmstadt (Germany), where annual happenings take place on the *Waldkunstpfad*, a "path" for traveling into "forest art" that has been curated by Ute Ritschel since 2002, for nearly twenty years. Imagine curating a pathway in the forest!

I tried to write about it in *Performance, Technology and Science* (2008: 101ff), not realizing back then that forests would soon inspire me as conceptual machines for installation art and cross-cultural/cross-species technological assemblages.

When I look at some of Schlemmer's bulbous, warped figurines, I think of mushrooms. Schlemmer was a painter and spatial artist, in the manner in which

I perceive landscapes by Robert Smithson, James Turrell and Olafur Eliasson, or choreographies by Dimitris Papaioannou, Meredith Monk, Mette Ingvartsen, William Forsythe, Christo and Jeanne Claude, and Lin Hwai-min. Landscapes that resemble forest art, airy and cloudy installations, compactions, liquid architectures, polytopes, scattered crowds. From a large distance (of nearly a hundred years), the Bauhaus objects and living machines may have become less legible.[1] Yet many of Schlemmer's forays that stimulated me were conducted in movement, in what he referred to as "metaphysical theatre" or "costume ballet" (Schlemmer in Blume 2014: 9) – where the "bodily" shapes and inventions of his "elementary-mathematical forms" (Scheper 1988: 49) dissolved into spatial volumes, and to some extent became dematerialized into abstracted geometries and non-human form, rhythms, refrains and arrays (shadows and light rays, in the case of the *Metalltanz,* and mesmerizing spirals in *Reifentanz*). Here I propose to take the *extended choreographic* some steps further, positing its virtual and homuncular potential expanding into constricted digital space (VR or VE) where artificial forests, plant lives and animals provoke varying kinaesthetic or perceptual responses. Constriction, as my design partner in the DAP-Lab also insists, is always enabling. Constriction, from a choreographic perspective, of course paradoxically means extension: it allows the movement to un-fold otherwise (if I were to suggest disalignment or asymmetry, or sliding-moving on the ground like a seal), to contend with obstacle.

At the same time, questions arise about *the things* in the real or augmented reality, and how things can contend with and choreograph others, and how this relationality (among co-participants) moves beyond traditional ontological or socio-political references points of nature and culture, of "plantations." Plants, planting, plantations – a string of associations opens out, from natural and industrial history to philosophies of vegetal life, botany, plant science, and science fiction. These associations have a colonial ring to them, referring to older economic systems of invasion and exploitation but also generally to industrialization and modernizing processes generating *plantations* as models for factories and capitalist expansion through scalability (Tsing 2015: 38). The scalability of design is at issue here.

All-body masks and deep-sea divers

The Bauhaus, which in 2019 celebrated its hundredth anniversary, was foremost an academy for the visual arts, crafts, and architecture/design. Schlemmer's stage workshop therefore appears to have been peripheral (just as his stone sculpture workshop was), since founder Walter Gropius' experimental model of art education focused on the development of prototypes to yield industrial commissions, furthering the mix of artistic inventiveness and industrial design/production. Initially still leaning toward an experimental occult spirituality (and Johannes Itten's influential Mazdaznan cult with its *Atemlehre*/breathing exercises, extended fasting, singing, hot baths, physical improvisations and mystical harmonization practices), the "plantation" under Gropius soon formulated a new credo (at the

1923 Bauhaus exhibition) as a demonstration of the fusion of *art and technology.* This proved to be path-breaking and complex, with all the queer edges that lined and irrupted into the Bauhaus aesthetic, with the female energies that pushed weaving, textile and object design, new techniques of photography resulting in breathtaking portraits of (trans)gendered bodies (Birringer 2020).

This concerns us today. We have not so many records of the stones from the stone sculpture workshop. But we have Florence Henri's nude portraits and the provocative "Mask-Photos" of Gertrud Arndt. Schlemmer, although not as strongly as Itten, had a metaphysical mindset and a strong interest in symbolic images, while at the same time exploring the precision logic of technological machinery, "scientific apparatus of glass and metal, the artificial limbs developed by surgery, the fantastic costumes of the deep-sea diver and the modern soldier, and so forth" (Blume 2014: 10). Records of the machinery workshops have survived, and here Schlemmer's passion also coincided with László Moholy-Nagy, Lucia Moholy and Marianne Brandt's interest into constructivist visual technologies of film, photography, photomontage and photosculptures (*Fotoplastiken*).

Schlemmer's stage workshop garnered a particular role in the interlinking of activities, combining art, craftsmanship and play. He had worked through painting, wood, stone and sculpture toward performance – this heterogeneous combination makes sense as a visual-movement art, in its modernist tendency toward abstraction. The *Triadic Ballet,* the dances for which Schlemmer is mostly remembered, had already been composed between 1916 and 1922 before being presented during the 1923 Bauhaus exhibition. Eighteen costumes were designed to be performed to correspond to the colors of the three parts (yellow, pink and black series) and encompass a total of twelve scenes (*Bilder-Tänze*). The costumes are well known from recent exhibitions (e.g. at Barbican Art Gallery 2012; *Human-Space Machine* created by the Bauhaus Dessau Foundation in 2015, and numerous shows during the 2019 centenary) and have evocative names such as "Sphere Hands," "Round Skirt," "The Abstract," "Dancer with Golden Sphere," "Dancer, Turkish," "Disc Dancer," "The Diver," etc. Schlemmer's work as a painter and sculptor fundamentally drove his experiments as a choreographer and his continuing concern with space and plasticity (Birringer 2013). The latter, akin with the architectonics Gropius imagined for the Bauhaus building design approaches (interior and exterior design of modern "life processes"), was formulated in the description of the main activities of the stage workshop. They comprised the investigation of space, form, color, sound, movement and light (Trimingham 2011: 18).

This scope of investigation drives my reflections on work I collaboratively developed with the DAP-Lab and other artists but also observed in a number of installations and digital art works.[2] DAP-Lab's current research links augmented reality and kinetic atmospheres – I call them *kimospheres* – with forays into Virtual Reality (VR).[3] VR is a method of envelopment or emplacement alluring you into an immaterial simulated world, inducing you to imagine being carried, transported, while at the same time also inducing you to imagine that you are doing the traveling and wayfaring, zooming yourself forward. You may also be induced into imagining the world moving at you, "as if" it were encroaching upon and swallowing you, and as if it responded to your motion and echoed back at you, as in the old song "My Shadow, My Echo, My Memory." These dives into virtual landscapes need to be examined carefully. There is of course drama beyond the human figure and gestalt thinking so apparently central still to Schlemmer's concerns. But how can landscape become protagonist? Dance installations as I had understood them over past decades were largely site-specific or geared toward an experience of sensorial or somatic tendencies: dance as physically interactive constellation requiring engagement, deep listening, touch, playful or ritual responses, attention, care, un-inhibition.

My knowledge of *body weather* techniques, which are in tune with such deep attention, is owed to butoh training and the Qigong method we practice in our ensemble. Body-Mind Centering and other somatic practices are directed at embodied perceptions and empathetic senses – in some instances specifically addressing our kinaesthetic, tactile and proprioceptive orientation in non-visual (blind) space.[4] The emergence of a more fully digitalized, networked and social media communications world seems to pose challenges to the real-corporeal, and to "being alive" in the sense in which anthropologist Tim Ingold describes interaction and moving with an animic cosmos, with sky, earth and weather (2011: 73). Digital, computational processes and immersion technologies generate new parameters, reductions and incompatibilities. They engineer new navigations through sensorial rewiring and perceptual adaptation; they may also re-connect to the more hallucinatory spiritual or spirit practices hinted at before. Moving rhythms, I believe, always modify the affective spacetimes of bodies; they enable excessive expressiveness and kinetic ecstasies.

In the following, I examine the transporting interfaces with digitally designed mixed reality and virtual realms, observing how wearable digital technologies are integrated or how they become somatic-experiential, and how, for example, the isolation of the VR headset – similar to the all-body mask of Schlemmer's figurines[5] – does constrain the wearer's ability to feel real immersion and to feel transported. It does not, however, prevent transport, and emotional and embodied connections and proprioceptive sensations. On the contrary, it may enhance dramaturgies of proximal and energetic, and even ecstatic, experience as it may cause neurophysiological dislocation experience, transfer illusions or an altered physical sense of selfhood. This is what I want to refer to as the *hallucinatory*: the design of a virtual environment can shift perceptual fields and generate a new viscosity of sensation through changes and disorientations of the mapping of one's physical body.

Sensory environments/choreographic objects

What are we to make of an all-body mask that constrains the wearers or even makes it impossible for them to see what they do? The notion of the costume as mask, given the abstracting, geometric figurations Schlemmer had built as wearable costume-sculptures for the Bauhaus dances, invites speculation on how a difficult costume (which hides the face and occludes vision) is meant to make the wearer move, prevent them from moving in any conventional way or compelling them to behave in ways they were not aware of previously. When I began to work with fashion designer Michèle Danjoux and noted encumbrances or challenges she increasingly built into dancers' garments, the performances we developed with the DAP-Lab gained a dimension that resonated with other "choreographic objects" (William Forsythe's term) created by contemporary dance-theatre artists who moved off the stage. Such installations were often participatory, inviting the audience to inter-act. Yet Danjoux's wearable encumbrance, or the threshold of animation she emplanted (Danjoux 2017), was unlike the spatial or object-oriented approach taken, for example, in Forsythe's *A Volume, within which it is not Possible for Certain Classes of Action to Arise* (2015) or *The Fact of Matter* (2009). Forsythe's *A Volume* is an architectural mask, and in this case indebted to the famously quirky triangulated design by Hans Hollein for the Museum für Moderne Kunst (Frankfurt). I will return to the concept of wearable space in a moment (Figures 5.2 and 5.3).

In Forsythe's *The Fact of Matter*, which I experienced in the exhibition *Move: Choreographing You* at London's Hayward Gallery, the "object" is not adopted built space but a participatory construction that invites visitors to climb through a very large number of rings hanging from straps at different heights – an kinetic obstacle course that seems enjoyably playful but turns out to be exceedingly tough. Almost

FIGURE 5.2 William Forsythe, *A Volume, within which it is not Possible for Certain Classes of Action to Arise*, 2015, MMK Museum für Moderne Kunst Frankfurt. Photo: Dominik Mentzos.

FIGURE 5.3 Visitors playing in William Forsythe's choreographic object *The Fact of Matter,* installation at *Move: Choreographing You,* Hayward Gallery, Southbank Centre, London, 2010–2011. Photo: AlastairMuir/Courtesy of Hayward Gallery.

no one is able to reach the other side without touching the ground or becoming too exhausted to continue. When I tried it I struggled for a long time, testing my strength and coordination while my body was constantly destabilized by the swinging rings. They toyed with me. I had to give up when it became impossible to sustain my swinging, and my pushing. The sense of gymnastic play was supplanted by an emotional affect of disappointment, or rather a realization about the limits of my physical sense of capacity, the capabilities of my body.[6] The things that choreographed me, on the other hand, seemed supremely confident, stable and impartial to my efforts. They were admirably unsusceptible to my emotional effort.

This choreographic object could be considered an initially indiscernible instrument, and also a technical object in Gilbert Simondon's sense of a machinic system which requires techniques of adaptation or produces them, indeed inducing a rhythm that moves with the machinic system's own force of organization. The machinic instrument or choreographic thing is an ecology of associated milieus that in-form the body or the participant perceptual emergences (Simondon 2012: 51). Bodies inside *The Fact of Matter* are altered or affected in a way that they are also opened environmentally – or atmospherically – to the tentacular: transduced to objectness, gravitational pulls, sweat and quicker breathing, heat and melting down, in the billowing and swinging of the furnishings, with colors, contours and sounds of others, contours, room of blurred permeabilities. The forest of swaying rings prevails.

Now is a good time to recall again the common and uncommon senses in wearable space – insofar as I want to draw attention to the phenomenological, sensorial-embodied perception approaches that underlie my writing here and point to the wearability of space, including space inside augmented virtuality design, which is three-dimensional and envelops you. VR worlds, I argue, can be like Forsythe's choreographic objects, entangling the subject while themselves being meshed and knotted of complex computational physics, graphics vectors and algorithms, the fact of virtual matter thus also becoming the subject wearing you and making you associate various uncouplings and re-accommodations.

A virtual tropical tamarind forest can play tricks on our minds, becoming an abstraction intensely impacting an unanchored body. I propose to look at different thresholds of perception – here in the context of VR performance installations – which take us across not only various materialities but also diverging modes of physical sensing-thinking. Such sensing, as I suggested previously, strays across registers not necessarily based on visuality alone. It includes subliminal and peripheral sensings, rhythms of sensation, vibrations, proprioceptive and imaginary relations, dreaming, and dancing as a kind of morphing, detailing the imperceptible, surrendering the curves and edges of others and other things.[6] The *un-common senses*, as I reported from a talk by neuroscience philosopher Barry Smith, are the ones we are much less conscious or certain about – thermo or mechanoreceptor nerves in fingers, arms or the spine, giving us tingling sensations; skin and hair sensing temperature and wetness or feeling textures, though not reliably; muscles and ligaments "hearing" how our anatomies, the bones, minerals and water in bodies, move along and stumble about; how organism and metabolism are comfortable or tensed, affected and afflicted by the environment as well as internal biophysical processes. I repeat these propositions here, as I continue to evoke artworks I find provocative. When I hung in *The Fact of Matter,* I also became intensely aware of the skin on my hands and arms, the ropes were rubbing against me, heating up and reddening my skin, leaving marks.

Another example jumps to mind, an idiosyncratic work by composer Yiran Zhao, titled *SSH – Solo for Head,* a one-on-one performance during which she asks the listener to close their eyes and wear ear plugs, while she touches their head and gently strokes their scalp, neck or ear lobes, gently rubbing fingers on hair and temples, fingering their echoing skull as it responds to physical impulses. The instruments, in this case, are the audience member's hair, ears, bones and the skin on the face (or of the beard). This reminds me also of Zhao's collaborator Kirstine Lindemann, a musician who taught me to listen to binaural audio recordings of ASMR (autonomous sensory meridian response) – whispering performances that tend to generate an experience characterized by a static-like or tingling sensation on the skin, typically beginning on the scalp and moving down the back of the neck and upper spine. It has been compared with auditory-tactile synaesthesia.

A further example is composer Frank Denyer's late work. Known for his delicate musical and ethnomusicological experimentations over several decades with a vast array of sound sources – new instruments of his own invention, adapted instruments, instruments of non-Western traditions, rare or virtually extinct instruments, and conventional Western instruments – Denyer created a string quartet in 2016–17 which impressed me with its almost devastating melancholic resignation, its seeming retreat into nothingness: the faintest tones of a violin or of a stone tapping very gently on a cigar box, sound receding into inaudibility. There are also tones held by two violins at the same time; together they appear, and eventually we hear overtones or undertones (something akin to the whispering that Denyer refers to as the "undervoice"), then they fall apart from one another and disappear to the intangible realms where we can only imagine sound waves still existing, to some kind of boundary realm, kinetic compressions within open-ended space. We strain our ears, and while doing so also become aware how hard it is, in today's permanent bullying soundscapes, to listen to the faint.

An ecology of associated milieus that in-form bodies or participant perceptual emergences, as suggested earlier, can provoke new cross-disciplined queries into deep listening, and indeed into imaginative possibilities for understanding performative roles of plants, animals, birds and biological lives of kin. How many kinds of undervoices are there, if we think of birds, for example? Such possibilities include inaccessible points of view of plant life, of trees and lemurs, but multispecies entanglement refers us to other qualities and agencies from which humans are excluded and yet tied into: barely audible undertones in plant-human relationships might receive re-valuations of what is now a serious concern of the Extinction Rebellion and creative activisms that seek to shift our anthropocentrism toward other experiences (to the biopolitical point where the notion of the "other" must be questioned as well). Or think of the bio-artist Bartaku's experimentations with berries he tries to grow in a small village in Latvia, his "seed scarification" methods trying to stir the dormant seed, Aronia Melanocarpa, with a sound poem which we hear in a recording of voices interpreting ("voicing cognition") the astringent black chokeberry in a choral manner. The berry, so to speak, animating the voices (Bartaku 2019). The choric materiality of an installation of the berry – which I curated at the "Things that Dance" platform in Karlsruhe (October 2018) – in all its mysterious and refracting obliqueness, is heightened affectively by the harsh vocals that hit us as we walk around the space. We feel the work viscerally, it flows through our bodies, muscles and tendons, it bounces off the walls of the underground studio, it tears us, and it also "cooks" us, so to speak (there is a steaming pot in the corner on a hot plate). We smell the berry; the steaming fluid is lignin for our roots and stems. Small memories: the vitalist impact of the installation also echoes through my bodily memory of an *arborescent movement* class I took years ago, its deep listening focus on skeletal bone nodes in the feet still fresh.

The be-holding of the faint or barely noticeable is a dynamic I want to re-connect to the hallucinatory and energetic sides of immersion and of an immersive aesthetics. The sensational, the forces of sensations are the nexus here, the connecting array of trans-sensory fluidity experienced and felt in a saturated environment.[7] Immersive atmospheres are entered into, and they enter into the receptors. Think of flying spores for which dispersal is not an issue. Atmospheres enter through surfaces of skin, through the bones (Figure 5.4).

FIGURE 5.4 Yoko Ishiguro performing "Flying Spores" scene in *Mourning for a dead moon,* created by DAP-Lab, Artaud Performance Center, London 2019. © DAP-Lab.

The under-ground of atmospheric, ambient interactivity is skin, porous and stretching. It is dark green moss sloping over bark, tree trunks, rocks and soil, again with elemental qualities developing sensori-kinetic "control" over newly emerging corporeal rhythms (in associated milieus). The manner of control or cohabitation and propagation varies, if we just think, for example, of patches of mycelial growth – do mushrooms listen when they exchange spores and propagate, or is their manner of listening a kind of chemical sensing, as bacteriologists have assumed? It is not too far-fetched to think of both fungal and human bodies in atmospheric immersion as behaving through multiple rhythms of chemical sensing. In the case of human subjects and their observed behavior in interactive installations, it is quite common to note sensorimotor habits, habitual reactions to situations that had been adapted to body schemas and to what embodied cognition theory would consider tacit knowledge. Sensori-kinetic control of one's reactions, therefore, would rely on how the organism tunes into an alignment with its habitual sequences or opens out into the errant. Derek McCormack, in *Refrains for Moving Bodies,* speaks of the possible superposition of disparate rhythms, one that

> engages the logic of sensation at the same time as it opens onto the incorporeal, the virtual, or the infinite. This is a matter of realizing how relations between motor, sensory, neurochemical, and other milieus are given a fragile consistency as rhythmic spacetimes.
>
> (McCormack 2013: 82)

Are the refrains of moving or orienting altered when inside a virtual environment, when knowing one's elemental body or surfaces are affected by shifting from outside to inside a forest generated by VR technologies? Do we move from Schlemmer's phenomenologically centered Euclidian geometries of the human figure to a meta-kinespheric realm hat Laban had not imagined either? Do we move inside a porous or tight envelope? Is the VR landscape infinite or bounded and conditioned? Is there such a process or thing as atmosphere-conditioning? These enveloping spheres are variable movement ideas, too, as forces of atmospheres and the potentials or constrictions of bodies unfold together. Spheres not only envelop but also extend, travel, if we think of sound waves and voluminous space. Thus they can open out or contract, physical movement attracted in relation to basic activating properties of expenditure and recovery, like the breath that animates, rhythms becoming movement, sounding themselves. We are enamored by this without knowing it. We love to breathe. We would love to spread our wings like storks do, gliding off into the winds, our thin legs dangling delicately. Writers like transformational practitioner Nicolás Salazar Sutil, communications researcher John Durham Peters, or cultural geographer Derek McCormack, as we have noted earlier, appear attracted to the elemental or ecstatic allure of atmospheric volume, and the vibrations and resonances lying dormant in our bodies (McCormack 2018: 47). In Salazar Sutil's *Matter Transmission,* he even exalts the notion of "grotesque immersion" (2018: 152), suggesting after his underground cave excursions that

> it is only through immersion in the dark, humid, dense atmosphere of the cave that the energy of the land can be internalized and channelled as sexual and sensual energy. To be immersed in cave space involves surrendering one's body to the cave's own erotic geophysicality.
>
> (153)

I am curious how this attraction to "freakish" cave walls and the ecstatic landesque immersion corresponds to my perhaps more irritating recent encounters with VR where I began to wonder how far the techniques in immersive VR push sensory conflict and disalignment (i.e. serious incongruency with body schema), and thus how far they can dislocalize sensation. Such dislocalization would run counter to the continuity or corporeal indigeneity Salazar Sutil aligns with landesque immersion and which I also, perhaps romantically, associated with forests, valleys and rivers of my childhood. I will attend to this more critically when I describe my experience of "The Plank." Regarding Salazar Sutil's notion of landesque immersion (157ff), lines from Charles Wright's *Black Zodiac* come to mind, where the poet claims that "all forms of landscape are autobiographical," but the scenery sometimes refuses privileged access of (his) life, as we learn from several lines in the *Oblivion Banjo* edition:

The river stays shut, and writes my autobiography... (71)
The silvery alphabets of the sea / increasingly difficult to transcribe (158)
aphorisms skulk in the trees, / Their wings folded, their heads bowed (506)

But what is such access? Laban's kinespheric architecture (for harmonious tracks of motion: high-low, forward-backward, right-left) does not necessarily determine movement but it seems to transcribe the orientations and directions, which can facilitate human movement possibilities. When I look up into the air, or listen to sound waves, bird song and crickets, I also improvise. The scales and intensities may be different: I sense vibrations of after-images, dreams, memories – expanded sensorium of the hyperreal, expanded volume of imagined exuberances. I see the sky and feel the fresh air. It could be a feeling of being high, the scent of the air and the grass transporting me – and why would it be so different from the hallucinogenic experience described by ethnomycologist R. Gordon Wasson and photographer Allan Richardson for *Life* Magazine in 1957 ("Seeking the Magic Mushroom"), after their journey to Mexico and an encounter with the *curandera* Maria Sabina, who gave them mushrooms to eat in the dark, reciting a ritual chant: "I am a mouth looking for you – but you are not paying attention. Come!"[8]

The mouth looking for you: homuncular environments

Come here, then, and inhale, kinetic atmospheres say. Lift off, dive in, let yourself fall, ascend, fly high, become buoyant, crawl into the tent, get on your knees, roll around, taste the flavor, bounce on the balloon, hang in the rafters, climb the scaffold, balance yourself on the see-saw, touch the screen, put on the earphone, smell the peat, allow yourself you to be pulled in, follow us ... *Immersion* takes on a growing significance as a category of attracting forces of affective experience, as the term is now often used not only in theatre and art installations but in conjunction with Virtual Reality, games and engineered atmospheres that range from the architectural, built environment, the urban spectacles of light and consumerism, to the various intimate aesthetic experiences designed by performance and sound makers, fashion and interaction designers, bioscientific experimenters. Immersion is sticky matter. It moves around bodies and their sensory desires. It moves inside bodies. The outside looms inside, the inside is an open wound, so to speak. It is swallowed out. Prostheses create affordances that point to their *Umwelt*, connecting body and world, enabling new qualities of existence, relationships, inhabitations. Immersion requires prosthetic techniques, and it can hint at vast imaginary lands and landscapes.

In a 2019 performance of *Formosa* by Cloud Gate Dance Theatre, at Sadlers' Wells in London, the designers used a highly sophisticated graphic projection technique to produce this effect – it was as if you were transported into the kinetic motion of poetry-writing, vertical streams of characters floating from the heavens. Digital-projected calligraphic motion scenographies began to interact, so to speak, with the dancers on stage. We had heard the poetry spoken before, by a voice in Mandarin, now the words projected began floating down slowly, turned more elemental, tilted sideways; they seemed to become clouds, black and grey rain clouds, altering their course and textures, raining down, streaking over the floor. Earlier, before we saw the tilted and diagonal projected characters, the Cloud Gate dancers, wearing coarse cloth in muted colors of the rainbow, ran and swayed rhythmically in similar coordinated lines, as if they were farmers working a field (Figure 5.5).

Refrains for moving bodies. The projections of words were very large and thus one could not avoid thinking of such stage projections as spectacular and overwhelming, but they had a clear consistency with Lin Hwai-min's choreography for the Cloud Gate dancers (and brought back memories of his *Cursive* series and the manner in which Lin, over many years, had integrated meditation- and martial arts-based vocabulary into visceral, abstract movement). What they projected were visual landscapes that were also "made up," as the program notes tell us, of names of mountains, rivers, cities and villages of the island, and gradually, near the end, these refrains fall apart and become dispersed into strokes and lines.

> Finally, one spring day
> Our children will read the following news:
> Migratory birds are returning north
> Drivers travelling along Tamsui River
> May not blow their horns.
> (excerpt from "Hope" by Liu Ka-shiang)

The recorded readings of poems were consistent with the melancholy mood of this dance work, which ends with projections of ocean waves. It is a sublimely introverted moment when the waters and the raw pitch of aboriginal singer Sangpuy Katatepan Mavaliyw's voice rise slowly: there is now only a white stage, with a single figure, as if a Caspar David Friedrich painting had come alive to be fused with the beautiful illusions of an island (Formosa), a place the Portuguese had sailed by, years before the Dutch, the Spanish, the Japanese and the Chinese would take turns in charge of this verdant Taiwanese island.

There are immersive spectacles we could mention that are perhaps not so sublime or overwhelmingly enabling but basically try to be beautiful and seductive. In exhibitions, they may require visitors to flap their arms like wings or gesture vividly with their hands when they discover that the motions of their extremities cause a small change to happen in the tapestry of digital projection: a change of color, a shadow, or a flying object may appear and this seems to excite audiences. I was observing this during a long afternoon at the Barbican Centre, London, during an exhibit on artificial intelligence, *AI: More than Human* (2019). The immersive installation I observed – teamLab's *What a Loving and Beautiful World* – had the opposite effect on me, compared to *Formosa*. It felt superficially pretty but also silly, with nordic ambient trance music trying to calm visitors down and lull them into a vapor. To my mind this installation had all the wrong attitudes – it tried to please me. I sat there for 90 minutes, in the middle of the room after feeling uninspired touching the walls, but had begun to feel bored almost instantly. Or rather, I felt it was offered as a self-congratulatory commercially entertaining 3D installation;

FIGURE 5.5 Cloud Gate Theatre of Taiwan in *Formosa*, choreography by Lin Hwai-min, projection design by Chou Tung-yen and Very Mainstream Studio. Performed by Cloud Gate Dance Theatre of Taiwan, 2019. Photo: HSU Ping.

tickets were timed, so visitors would walk in at their appointed time, take a quick look and a selfie, touch the walls, and leave again. In a review, however, I somewhere read that it was

> an endless immersive digital installation in which tumbling calligraphic characters transform into animated images when touched by a visitor's shadow. Flocks of birds, mountains, thunderclouds, cherry blossom, sparks of fire, trees and raindrops leap out from the shadow of your fingers, skitter across the wall and interact: it's enchanting and mesmerising.

Not sure what shadowy fingers the critic applied, but it did not work with mine, no flocks of birds tumbled. This of course is also interesting, if indeed the wall responds differently each time, and is displeased by the mindset of the toucher. A similar interactive 3-D installation, *The Worlds of Splendors* (瑰丽—犹在境), premiered in Beijing, produced by the International GLA Art Group, and the images I have seen from visitors interacting with the walls are almost identical to the ones I witnessed in London. teamLab now lead this branding of immersion spectacles, having recently opened their permanent exhibition *Future World*, sprawled across the ArtsScience Museum in Singapore.

When I mentioned my reaction to teamLab and GLA Art Group to colleagues in a discussion on how to exhibit Virtual Reality work, not everyone of course agreed with me. It was pointed out that teamLab's work spans a whole spectrum, from the straightforward commercial to a reconfiguring of spatiality, to some more political pieces. Christiane Paul, who made this comment, also mentioned that Japan makes much less of a distinction between art and design, which may be problematic in other contexts. Paul argued in her posting to the New Media Curating listserv (www.crumbweb.org) that teamLab's work is also very much embedded in aesthetics of Japanese art and culture, which might require some translation (e.g. their work on what they call "ultra subjective" space is a continuation of the depiction of people and scenery as all relative to each other in traditional Japanese screens or scroll paintings, and it tries to play with agency through multiple viewpoints within one virtual space). Interestingly, and perhaps telling in regard to how diverse audience reactions can be, a few days after Paul's posting, another subscriber to the New Media Curating listserv argued that "teamLab are visionaries! The work is breathtaking with no need for a Japanese cultural aesthetic explanation…" (Yvette Mattern, 24 June 2019). There were two contrasting reactions that subsequently opened up in this debate on curating VR and immersive digital art. Mattern's praise was followed by Ashley Lee Wong's comment that he had been looking into teamLab's work realizing how easy it is to dismiss it as decorative and populist. But then Wong argued that teamLab are not the only studio working across art and design in this way, that there are in fact numerous studios that tread this line (e.g. onformative and FIELD), not particular to Asia.

> I'm more interested in how they support their own artistic practice through their commercial work. teamLab's studio is organized horizontally and

they work collaboratively in groups to explore ideas in a realm of the 'unspeakable' in our relationship to our surroundings, nature, city. Interestingly teamLab never thought of themselves as artists and never circulated in the gallery world until 2011 when Takashi Murakami (who coined the concept of 'superflat') invited them to do a solo show in his gallery in Taipei. They also met an art advisor Ikkan Sanada, a Japanese gallerist, who knows people at PACE and helped position their work in the art world. For me it was the attempt of the art world to capitalize on tech innovation by promoting 'new forms of art.' What they do is not new but in the narrow realms of the art market it is. I'm not sure they really need the art market validation since they have created their own economy to support their work.

(Wong 25 June 2019)

Wong concluded that he was impressed how teamLab were able to reach a wide audience, especially young people, arguing that they work at scale with high production which is unprecedented, and generally unavailable to independent artists. "Only a large studio could achieve such scale. They want to make a kind of work that is universal and not specific to a culture or geographic location and can be enjoyed by anyone living in the contemporary age."

Wong, who is a researcher and artist based at the School of Creative Media, City University of Hong Kong, was then contradicted by Simon Biggs, a well-known interaction designer now working in Australia.

I wasn't thinking of the art market when I came down on teamLab, but art. I hold the art market in the same suspicion as I hold any form of commerce. I think teamLab make an excellent fit for both the design world and the art market. When I say I think their work is trivial I mean that I don't see it addressing any key concerns about the human condition. I recognize the skill that goes into their work – they have the resources to draw those skills in and create the kind of spectacles they do. We see this in a lot of popular art at this time. Spectacle seems to have taken over from thoughtful attention to the interstitial and liminal, where what is interesting about being human is usually found lurking.

Biggs' argument is understandable, if we follow his critique of the commercial veneer of these large scale 3D immersive installations that seek to overwhelm or please their audiences. His critique of Wong's idea that VR design aims at being "universal" is worth quoting in full:

Suggesting that teamLab is seeking to make work that is 'universal' reveals what might be considered a naive understanding of what makes something interesting to people. There is no 'universal' aesthetic or effective form of communication. All representation, and its reception, is subjective and culturally contingent, the product of numerous factors (cultural studies 101). People are receptive to things for a lot of different reasons. I find teamLab's work kitsch and ugly, which I admit is a subjective view and only my opinion. If somebody else enjoys it that is fine. But I am troubled when the people making such work, or those interpreting it, somehow think it is universally 'good.' That betrays very lazy thinking, which is why I think the work is trivial.

(Biggs, 25 June 2019)

In this discussion, which went on for some time, we could not find agreement. Those who argue like Biggs stand closer to a participatory and interactive performance aesthetics which implies working with a multimedia environment (including immersive projection) in which everyone can interact in the environment at the same time (whether with other people or with the synthetic elements in the environment) or observe such interaction analytically, without self-disciplining or satisfying the self-entertainment requirement, but in the Brechtian sense of noting contradictions, uncanny irruptions, and the environment's autonomy since it will be based on algorithmic programming. Thus the constructed space transmits a shared experience that focuses on forms of *agency* rather than *spectacle*.

My own sense of immersion as a technique is also closer to such forms of performative interaction, and thus I keep returning to *movement* and *kinetic objects*, and how they might affect us autobiographically (to pick up on Wright's subterranean and transitional poetry). VR designers, such as Jaron Lanier, may also dance synaptically as/with their avatars, but their initial questions are very different: they point to *mapping*, i.e. controlling motion in virtual environments, making measurements on the body of someone wearing a capture suit, calculating an aspect of the flex of a wrist to be applied to control a corresponding change in a virtual body. Immersive virtual reality allows persons to inhabit avatar bodies that differ from their own, Lanier argues. This can produce significant psychological and physiological effects, and Lanier's concept of *homuncular flexibility* (Lanier 2006) proposes that users can learn to control bodies that are different from their own by changing the relationship between tracked and rendered motion. In later research at his Stanford lab, Lanier and his team examined the effects of remapping movements in the real world onto an avatar that moves in novel or unusual ways (Won, Bailenson, Lee and Lanier 2015) – thus shifting the mapping of one's physical anatomy (occurring in the motor cortex) to rather different homuncular and eccentric "body schemas," avatar creatures that are indeed virtually different. Lanier mentions a lobster with a trio of little midriff arms on each side of its body. What he does not fully explain is how an "alien" body schema can be neurophysiologically and neuroaesthetically incorporated; how it alienates and traumatizes; how it associates injury or delusional states;

how it might cause exciting confabulations and morbid distortion to the human perceptual experience.

During the development of our *kimosphere no. 4* installation, and in many of the subsequent international workshops, the virtual differences that interested me were atmospheric, not avataric. Thus I turned to building habitats of certain colors, textures (red sand, silver foil, cling film, tree branches, moss, pebbles) and tropical vegetation (coconut, tropical fruit, palm tree fronds), with dynamic environmental elements that incite immersants to bathe in them, feel the swarm of fluctuations, crawl into them, lie down, dive, surf and fly across the currents and mutations of such kimospheres: to imagine becoming bird-like or amphibian, participating with flow and uncanny connections emerging from a geomorphological base. The immersant imagines this, rather than delegating agency onto something other or alien. The immersant also feels rawness, affected by physical sensations and tendencies. With tendencies I mean ambiguous potentials that are plural, virtual and real, with sensorimotor capacities still quite active. The immersant is not still. The immersant arches backwards, rolls around the floor, stretches out forward, bends down, thus even becomes dizzy, motion-sick, noting imbalances and paradoxes – namely contradictions in the perceptions of their own movements as they are implicated in a VR world.

This is the trans-sensory, hallucinatory atmosphere I want to conjure up here. The aesthetic theories of atmosphere (cf. Böhme 1995; Zumthor 2006; Pallasmaa 2014), derived from philosophy and cultural geography, now also relate to stage design, architecture *as* stage and installation design, to questions of how designed space surrounding our bodies affects our emotions and moods. These theories are centered on human response, thus less capable of taking into account other material propensities in wider ecologies, i.e. in constructed environments that could be planted to be left (alone) more permanently, to grow out by themselves, to wake up like Bartaku's seed. I tried to evoke this "growing out" once at a lecture I gave to the Bartlett School of Architecture, University College London, in 2017 ("Kinetic Atmospheres: Performance and Immersion"), but realized that many of the architecture students were already designing models in VR and exploring conversions of material-immaterial space, without necessarily taking external ecologies, thresholds into the raw or the geophysical, into account. I showed them some photographs from an astounding book I had discovered – Wolfgang Meisenheimer's *Choreography of the Architectural Space/Choreografie des Architektonischen Raumes* (2007) – to draw attention to ways in which spatial immanence can be imagined also through dance and movement. Meisenheimer addresses passageways and entries, and implies also a way of surrendering (the body) to physical and architectural anatomies. He displays some very evocative black and white photographs of butoh dancers in a chapter on "Thresholds," and in this context evokes the idea of atmospheres as well:

> constructed spaces have challenging or calming effects on us…This is due
> to the fact that formal and proportional composition tends to correspond

with the way we perceive our own bodies, i.e. with our immediate idea of ourselves, our human body, its shape, and its range of possible variables of movement.

(Meisenheimer 2007: 27)

He still echoes the anthropocentricism of language that links (proprio)perception of movement to human movement, yet the photographs in the book point to further pluralities of bodies, extending Platonic ideas into the geological and geophysical, especially when he elicits images of the disappearance of space in time.

When speaking of *augmented reality*, say in theatre, music concerts and art installations, or in the "Constructing Realities" workshops I have conducted, it is implied that what is planted to be physically affective is also extended and *amplified* through technical means (sound diffusion, digital projections, lighting) to be experienced somatically by immersants. As audiences go, most of them are human, but we have had animals attend (dogs guarding our blind visitors; birds and bats flying into our installations, critters showing up). An expanded sense of the choreographic as raw and wearable space, as becoming-multi-perspectival, grows out. The material is co-extensive with the immaterial, it stretches. Amplification is always a psychoacoustic form of choreography, of orchestrating rhythms, volumes, and propagating reach, sonics' agencies, as Brandon LaBelle affirms in his political-musical study of diffusions that can generate joyful and erotic passions, and a "vibrant and tensed bridge between the spiritual and the political," echoing with what I earlier called ritual force, a "commonality of feeling" (LaBelle 2018: 127; Figure 5.6).

FIGURE 5.6 Augmented reality/digital forest. "Constructing Realities" workshop. CNDP Bucharest 2018 © DAP-Lab.

We are returning to the forest I evoked in the beginning, approaching the ritual force through a growing prehension of these spaces as *latente Allmende*, or latent commons (Tsing 2015: 255). Amplified and augmented space enters us and our receptors receive many (often ambiguous) clues. But this is not entirely true; one would probably have to carefully describe each stimulus and atmospheric condition to ascertain what is affective, and how, and in what variation that indeed may exceed any technology of measurement (how do you measure the wind, the cold and the heat in an environment where fifty or a hundred visitors or more have gathered? How do you measure affect in a stadium during a sports event or a rock concert?). In the workshops and installations I am referring to here, the physically affective is composed of various organic and synthetic materials that are embedded in the space. The space is a wider ecology. The materials and aesthetic occurrences are emergent, dynamic, and thus may generate different kinds of attunement. Yet I am still describing it all in terms of affective values (atmospheric qualities) that concern human perceptions, perceptions of a limited amount of visitors, say between ten and one hundred. I wish I could go beyond, and atmospheric values indeed point beyond to the elemental, the meteorological, the viral and bacterial, scaled up and scaled down (Figure 5.7).

The fullness of the real – the physically affective that is expanded by the virtual and the virtual that is augmented by the real – heats up such densely sensorial atmospheres toward a ritual aura, a ritual sense of performance events that may

FIGURE 5.7 *kimosphere no. 6.* Visitor touching branch in virtual forest, inside augmented virtuality space where the sound of the suspended silver foil creates a rustling (of leaves). CNDP Bucharest 2018 © DAP-Lab.

indeed be "meta" – moving across varying trajectories and speciated desires. I believe there is a ritualistic, elemental quality in such environments. Immersion is craved because it brings participants a little closer to the sublime, the erotic and the rapturous.[9] As I described in the previous chapter, *kimosphere no. 4* was our first environment that included 3D video and VR stations where visitors would be given head-mounted devices to enter and become absorbed in an immaterial world. Before visitors are given the headsets, they have already been heated up, immersed in the physically affective space, drenched in blurred red light. Their induction into the environment has sensitized them to a more unconscious, or a more dreaming, state of mind. The craving for ritual of the latent commons, that I suggest, seduces the wearer of goggles to worry less about the visual/optical VR experience but to "touch" space and be transported to other perceptual realms – and this is of course what "teleporting" in VR induces – jump-cutting, flying, dropping, perhaps therefore even enjoying the sense of uncertainty in restricted virtual space.

Green grid lines pop up when the immersant with the VR-HMD (Head Mounted Display) walks too far. Suddenly, there is a fence, so to speak, a barrier. This could be a jolt, reminding the wearer that they are in an artificial world. But the jolt is short-lived; the green lines vanish and the immaterial virtual landscape pulls them forward, enmeshes and animates them. Exploring such imaginary limits and contagious material conditions, our DAP ensemble tries to fabricate hallucinations, also preventing the immersant from noticing borders and fences, using very simple technique of "safe-guarding" the immersant, yet also adding effects of *augmented virtuality*. When the visitor inside the VR tamarind forest is pulled forward, our guides carefully prevent them from leaving the virtual realm (stepping outside the grid), using physical affective means to make the VR immersant feel the immaterial suddenly have real tangents. The "immaterial" has surprises in store. A tree branch touches the arm, cool wind caresses the cheeks, a smell of leaves and moss enters the nostrils, temperature changes, and twigs and soil on the floor make the visitor imagine that they are in a real (virtual) forest, make them be more attentive and more self-conscious. They touch an other reality, and their own somatic pronunciation of their senses, their lives.

Materials move. Fabrics, the smell and aroma of cloth and of organic materials (soil, leaves), the quality and temperature of light, the character of sound, all these sensory generators fill a space with mood and *tune* it, as Böhme would say, and the projection of film, graphics and light can also create rhythms and temporal sensations, kinetic characters, masks, a patois of oscillations, echoes, shadows. All of these sensations are built into a simulated tamarind forest – and the strangeness of our virtual forest should be completely apparent to the visitor if they observe the VR world carefully. They will then notice that it is incomplete, so to speak, it is actually "being built" as they watch, the illustration/design meta-process is visibly ongoing: one can see how the trees and the branches are composited by brushstrokes.

And yet, this very compositing process, which is a Brechtian device to undermine illusion, can be mesmerizing, as the forest grows and grows, becoming more and more populated and complex. The sensory atmosphere impregnates aural and tactile experience while it implies kinetic movement, a "trans"-motion across. The immersant can fly up and across, in this forest, the controller allows the visitor to make quick changes in location and perspective. Performing (with) this forest architecture, then, is one of the sensory challenges I propose here for embodied VR scenography. It is a 21st century step further ahead from the Bauhaus and the all-body mask in real theatre space, mentioned earlier. How does embodied scenography enjoin with spatialities both material and virtual? A haptic feedback relation seems inevitable, when we speak of feltness of these different materialities.[10]

In *kimosphere no. 4* (2017), first staged at the Artaud Performance Center in London, we tried to test the haptic feedback by physically sensualizing the performers' imaginative processes. We wanted them to become tangibly engaged in the virtual with their common and uncommon senses, touching things that were not there. Thus, with our dancers acting as guides (and all of them wearing masks), the audiences become the performers invited to step inside the physical landscape and touch, listen, move around and at some point put on the VR-HMD to enter into the virtual environment. With the goggles, the wearer dives into abstractions, i.e. a tropical rainforest that is clearly (being) digitally fabricated and not real. They cannot see their own extensions (their hands or feet) in the layered environment, but they can feel them or compensate (when they touch a branch, the sand, or leaves on the floor) by trying to balance their bodies "floating" in destabilized architectures. They reach, as they feel urged to touch the tamarind trees or the grass. Some of the visitors, perhaps unaccustomed to VR, choose to sit or lie down, yet move their bodies around, turn, twist (Figure 4.12). They appear to hover in the tropical surreal, and at the same time are surprised by events they may not suspect (wind touching cheeks, smell of moss and tree bark, crackling sound of branches, whooshing sounds, etc.).

Hovering presences – where atmosphere also appears uncontrollable, emergent – evoke complex ontological and spiritual questions. The wildness of nature – if we have grown up as children of dark forests and steep hills – is perhaps harbored deep inside our skin and bones, the underlying muscle memory, internal perception and emotional conditioning. The wildness may also be imaginary, pulling us into the past, the back beats of subterranean ghost stories and fairy-tales. How are we to think, then, of trans-sensory hallucination as other than an effect of elemental materiality in contagious synaesthetic constellations? The forest touches us.

The production of atmospheric conditioning through design, with sensorial impact on perception and also ethical perspective (namely how to react to affective presences and lurking environments), thus points to material assemblages. Augmented virtuality – where a real branch reaches into an unreal simulated tamarind forest – transforms the destabilized, solitary proprioception into more collective "seeing" that no longer arises just from eyes (looking at the VR world through goggles without peripheral vision of one's own body and limbs) but slides into audible-visceral kinaesthetics. The homuncular innervation and the "seeing" with phantom limbs are felt sharingly: audiences inside VR also touch and are touched by guides and objects outside of VR. The cortical plasticity implied by such innervation and alteration (first a dry branch, then wet algae are grasped) allows the wearer perceptual rearrangements. "Slimelight" – the name of a goth metal club in London – might be a pertinent ascription of the strange combination of physical sensations. Sensory feedback of new affordances overtakes the perceptual arrangements, and *kinesis* reverberates with ritual force. It makes us dwell in a shared circle of continuous community. Visitors are aware of the co-presence of others wandering around the milieus, before each of them mounting one of the HMDs to slip away into the forest. They stray off.

Hallucinatory scenographies

These shared experiences stood out to me at the open workshops for *kimosphere no. 6* at the National Center for Dance Bucharest (2018) and for *VR Lost and Found (kimosphere no. 7)* at Verkstedhallen in Trondheim (2019), where participants both guided and guarded each other, observed and followed one another, a single participant at a time wearing the VR-HMD. Since there was only a group of 18 participants, and each of them exploring the virtual world inside the augmented reality design of the installation as well as helping the others be-holding their immersive experience, one could speak of a collective sense of threading through the envelope, grasping the scene from inside, perceiving. Clearly, there is a very transitive sense in these enactments, the "be-holding" already mentioned in Chapter 4 when I discussed d'Evie's recuperation of the older meanings of be-holding as not focused primarily on ocular relations, but on other sensorial grasping of tenuous threads and peripheral distractions while affirming "wayfinding through blindness" (2017: 43). Indeed, the threaders who are holding the cable of the headset – protecting the VR immersant from falling or moving outside of the grid – resemble puppeteers in some kind of somatechnic Bunraku theatre. They delicately protect the scene, encouraging the visitor to step up to the VR station, and help them to put on the headset, fasten it and make them feel comfortable; then they nearly unnoticeably guide, lure or gently caution the wearer, holding the string behind their head. In one moment, the handler extends a small rock to the hands of the immersant, letting them touch it and grasp it while they are wandering inside the tamarind forest, looking up to the tall tree crowns or spotting the lemurs.

In this interactional scenario, the immersant has slipped away into the virtual environment but can still feel safe and safely stimulated, knowing that the guardians are there and that they can now let themselves fall, just as they probably noted the acrobatic dancers, when foraging in another area of the installation, hanging upside down and swinging from the rafters, enjoying a carefree sense of joy similar to what children experience in a playground (parents watching from a near distance). I use this comparison cautiously, as I want to avoid making any facile connections to playgrounds or insinuating that immersive art installations infantilize the audience. On the contrary, an invitation to play or to let go of control – it is a serious matter (cf. Alston 2016b). The immersant is on the up-swing. Yet such absorption can lead to carelessness or ambition, in gaming terms one might even think of competitiveness – the virtual world is a challenge or obstacle, in-game choices woven into the thematic experience of "lives of the forest," level up. In all encounters with visitors in *kimosphere no.7*, an installation that has a very wide range of real/physical, projective, acoustic, somatic and digital/3D-immersive components, I observed that initial apprehension or self-consciousness tended to wear off very quickly. Participants "played" the tangible objects we gave them to handle, carry, intone and listen to while they also let themselves drift into intangible spacetime, once the immersant felt secure and enveloped inside the virtual environment. "Landesque immersion," as Salazar Sutil calls the physical-material transmission felt in a cave (2018: 157), can be overstimulating and overwhelming, as I noted in some of my descents into natural caves. Salazar Sutil warns us to make easy analogies between cave and VR immersion; in his view, transmission is a physical-material mode of mediation, sustaining an immanent ethical and ecological connection between land and bodies. The virtual forest in *kimosphere no.7* affords a different quality of attunement that cannot have the same thermal, kinetic, geodesic and frictional energy that a limestone cave delivers. The virtual forest cannot pretend to be immediate, yet neither is it instantaneous ocular spectacle to be gazed at. The atmospheric quality to be grasped in augmented virtuality depends on the corporeal awareness which the immersant develops during the interaction with palpable and intangible objects at the same time, adjusting their bodies – which are now also bodies in code (human-computer-environment-interfaced) – inside a particular volume and voluminous prehensive play space.

Each in their own way, immersants are in-formed or attuned to let go of their fears or apprehensions and to become captivated. This is a conditional extended choreography, largely unpredictable, and I can only surmise from the participants' bodily behavior, from their "release," that an alluring experience was precipitated. In somatechnical terms, for example, release technique, one can feel the effect, working with compression and release (beyond tensegrity) to freeing up the scapular movement, the spiraloid, fluid morphology of the body. I witnessed such release in the immersants' opening up – the sun coming out (we could call this the weather of bodies). Some of them became much more inventive, exploring the altitudes and latitudes of their VE, reaching up, slouching down, even rolling over (here the guardian is very important in their Bunraku role). From an observer point of view, this is very fascinating, to see the almost delirious somatechnical responses, the immersants also learning and improving their skill with the controllers (which allows them to zoom into locations, tele-transporting themselves, flying like kites).

I conclude with a couple of other examples of Virtual Environments, exhibited in the United States and in Germany. The latter left me puzzled for many weeks after I experienced it, making me question any easy assumptions about the multisensory or trans-sensory when applied to immersive AR or VR installations, and alerting me to the more dissonant, disturbing side of technological intervention. The former, as a "stand alone" choreographic object, so to speak, appears now in a more conventional light and customer friendly way: a soothing experience of having been lured inside a simulated "planet." Japanese artist and filmmaker Momoko Seto's *PLANET ∞* (2017) was displayed at Moody Center for the Arts (Houston) in early 2019, in a small rectangular studio-gallery which you enter through a door and then a small corridor of black curtains, after which you reach an enclosure with four swiveling chairs. Four Oculus Rift HMDs lie on a small table at the rear. The room is empty at the time I arrive. I pick up one wireless headset from the table, unplugging the battery charger cable, and press the button, then sit on one of the chairs. The chair turns out to have been a poor tool for decompensation.[11]

And off I slip inside a seven-minute world of underground ocean ecosystems where strange fungi, insects, fish, giant carnivorous tadpoles and other biomorphic creatures float in an aquatic space (https://vimeo.com/220965048). The oceanic world is swirling around me accompanied by a familiar cosmic space sound track. The technical set up is easy and self-explanatory (especially for gamers it would present few issues): a small card is placed on the table giving me instructions. The gallery did not provide a guardian or guide. I stay on for a while, to watch visitors come and go, experiencing the aquatic work for this short 3D VR film, then they leave, after taking a selfie. While the VE runs inside the HMD, it also is also projected, via multiple video projectors, onto the four walls of the gallery, thus creating a CAVE-like environment with digital surround projection (including the floors). Having stayed a while, I note that the only problem might be the re-charging of the batteries of the wireless sets. If a visitor forgets to put the Oculus Rift headset back on the charger, subsequent audiences will have to wait, and figure it all out by themselves what to do.

It occurred to me that Seto may be using the medium in a way that is similar to 3D cinema – pass out the 3D glasses and your audiences know what to do, it has become natural. In our first *kimosphere* performance with VR, we did not think of participatory immersion with a VR-HMD as natural but as potentially destabilizing, a threshold needing to be crossed with the help of a "conductor." Designer Doros Polydorou acted as this conductor to make sure that all programs were running; he held the cables of the VIVE VR set (which was not wireless) to make sure the immersant would not get entangled or strangled, and encouraged movement with the headset and hand-held controllers. The entanglement side is sweet of course; I gather there are some stranger things one can design to make your heart stop and trick your brain, and perhaps also lead the visitor astray, so to speak. When I participated in the "Digital Materialism" workshop at Tanzhaus NRW, Düsseldorf (May 2019), I experienced such a form of precarious immersion. Although I was quite relaxed entering the scene when it was my turn (there were six groups of two people each, taking turns), I was not at all prepared for what was about to happen to me. I describe it in the following.

I am in the last group in line, but since there are other installations happening in the large performance space that take up my engagement, I do not pay attention to my predecessors. I also want to have a fresh mindset before straying inside. (The other workshop-installations include a holography piece; one with smartphone live feed and sampling of real-time improvisations, using a special app which drives me crazy; and one with a young 16-year old influencer who works on the TikTok social media platform where she creates 15-second dances and has about 1.4 million followers. She baffles me with her calm self-confidence, whipping out a new video on the spot, teaching us how to use silly special effects, all of which I find somewhat disconcerting.)

When it is my turn I move over to the corner where the scene was set up for *The Plank,* created by Michael Bertram, Stephan Meyer, and Christoph Vogel, designers from the Hochschule Düsseldorf (HSD).[12] They also work with a young dancer, Malila Ali, who acts not as a guide but an interloper. The set-up is simple, it is more of a technical workshop installation than a full theatrical scenography: VIVE goggles with headphones are placed on my head, two sensors are attached to my feet; the action space is small. On the floor: two wooden planks, one white cardboard box. That is what also what we see in the VR world: a narrow white strip, and the box over there. The immersant has the controllers attached to the feet, which is unusual and immediately arouses my curiosity. With my VIVE goggles and headphone I am sent off, having stepped onto the wooden plank, which is raised 10 cm from the floor, and is about 25 or 30 cm wide. You know this when you step on because you have already seen these two flat planks. In between the two planks there is small gap, maybe 30 cm wide. You imagine beforehand, ah well, you will be walking on this small plank, perhaps a bridge?

Inside the VR: I admit, the first minute or two are uplifting, I am high up, on the rooftops in Manhattan or Düsseldorf, I see the rooftops, chimneys and smoke rising, above me I hear and then see a helicopter fly by. I wave at it and am in a good mood. Then I look down and see the plank, it is not a bridge but a very narrow roof ridge, I assume it is metaphorically like a high wire act, to get me over there, forward, to the second ledge where there is that box. This is the box I must find and pick up – I am now in game mode, having to fulfill a task. In front of me, on my tiny plank, ah, some kids left their beer cans, there is also a bucket. They are in my way. I try to kick them with my left foot, the beer cans fall off, but I have to try a few times, since I am clumsy and am losing my balance. Now I slowly notice and realize, I am very very high up, maybe the 40th floor, there are courtyards down below, in the abysses, left and right. I stare down and see cobble stones very far down at the bottom. I inch forward, finally manage to trip off that bucket with my right foot. I now get scared.

At this point, adrenalin rushes through my veins, my blood pressure rises, I have panic attacks. I try not to look down. I am on the edge between first and second plank, need to step forward but I freeze, I am scared, cannot do it. I decide to give up. But that would be shameful. I cannot possibly go backwards. So I become angry and agitated, I take a deep breath, I imagine I want to hold on to the rooftop on my left, but suddenly everything has changed in the atmosphere: it starts to rain, a cold wind hits my cheeks, it gets quite cold. I am startled. I swear, and try to catch myself/my composure. I then quickly make the step to the second plank that gets me to the edge of a roof, where the box is further ahead. I succeed and for a moment I am relieved. I decide to crouch down, touch the plank with hands. My hands are invisible. I want to lean against the roof top, there is no roof top. I see pink and purple balls swooshing by, like the trail of an insect (later I gather that was the dancer – google-tiltbrush-painting something or other, popping these red balloons into my virtual environment, distracting me). I must not be distracted, if I make one false move I fall and die. I begin to sweat. I slowly, very carefully pick up the box, I decide to return with the box, I turn and rather than go through the slow scare and hesitation again, I try to speed up this time, make three very quick steps. I am back on the first plank, near safety, move forward, someone takes the box out of my hands and says "well done!" My plight is over.

Is this not rather ridiculous? My brain fell for the illusion, my whole nervous system went haywire. I should have known, my brain should have known, from the first moment, before putting on the goggles, that the planks were 8 or 10 cm off the floor, so I cannot possibly fall anywhere. That white box – no danger really, so unlike most video games with role play heritage, I did not expect this banal looking setting to be based around an agonistic scenario of combat and mastery (in the militarized masculine culture of technicity that is pervasive in mainstream virtual landscapes) or a complex ludic environment where the immersed player is invited to experiment and assess reality through fantasy. The brain, with the goggles on, took visual information at face value or switched into a naturalist mode of perception. It imagined to be on the 40th floor roof top in Manhattan or Düsseldorf, it took the helicopter for real. The psychosomatic response patterns amaze me – and I am slow to admit or understand in retrospect: were they automatic? I am sure there are other clever VR designs that are meant to induce hallucination or this high pressure of simulated danger (presumably all flight or military simulator VR programs work this way). Perhaps first-person shooter games work in very similar modes, and that is their exact allure, causing full excitation and adrenalin rushes. I am not too familiar with such subject positions and the codes which rule fantasy, game play, quest, multiplayer strategy, narrative geography, interpassivity, and so on, as I am not a gamer. (I did like the irony that seemed to resonate in the term "Ego shooter" used by friends in Germany for first-person shooter games; and I did for a while visit Second Life and entertained an avatar there.)

After the experience of *The Plank*, we are asked by the designers to sit down for a debriefing, each of us invited to describe our affective response and feelings. My partner in the 6th round, and some of the others, speak about the reactions they experienced, how they felt excited and stimulated to do the high wire act, or how they froze, how they panicked or how they enjoyed tossing the box into the abysses. In this discussion, I mostly express my anger; I comment on my astonishment having fallen for the illusion, having been tricked. I also now can recognize the hallucinatory effects machinery for Act II, when I step onto the second plank. The designers use two huge wind machines to create the cool air; their soundtrack simulates rain and wind sound. So while I try to explain my anger, I look at the wind machines. And then I cannot help it but smile – these machines are of course similar to the analog techniques (branches and towels) we used in *kimosphere no. 4* to generate an *augmented virtuality* – producing such disorienting effects of elemental materiality, recompensating for the lack of real raw sensation in the anime.

The Plank, then, is not asymptomatic at all. It falls squarely into the potential range of kinetic atmospheres designed to affect viewer's psychosomatic faculties, to transpose challenges and even risks to the immersant by provoking perceptual dissonances or psychosomatic excitations that can be hyper-illusionistic. The hyper-illusions, in this case, made me (or my psychogeographical movement inside the augmented virtuality) succumb to a phantasmatic experience, an "as if" scenario elicited through virtual embodiment. I literally feared to lose my balance and fall to my death, a logical paradox as I had seen the scenography beforehand and noticed there was no danger; my body schema (the conceptual image of my own body) was obviously intact. But the VR design fooled me into an almost sublime allure, extreme height and danger – it fabricated a certain excess which was however still based on realism, on a topological comprehension. This is where *The Plank* differs from the more ambiguous poetic "forests" and underground spaces that I had integrated into my dance environments – those do not evoke aerialists and highwire-walkers, nor danger and catastrophe, nor a FPS perspective on shooting down zombies or enemies. They are more directed at "recompensating" expressive vibrancy verging on a synaesthetic sensorial imaginary, and on working with expressive articulations for elemental conditions of an atmosphere. This points to the core aesthetic interests of my writing here, and grasping such an imaginary clearly and palpably is not easy – affective atmospheres that move us and immerse us remain ephemeral and unstable, difficult to pin down in their affective capacities as well as their emotional impact. What are forces of things or projections, raw or cooked, more or less tangible – if indeed kinetic atmospheres are neither things nor projections but processes and events that hover between and beyond, that lure and withhold?

The Plank is an interactive VR design installation, probably with an eye toward the artworld or commercial media world, thus less inclined to be used in medical laboratories or scientific tests studying first-person visual feedback to body ownership, full body illusions, phantom limb misperceptions, and perceptual or behavioral disorders in exteroception and interoception. The virtual environment illustrated by the *Plank* designers is an urban topography; it does not simulate a game world into which my avatar enters as an extension of my physical sense. There is no replacement of self or my own body schema with a homuncular creature or polygonal abstraction. I see a white box where there is a white box in RL and VR, and I feel my feet where I assume my feet to be when I walk the earth or, in this case, a ridge on a roof. I am choreographed through implicit instructions (to retrieve the white box), and I noted there was a dancer in real life moving around me, though inside my goggles she is gone and nowhere to be seen. This is not a multiplayer environment, unless one sees the helicopter as one, or takes the beer cans as signifiers of other folks having been on the precipice before me.

The immersive reception that I have likened to hallucination is triggered by the confusion of the factual and counterfactual, apparently a thin line that such engineered atmospheres help to highlight. Stepping inside the urban rooftop scenario I inhabit a virtuality and am still embodied in it as I have not left my bodymind. My sensorimotor functions and reactions are smooth. I become a presence in somewhere else (as we used to consider telepresence to connect us to another location), an other present realm generated by VR simulations (and later augmented by "meteorological" manipulations affecting temperature, climate and thus the properties of the atmosphere). Neuroscientific studies, such as the immersion tests done by Mel Slater or Olaf Blanke (at the Brain Mind Institute in Switzerland), confirm that the human brain appears not to draw a sharp line between reality and virtual reality – our nervous system reacts very quickly when it perceives danger and experiences arousal. As a form of self-protection, the bodymind reacts with signs of acute body stress responses, no matter if it knew beforehand that the plank was only 10 cm high and presented no danger.

The rooftop, forest or underground tunnel (the latter part of a new VR dance installation I am developing titled *Hannibal in Afar*) present opportunities for the imaginary and for telling stories; these kinetic atmospheres become aesthetic vehicles for trans-sensory modulations, and for looking at behavioral responses of the participants. Inviting them to share their feedback, we can learn about variable dimensions of participation (social, ecological, emotional and ethical) that instruct our forms of life. The discussion about curating VR Art, mentioned earlier, was initiated by Adinda van't Klooster (a Creative Economy Fellow at Durham University) in June 2019 after she had attended "DIS:SOLUTION" – the 2nd Virtual Reality & Arts Festival Hamburg (VRHAM). She was intrigued by the problems encountered, namely, that the twelve featured VR artworks had to overcome the fact that each work was for only one person at a time, involving a queuing process and long waiting lines, frustrations and disappointments. It was also mentioned in the discussion that queuing to have a solo experience, while everyone around you is looking at you doing it, might create discomfort. Simon Biggs argued that one can choose to turn the user into part of the show, although possibly at their expense as most people do not enjoy being on "show" like that.

The ethical implications are obvious, as they also become virulent in immersive theatre productions (by Punchdrunk, La Fura dels Baus, De La Guarda, Anne Imhof) which cast audience participation and witnessing, or delegate performance and procedural complicity to the immersed participant. I do not think one can generalize how audiences react; I have not seen many people refuse to use VR when there are other people around watching, although I agree with Biggs that some participants will have an innate aversion to being the center of attention and not being able to see those watching you.

Another curator and digital artist, Simon Poulter, suggested that for him it is perfectly all right that thousands of people may drift through a large scale immersive productions (such as teamLab's *What a Loving and Beautiful World*), but then the same would be true if it were a handful of people dressed in black. Poulter may have thought of the arguments made against commercial art, yet it is hard to imagine how an experimental arrangement such as *The Plank*, with a dancer working as interloper trying to distract the immersant (when I talked to Malila Ali afterwards, she told me that she was in fact trying to help the immersant *not* to fall off the roof!), could be reproduced for a large exhibition, with thousands of visitors. What I do agree with is the expectation that inevitably we shall witness much more porous processes of devising where installation, storification, live music, game playing and theatre are co-authored in technological environments, as Poulter proposed. He added:

> I'm less interested myself in being surrounded by sharks underwater or standing in a multi-coloured forest. There is a specific point though of interest, notably scale and ambition. I can't make *Toy Story* in my studio, I don't have the industrial means of production or the processing power but I can make 360 watercolour paintings for people to look at or experience in open source VR.
>
> (Poulter, 25 June 2019)

His not craving to stand in multi-colored forests made me pause, as we had in fact discussed scale and immersion and probed issues of curating VR (single experience in a booth, interactive experience in large space AR and VE), but rarely did discussants offer a more detailed critical response to aesthetic content of actual works at VRHAM, or of Simon Biggs' interactive VR installations with choreographer Sarah Neville, or Poulter's watercolor paintings. But I did come across Sarah Lucie's review article in *PAJ* critiquing *Programmed: Rules, Codes, and Choreographies in Art, 1965–2018*, an exhibition at Whitney Museum of American Art, New York (September 28, 2018–May 14, 2019), co-curated by Christiane Paul and Carol Mancusi-Ungaro. Lucie's response is helpful in this respect, as it looks at software and algorithms as new forms to create with, but keeps in mind the notion of a "choreographing" of materialities and concepts, trees and toys (linking, say the current VR art that I discussed here, to older conceptual art aesthetics that used ready-made video graphics or loops of various imaginaries, from abstract spectral videograms to clown tortures and infinity rooms). Lucie proposes to look rigorously at the programming of today's trans-sensory hallucination artworks, anticipating a "post-human" trend in the cybertyping, algorithmic and AI discussions, yet asking how "human" is defined or how the still existent binary divisions between human and post-human are retooled, how the elemental and the animic are dislocated/relocated.

If one were to think historically and remember earlier programming of video installations, such as Nam June Paik's massive and hypnotic screen work *Fin de Siècle II* (1989) featuring a continuous re-sequencing of music video clips, the conceptual algorithmic lineage is apparent. But the material presence of Paik's tall multi-television monitor wall is of course overbearing, and at quite a remove from goggles strapped around one's head. Like with Paik's *TV Garden* (1974–77), we face a hypnotic machine; and in the case of the "garden" it is obviously a constructed environment that invites the visitor to become engulfed by the surrounding density of foliage and the flickering lights (and sounds) of the forty or more television monitors, global grooves and voice overs nestled among the leaves, the manipulated music and modern dances mixed with Japanese Pepsi commercials and some Stockhausen. Gardens and plants need tending, mediate matters.

I conclude by invoking atmosphere as such an undefined aesthetic of physical-material, immaterial and imaginary mediation. Atmospheric forms, as I discussed them in the *kimospheres* and the other VR examples, involve somatechnical reactions, transforming relations between somatic bodies and extended choreographic objects. Entering into the kimospheres of augmented virtuality raises difficult questions about the indigenous and neurophysiological response patterns of bodies, their transportations, falls or ascensions into poetry or the spacetimes of the imagination. The material and computational sides of constructed realities force us to admit to mutual interrelations, to lives following codes and coding. Experimenting with kinetic atmospheres points to the extended choreographic also in terms of programming, rules, instructions, procedures and algorithms, as much as it invites and speculates on living responses to poetic combinations of practices and technologies.

Notes

1 For a cautionary and critical introduction to the "moving objects" (the works, pedagogies, legacies) of the "disappearing Bauhaus," see Saletnik and Schuldenfrei 2009: 1–9. See also Birringer 2013.

2 The lab was mentioned before, but I sketch its history more fully here: The Design and Performance Lab (DAP) was founded in Nottingham in 2004, and moved to London in 2006, after NTU's Live Art program was discontinued. It has produced a number of telematic/networked and onstage performances, most notably *Suna no Onna* (2008), *UKIYO (Moveable Worlds)* (2010) and *for the time being* (2012–14), the latter a suprematist homage to the early futurist opera *Victory over the Sun* created by Kruchonyk, Matyushin and Khlebnikov, with designs by Malevich. Since 2014, DAP-Lab cooperated on the European METABODY project (www.metabody.eu), with other arts organizations

and research labs including Reverso, Hyperbody, STEIM, InfoMus Lab, Stocos, Palindrome, K-Danse and Trans-Media-Akademie Hellerau (Birringer 2017). After a joint production in Madrid with the Hyperbody architects from TU Delft, DAP-Lab staged a series of six *kinetic atmospheres* between 2015 and 2018, in Madrid, Paris, London, and Durban, South Africa. A further installment was developed for ISEA 2019 in Gwangju, S. Korea, to be followed by a dance work addressing the climate crisis: *Mourning for a dead moon* (2019–20). Filmic excerpts of many of these works are available online. DAP-Lab's website is here: http://www.brunel.ac.uk/dap. On the website are numerous links to publications, conferences and exhibitions that featured our innovations in wearable design created by fashion designer and co-director Michèle Danjoux (http://www.danssansjoux.org).

3 Some researchers prefer the term VE (virtual environment) to draw attention to a body's remaining environmental/spatial connection, even if the VR head-mounted display (HDM) is restrictive like a blindfold and blocks out all visual information from the surroundings. The VIVE and Oculus Rift headsets are the most commonly used ones. Sensory navigation and orientation continue in the virtual world that is layered over the concrete one. Cf. Thomas and Glowacki 2018. The goggle systems so far were mostly cable-connected to the computer hardware; the new VR generation, with Oculus Rift, will use wireless system so that movement will become less restricted. The term *augmented reality* (AR) refers to interactive experiences of a real-world environment where the objects that reside in the real-world are *augmented* by computer-generated perceptual information. When we began to use VR headsets in our installations, we tried to invert the ocular-centric headset information with supplementary real world organic objects and tactile occurrences (air/wind, smell, temperature, pressure, contact, etc.), thus the somewhat paradoxical or ironic term *augmented virtuality*, which probably was known in the interaction design, human factors engineering and 3D visualization communities for some time; see Milgram and Kishino 1994.

4 In Chapter 3 I briefly described an extensive outdoors "blind-folded" exercise I did with eyes closed during the "E/motion frequency deceleration" Choreolab in Krems, Austria. In the same laboratory we also practiced butoh and body weather techniques (the latter derived from the field teachings of Min Tanaka). For my forays into choreographing moveable space (interlinked with telematic space), see Birringer 2010. My work with networked performance spaces dates back to the early 2000s when I co-founded the ADaPT (Association of Dance and Performance Telematics) collective while conducting digital and motion capture research at Ohio State University. The ADaPT collective was a very energetic group of performers, filmmakers and digital artists who worked together for some years (across the United States, Europe, Japan and Brazil) and even achieved a nomination for PRIX ars electronica 05 based on its telematics research and development of international digital community. This telematic work I know see as a precursor for my later experiments with VR and VE.

5 It was fascinating to find out that Schlemmer, during the early 1922 Stuttgart performances of the *Triadic Ballet,* was one of the three performers wearing the constrictive costumes himself, discovering that unfortunately he "could not see the scenes in which he performed and hence did not have an overall picture of the performance" (Schlemmer cited in Cramer 2014: 22). He also comments on the differences in techniques, for example, how trained dancers like Albert Burger and Elsa Hötzel might wear the elaborate full-body sculptures and oversized masks, compared to non-trained performers.

6 Forsythe's choreographic objects as entanglement of the participant: this was also beautifully demonstrated in *Films* (camera: Dietrich Krüger), an extended improvisation with ropes exhibited during *William Forsythe: Suspense* (2008). A very different, exuberantly programmed animation of ragdoll-like figure simulations (driven by motion capture recordings), floating as kinetic cloth inside a fake curtained "theatre,"

appears in Canadian digital artist Mike Pelletier's *Constant Iterations* (2020): https://vimeo.com/399816991.

7 See my earlier comments in Chapter 4 where I refer to my public engagement lectures on sensory techniques and somatechnics. This work on trans-sensory fluidity is indebted to collaborations with colleagues in social work, disabilities studies, and the Welfare, Health, and Wellbeing Research Group at Brunel University, and in dance (with collaborator Robert Wechsler/Palindrome Dance Company). I have also been inspired by Fayen d'Evie 2017.

8 Quoted from the exhibited books and photographs in *Mushrooms: The Art, Design and Future of Fungi*, Somerset House, London, January 31–April 26, 2020. I'm indebted to Matthew Bevis for alerting me to the particular lines of Charles Wright's landscape poems in *Oblivion Banjo* (2019).

9 Haein Song, a dancer/choreographer who practices *kut*, the traditional Korean shamanic ritual performance, recently completed a series of works that intermesh the traditional and the digital. She describes the ritual techniques (*mugu*) deployed to achieve the desired collective healing and well-being effect of the practice in *Ecstatic Space: NEO-KUT and Shamanic Technologies*, PhD thesis, Brunel University London, 2018. This stands in a complicated, not easily resolvable relationship to the "wellness" effect that teamLab's large-scale immersion environments might seek to generate, as in their *Borderless* "world of artworks without boundaries" (https://borderless.teamlab.art/) now installed at the Oidaba MORI building (Digital Art Museum) in a 10,000 square meter environment equipped with 520 computers and 470 digital projectors. It is not an augmented virtuality but completely artificial. As I have not been there I cannot comment; then there are the 360 degree all-around 3D digital projections of virtual worlds in the newly opened Nxt Museum in Amsterdam. The first show at the Nxt Museum, *Shifting Proximities,* opened August 29, 2020, and the critic Hanno Rauterberg comments on the intoxicating totality of such immersion that aims at "delimitation" – dissolving the ontological division between the world of oneself and the world outside, between inside and outside, sub- and object. It is not about wiping out or even overpowering the individual, Rauterberg suggests, setting this all-around immersion apart from the old romantic Sublime, which wanted to show the individual their own nothingness. In Amsterdam, the illusion is always aimed at "participatory harmony" (*partizipativer Einklang*), to grasp something like "the cosmic connection between black holes, dying stars and our own existence" as the museum promises (https://nxtmuseum.com/). See Rauterberg's skeptical review, "I'm amazed, therefore I am," *Die ZEIT*, 37 (2 September 2020).

10 Cf. Danjoux 2014, 2017 and Birringer 2010. For the *kimosphere no. 3* staging, arranged for vision-impaired visitors from the Ealing Association of the Blind, see https://www.youtube.com/watch?v=5DdAcv37jmc. For *kimosphere no. 4*, see https://www.youtube.com/watch?v=0aIW6Klfm1g.

11 In his thought-provoking comparison between raw cave and CAVE (Cave Automatic Virtual Environment or wearable digital VR environments that provide technologically mediated immersion), Salazar Sutil points out that unlike the multimodal experience of "landesque immersion," VR environments produce sensory *decompensation* as they are unavailable to touch, taste or smell, resulting in a disconcerting discrepancy between

overstimulating audiovisuals and understimulated tactile-kinesthetic, olfactory, and gustatory inputs. Decompensation is a medical term that denotes the failure of an organ (for instance, the heart) to compensate for the functional overload resulting from disease... In the case of iVR technology, sensory decompensation is caused by the overload of visual and sonic stimuli in surround screen and surround sound interfaces, and a lack of full-body and raw material sensation, which causes an imbalanced perception of reality at the sensory level, quite often prone to feelings of uncanny disconnection and alienation.

(2018: 157–58).

12 The installation of *The Plank* was developed by the working group MIREVI (Mixed Reality and Visualisations) in their mixed reality research lab at the University of Applied Science, HSD, Hochschule Düsseldorf. See: https://mirevi.de/. The prototype intends physical (the wood plank) and digital objects to co-exist and interact in real time, in order to raise the level of immersion. The user is provided with a small task (carrying the box from point A to point B) that raises the level of immersion even more. (An older version of the plank, installed in the University's lab, is here: www.youtube.com/watch?v=DnLDCz2Ppjg) As Christoph Vogel told me in an email, the team basically intended to provide the *Plank* installation for the curators of the Digital Materialism workshop, for them and the participants to get creative; the dancer got involved in the last minute, her role was more an artistic choice made by the curator of the workshop, not the design team.

6

COMPOSITION OF ATMOSPHERES

Floating islands

> The croaking of toads in the fields flooded by clear waters: I go on living thinking day and night about the moon that, unfaithful, floats from field to field.
>
> (Hanaogi IV, print by I. Koryusai, 1785–89)

Floating

The architect Zaha Hadid is reported to have said "why stick to one when there are 360° possible" – and this has always seemed like a good idea when you start out composing a dynamic architecture for multimedia dance and performance. For any performance, in fact.

In this chapter, I associate the idea of "floating" with digital intermedia scenography and the particular tasks we face when composing with mixed realities and mixed temporalities, the *time-space* of recorded action and the *time-space* of interactivity and real-time synthesis. As I move along, I shall question or modify my ideas on *composing,* and reorient assumptions about environmental architectures toward other notions of ecological experience and body weather, atmospheres that can perhaps not be *designed.*

This reorientation now appears imminent, more drastically than I ever could have imagined. It was not something I had planned for this chapter or the book. Therefore, I want to pause for a moment, right at the start, for a reflection on *slow space* and *slow time* precipitated by the 2020 lockdown experienced by peoples across the planet as an emergency response to COVID-19, the coronavirus disease emerging in late 2019 in China and then becoming a global pandemic in early 2020. It has been a crisis of unprecedented proportions, certainly in my life time, as I have no recollection of the cholera, the Spanish Flu pandemic of 1918, or earlier plagues such as the Black Death (considered the most fatal pandemic recorded in human history). I have no recollection of a lockdown.

DOI: 10.4324/9781003114710-6

Nor am I comfortable hearing comparisons being made to war time. This global pandemic cannot be fought or comprehended as a war, but it may well be the first pandemic that generated overarching systemic effects of suspension of activity and economy, the quarantine of entire countries and continents. This suspension will have severe and distinct effects on almost every country, immediately as well as in the months to come, after the emergence of the virus and its disease. The plague and its human costs thus create a barely understood complex scenario that overshadows everything in our global ecologies and our being in the world.

This plague is not something Antonin Artaud had in mind when he wrote his 1934 essay "The Theater and the Plague" – and yet it is not impossible to imagine Artaud deliberately theorizing the notion of the theatre as epidemic with awareness of early modern medicine's radical reconceptualizations of the body and its boundaries, as well as of the ideas of health, illness and immunity. The theatre's boundaries naturally proved to be porous, and those discourses of theatre, medicine, and immunology may need to be returned to, in due course. I already mentioned earlier that another crisis, after 2015, was perceived through scapegoating boundary crossers, the migrants and refugees in the Mediterranean who began to be understood as a major political and socio-cultural challenge to the EU and its open borders. Global migration at times was labeled a kind of "contagion" by populist, nationalist discourses, a threat to the common wealth. At the same time, global migration of course also was an inevitable outcome of globalization and neoliberal market policies.

I grew up, after World War II, hearing innumerable stories from parent and grandparent generations: about hunger and fear, the devastating experience of cities being destroyed through bombing air raids and fire, rubble on the streets, armies moving through, family members returning home traumatized and mutilated, then the clearing of the rubble (my mother belonging to the generation called *Trümmerfrauen*), the slow post-war reconstruction and longing for a new normal. Migrations and innumerable relocations due to the effects of the war were commonplace after 1945. Concentration camps and the Holocaust left a most profound wound and trauma, continuing to exist and affect the families of survivors and the Jewish people. More recent studies show that war and Holocaust traumata are not limited to survivors themselves, but can be passed on to the next generation born afterwards, raised in their shadow. The connections between these shadows of trauma, ritual and immersion are tentative, at this point in my thinking, but I am grasping at something that is extraordinary, perhaps threatening but surely uncertain and uncanny. Connections that are also disabling: undermining the sense of theatricality that one knows to be underlying the ancient and modern apprehension of ritual. We recognize rituals through the repetitions we enact, just as the seasons repeat themselves and religious festivities follow their calendars. The significance of ritual rests in its ability to measure our repetitive lives, and thus fresh performances of rituals will remind us of earlier performances and how we might remember them emotionally, intellectually and spiritually. Recoveries and survivals, for example, of grave illness or surgery, are also part of such refrains that we know, patterns and psychosomatic practices required by healing. My first steps will be to approach contagion through the notion of ritual.

The 2020 pandemic has brought a range of new terms into currency that we had not known or experienced in this sense beforehand, such as self-isolation, social distancing and dissociality. The idea of a curfew or quarantine of course was comprehensible; I felt it was socially acceptable for the time being, even if legal-political debates started soon after. Psychosomatically, quarantine begins to resemble a private ritual. It also of course involves the repeated self-cleansings that are ordained upon us, ordered as a new protocol of self-sanitation and autoparticipation in a (presumably) collective ritual cleansing and self-protection. This ritual is politically sanctioned as the protection of all others. That is the mask. Medical advice I receive tends to clarify that the mask is less helpful to protect myself than it is helpful to protect others.

Elias Canetti, in *Crowds and Power,* proposed that masks are fascinating since they attract us and at the same time force a certain distance. Alongside this attraction to masks and masquerade, Canetti was fascinated by the "mass" (a better translation for the German *Masse* than "crowds"), the masses and multitudes in civilization, a notion that he saw as non-confined by a national attachment, as non-organized. And it is beautifully paradoxical, in the current context, to hear Canetti argue that being immersed in a mass, the individual is relieved from a phobia of *contact* that so often regulates their relations with others or influences their fear of touching/being touched by the unknown.

> It is only in a crowd that man can become free of this fear of being touched. That is the only situation in which gear changes into its opposite. The crowd he needs is the compact crowd, in which body is pressed to body; a crowd, too, whose psychical constitution is also dense, or compact, so that he no longer notices who it is that presses against him. As soon as a man has surrendered himself to the crowd, he ceases to fear its touch. Ideally, all are equal there; no distinctions count, not even that of sex. The man pressed against him is the same as himself. He feels him as he feels himself. Suddenly it is as though everything were happening in one and the same body.
> (Canetti 1984: 15–16)

Canetti distinguishes between open and closed masses. There are open masses that arise out of nothing and can only exist as long as they grow. They are emergent. And there are closed masses that are limited externally: they do not grow, but they do not disintegrate as quickly as the open ones. Closed masses, he suggests, consist mainly of repetitions, for example, in the form of regular meetings or rituals (e.g. in religious worship). Crowds need "discharges" and these are moments when all members of a crowd feel the same, even though in reality they are not. Nor do all lives matter equally.

One might think that most readers of Canetti, including Toni Negri (the author of *Multitudes*), recognized the critique of Freud's emphasis on the individual psyche and Canetti's effort to explain the phenomena of mass movements (as he experienced them in the 1930 during the rise of fascism), while they also often thought of Canetti's "mass" as a formless and floating, swarm-like phenomenon. A few years ago, when "locative media" were all the rage, swarm behavior was studied in various quarters, including the dance world. Theatre scholar Gabriele Brandstetter co-edited a much discussed anthology (Brandstetter, Brandl-Risi, and van Eikels 2007), which she introduced by using the term *Übertragungen* (transmissions), referring to two types of transmission phenomena that she considers a challenge to movement research. On the one hand, the swarm is a dynamic collective (similar to the way Canetti understands masses) whose movement organization is particularly fascinating because it combines a free play of forms, at the same time, with a relative degree of control yet manages without a central control or supervisory authority. Brandstetter transfers observations from flocks of animals to human behavior (and dance improvisation), thus offering insights into scientific and kinetic research. But, mostly, her ideas on political, social and artistic actors/agency open up provocative questions about how we think collectivity (and thus ritual), how we imagine it when it is an emergent effect or freely dis-organized. When Brandstetter uses the notion of *(e)motion* for the analysis of swarm behavior, she is drawing attention to the dynamics of transitions and interactions between physical movement and emotion, "outer" and "inner" movement, voices and voice bodies, somatic knowledge and movement knowledge in dance and its relationship with medicine and physiotherapy (Brandstetter 2007).

Yet most interesting to me seems the underlying idea I sense in Canetti's figure of the *mask* that cannot be properly seen, as changing or transforming or emerging, and yet impactful, how it affects metamorphosis. But who is metamorphosed, the wearer or the one who looks at the mask? How is (e)motion transformed? A certain relation of distance is necessary in the closeness of the discharge/change, to see the immutableness of the mask which resembles other masks, so I take it that Canetti is discussing the effect or affect of the dancing mask on others:

> The working of the mask is mainly outwards; it creates a figure. The mask is inviolable and sets a distance between itself and the spectator. It may come nearer to him, as sometimes in a dance, but he must always stay where he is. The rigidity of form brings about distance: its immutability produces its fascination. [...] Charged with a menace that must not be precisely known – one element of which, indeed, is the fact that it cannot be known – it comes close to the spectator, but in spite of this proximity, remains clearly separated from him. It threatens him with the secret amassed behind it.
>
> (Canetti 1984: 375–76)

The figure of the mask thus can be connected to the ritual sense of performance when audiences gather together or move together with the dancers or the music and rhythms of the kinetic objects that animate the environment. I also wish to connect this figure of the mask to the ambiguity of distance and proximity in the current world in which we must keep a certain distance. The mysteries inside rituals have to do with the inscrutable, and the mask reminds us of the visible and audible strangeness or foreignness that attracts (invites) the hiding of mutability, in the phantom of the mask. The mask protects and allures while also "threatens" both wearer and the perceiver, I imagine. It blurs the wearer's vision.

The pragmatic collective social choreography, as I experienced it in London in March 2020 and in Germany during the following months of the lockdown, was coherent if one argued, and believed to adopt one's actions, on behalf of the common good and the protection of others and self. Argumentation was necessary, as the lockdown of course meant restrictions of one's civil rights and the accustomed freedoms one had taken for granted. The idea of the common good and welfare became an important corner stone for reflecting on the ethics of sharing public space, economic and ecological space (*latente Allmende* or latent commons, as Anna Tsing calls it), ritual space, spaces of health care, care for the elderly, food care, special needs, care of animals and nature. Shops and factories, schools and kindergardens were closed. Theatres and museums and all arts activities shut down, festivals canceled. The compulsory wearing of face masks became the norm (no reason here to make any further theatrical allusion to masks and what Brecht, in *The Measures Taken*, calls the *Auslöschung des Gesichts*, but it is of course a strange irony also in regard to the laws that were much contested in some European countries regarding Muslim women and their wearing of hijabs).

We all had to adapt. By the time I internalized the self-isolation, I was prepared for a long sustained period of solitude, a slow time to listen to nature, the trees, plants, and birdcalls in my environment, and also the silence in my inner space. Violence and civil strife over racial oppression, as it erupted later (in late May 2020 in Minneapolis) and then raised a storm across many cities in the world, seemed far away and unwarranted. Yet assaults on the body, from a dangerous virus, and from dangerous racist ideologies and practices, came to be an overbearing trauma by midsummer, a trauma for which people needed proper tools to recover from. Existential questions not only moved to the foreground for most; the complex interrelations between quarantine, oppression and social accountability were suddenly immensely discomforting for all who felt trauma, experienced loss and a deep sense of fear, or yearned for intimacy and the accustomed lightness of interpersonal being. In regard to artistic practices built upon embodied expression, it quickly became apparent that all of us had to re-imagine breathing together, the breath in dancing and theatre and music making that is absolutely vital and reassures us of the simplicity of being which also is intimacy. The intimacy of a choir singing a chorale, or a congregation sharing a hymn at a funeral.

New forms of intimacy, or strategies of imagining the future of theatre and ritual, are required. One might argue that creative and ritualized compositions, for example, in the creation of kinetic atmospheres as I discuss them here, are now *more necessary than ever*, at this time when our vulnerabilities have been exposed and we need to keep exposing them further to examine them. We wrestle with such intimacy and what it will be. We also learn about behavioral modifications, surprisingly quickly implemented regulatory behavior that alters our "tacit agreements." Norbert Elias had assumed civilizational habits and social gestus evolve over a very long time – a slow process. But in our current globalized digital era, the pandemic has led to a highly accelerated pattern of adoption of changed social manners due to external constraints and self-restraints practiced by most. The notion of "tacit agreements" I take to be a cultural set of unspoken manners or etiquette that we know intuitively when we encounter art works, rituals or performances. I remember writing about it when British sculptor John Newling, after having created a significant body of participatory public art words, collected his notes and essays on these *public transactions*. Since we were neighbors in Nottingham, he asked for my commentary.[1]

When I meet John at his house, we sit at the kitchen table to eat from the fresh bread he has baked. He tells me of the bread machine, and while we eat we are reminded that as human beings we first draw on small, even mundane rituals that allow us to enter into a conversation, a sharing, a communion. The arrangements for a meal or a conversation in a home carry echoes of other arrangements, more public and formal perhaps, or burdened with tradition and socio-cultural values (or distinctions, as Bourdieu taught us). As members of social communities, Newling believes, we need such common agreements. They are a necessity even if, as he also suggests, they are "tacit" – unspoken conventions we understand as necessary as they allow us to function within the social bounds of tolerance, acceptance and respect, and according to the rules we have learned in order to understand or manage the business of "transaction." Transactional behavior, conduits of communication and communion (and there are religious or sacred undertones involved too, apart from legal, economic, social and political connotations), thus also involve faith in the currency, in the breaking of bread as much as in the uncertainties of site-specific art or installations, the public reactions and participation.

Floating colors or, a wop bop a loo bop a lop bam boom

I am invited to join a website forum on the corona crisis where I have to choose a color for my name. Light red 3 is my color. After a while, I realize I am not isolated and in distress as much as I am tired of reading panic news, blogs or diaries on the COVID-19 crisis. I am retooling, partly to avoid being overwhelmed by news that could be traumatizing and damaging to one's mental health, as there are so many conflicting viewpoints floating around the ether. I am retooling also in the sense of adjusting to a few newly configured transactions. No one shakes hands with me anymore or wants to be hugged or kissed.

And still I get asked to join this or that forum, to enter my opinions or relate what I have found, from the rhythm and blues of popular debates, the wild songs and *Werkstattberichte* of artists, to the statements by virologists, pulmonologists and medical researchers, criticized and questioned by the admonishers, master thinkers and prophets (Sloterdijk, Agamben, Gumbrecht, Žižek, Ferguson, Finkielkraut et al.). It is hard to keep up. May Day approaches and the weather forecast for demonstrations is bad. No public assemblies are permitted yet in the county where I am locked down. Little Richard dies and I read long obituaries. Over the weeks to come, many more obituaries. Bob Dylan sings an interminable ballad, *Murder Most Foul,* that the music critic in Germany's weekly *Die Zeit* considers a funerary playlist for the end of the United States.

Preparing for my eventual next appearance in the theatre, I keep up my bodily practice. I train outside, in fresh air. Of course I do not know when our next dance performance will be. They say everyone has a bit more time now to write and reflect, naturally. Manjunan Gnanaratnam, a composer friend from Minneapolis, mentions to me that he feels able to listen to himself better now. I find this is not necessarily true in my case, although the assumptions, during the general lockdown, make sense. Many assumptions also change on a daily basis. Another colleague, dance researcher Michael Kliën (Duke University), suggests that I join the weekly sessions of the Social Dreaming Matrix arranged by his Laboratory for Social Choreography. The Matrix promises to enable associative thinking (based on the dreams we share) – for new thought to emerge. I decide this will be a good exercise.

I also join a COVID-19 "archive" to which a young Swiss writer had invited me.[2] I choose my color for my writing. When I look at the archive for the first time, I decide it will be impossible to catch up and read everything. I already give up. Then I wonder whether my writing could be like a small dance, something that moves outside and blossoms like a tiny flower, peeks through the asphalt cracks or makes a small noise like woodpeckers do. Can one bring photographs and sound files here? Or just ask questions? Or not participate, but acknowledge that this – asphalt cracks and archives – exists? The archive has an index and categories for the entries (News, Agamben Debate, Fascism, Libertarianism, Climate Change, Animals, etc.). Can one not be sophisticated, and remain also uncategorized? Perhaps one's writings, like small movements in the outdoor areas we are permitted to walk or run to, are really only interesting to the one who writes, as a mild self-confession, an excuse. "Dark writing," as playwright Ngozi Anyanwu calls it during one of the Segal Talks broadcast by Segal Theatre Center and HowlRound Theatre Commons[3] during the first months of the pandemic.

She means it as method to drive away dark thoughts including the guilt one might feel for not knowing how to engage, how to share one's isolation, how to not feel abandoned. But small movements should never need an excuse. I realize of course I am not Little Richard, but I imagine, in my small movements, the infinite potentials of funk, the little ecstatic falsetto shouts and restless flamboyance, the woos of *kinetic flamboyance*. It ain't the ocean, it's the motion, he sang for us. In the second half of this chapter, my evocation of *UKIYO (Moveable Worlds)* is composed as a kind of libretto, replying to this flamboyance. In that sense, it is also an homage.

Resurrection

Two trees were damaged, in the garden, since before and after the outbreak of the epidemic. One was the crown of a blue fir tree, decapitated and thrown across the land by the storm. The other: a slender beech, it fell sideways and was caught by another beech, and now they formed a strange duet. I tried to re-erect the fallen one, even though the root seems rotted. When it happened I was in London, far away. Then I crossed numerous borders, and am back in the countryside forest areas of the Saarland. Polish writer Olga Tokarczuk, author of *Drive Your Plow Over the Bones of the Dead,* said the other day that borders are alive and well, they have their "second coming" in the EU. This sounds humorous but of course it is not. I crossed back over the channel from Dover to France, Belgium, Luxembourg, then on to Germany. The last crossing was not the bridge I usually take in Remich (barricaded and closed), but via the highway, at a location called Schengen. The police was polite and firm, checking my reasons for crossing the borders, sending me off into quarantine (Figure 6.1).

I ponder the increasingly devastated garden I find upon my returns – some trees reaching upward to the sky like ghostly torqued skeletons, some buds already dead before blooming, the earth a strange green-yellow, fallen branches strewn all over, some mighty green fir trees waving in the winds nearly ready to fall, too. The little solar lamps stopped working. There is a new ant hill,

FIGURE 6.1 Tree resurrection, with *Majestät des Schreckens* paper clipping, 2019 © Johannes Birringer.

which I welcome, although my Russian neighbor, waving at me from his garden, tells me they are bad news and need to be dealt with. I find holes made by field mice, and mole hills; in early April this is rather unusual, but I cannot remember now whether I have ever been here in April and May. Herbert Blau, one of my theatre mentors, some years ago taught a whole series of rehearsal workshops on Kafka's *Der Bau* (*The Burrow*), associating the traumatized and paranoid behavior of the burrowing creature with actor training. Or rather, treating the Kafka story as a parable for an acting theory – a speculative theory of an architecture – of vanishing, a disappearance and re-appearance on some stage or frame of consciousness. Most likely, Hamlet and the ghost father were on his mind, with the actors wandering in the fog and asking, what's there? Who's there? Anybody here?

On the north side of the garden, my resurrected tree gallery is holding up well, in spite of the storms. The barks are peeling off, something I did not see coming. The newspaper clipping I had attached to one of the tree sculptures ("Von der Majestät des Schreckens" – a review of Romeo Castellucci's staging of Mozart's *Requiem* in Aix-en-Provence), which hung there for months, even in midst of winter, has disappeared. Perhaps eaten by magpies? Resurrecting trees, or composing statues with the truncated pieces of the stems, is a form of scenography too. I also use some of the trunks in our dance installations.

The chapter subtitle of "Floating Islands" now gains a bitter resonance that I had hardly anticipated. The confinement of "quarantine" was probably not intended by earlier uses of the term *isolation*, which Carlyle (in the mid-19th century) likened to the sum total of human wretchedness. Surely he was referring to social isolation and perhaps even pondering a religiously motivated critique of the emerging capitalism in industrialized societies. The term "isolate," however, goes back to the Latin *insulare,* literally meaning "making into an island." And as governments were issuing commands to self-isolate, slowly and gradually (after inexplicable delays), when I first heard the regulations mentioned I thought the term was *self-insulate*, to become an island unto oneself or create protective boundaries around oneself.[4]

This left a strange taste in my mouth as only quite recently, in 2018, I had organized two workshops on "Shadows of the Dawn: Migration and the Indeterminacy of Community and Immunity" targeting the so-called refugee and migration crisis that had taken over the news and political maneuvering in the Mediterranean and the larger EU. I wanted to protest the creation of protective borders. The migrations of political and economic migrants, from conflicted war-torn and impoverished regions in the Middle East, Afghanistan and Africa to the North, caused a number of debates that frequently touched upon the language of contagion and immunity – and the logics of encampment, states of exception, biopower, immunization and "thanatopolitics," as philosophers like Robert Esposito or Giorgio Agamben had theorized them. The indeterminacies of community that I sought to explore in our workshops are now, at this point of writing in 2020, overshadowed and dramatically heightened by a pandemic which has exposed, more than any other event in recent history even exceeding the expanding doom of the climate crisis, the entanglement of all, between everyone, everything and the world, both at molecular and global scales, revealing our intensive fragilities.

In Chapter 4 I mentioned these workshops, and the invitations extended to artists from Iraq, Syria, Lebanon, Iran and Greece, as I hoped we could learn, in a collective physical workshop that also included performances by Iraqi Bodies, BADco, and Lambros Pigounis, from one another's approaches to movement, transitions, voices in exile. Having evoked architecture and scenography means to think of them as transitional transactions too, as *interactional* – and to propose composition as a design process that understands spaces (settings) as changeable, changing, unstable, moving, moveable, multi-perspectival, multidimensional. If built environments had once been considered stable structures (buildings, houses, rooms) with foundations, floors, walls, columns, ceilings, roofs, etc., organized into spaces for living and working, giving form to civic or individual aspirations and providing the stage for social infrastructures and urban life, we have become increasingly aware that fixity or duration in architecture are not absolute values. Technological change has reconfigured the typology of buildings, and newly recombinant forms are emerging all the time, reshaping older ideas of functionalism or attention-seeking design invention as well as reflecting greater sensibilities toward social and environmental contexts, toward proximity and distance (*Abstandsregel*, as it is called in German building code). In theatre, even the most minimal stage set is constructed for (inter)action, for potentials to unfold, for the evocation of imaginary scenes – whether naturalistic or abstract-expressionist, mundane, poetic or absurd – to be performed in such settings. Perhaps it is true today that architects do not only draw plans of rigid geometries and structures but imagine movement, folds, and flows of people, unfolding possible ways of inhabiting spaces, and of returning home (from global mobility and the desire for constant, accelerated mobility). They have to be concerned with people's behaviors, and cultural values – and memories of such values – particular to a specific community.

In the introduction to *Machining Architecture,* Lars Spuybroek (2004: 6–7) uses a neurophysiological approach to architecture's fundamental need to be "plastic, topological and continuous," recalling Merleau-Ponty's philosophy of perception which posits that action and perception both form a system which varies as whole, through the *Gestalt* of a body-schema of body-image. Spuybroek interprets such variation to imply the system's mobilization that combines actual structure and virtual organization, behaving like a human body-schema that is plastic and dynamic. Neurological studies confirm this: our body-minds are updating and remodeling all the time, allowing us to reorganize our action-perceptions according to the contingencies of experience, namely how they match the contours of our bodies and extend into space.

A "floating" architecture, as I call it, therefore suggests an intimate connection between activity and perception, between spatiality and the viewing/

coordination of objects. Rather than thinking through theatrical production from the points of view of the actor or dancer, I propose to take a different approach in the following, concentrating on composition of atmosphere in terms of this aforementioned plasticity and formative procedure, as a design art that *situates* interactions, preparing for active embodiment in settings both analog and digital. Contextual design propels relations and negotiations, preparing the ground for movement and action in performance, building the dispositions for engagement with the real and the virtual, with changing spaces and changing temporal frames. These changing spaces are living spaces in a living theatre, and therefore the movements of actualization and virtualization (the "machining architecture" posited by Spuybroek and his innovative Rotterdam-based studio, NOX) require attention to body-schema and the rebuilding of the schema.

At the same time, evoking architecture and design does not mean that we can easily separate the setting (*mise en scène*) from directing and performing, nor does the design approach easily provide conclusive answers about participatory models of art production, about spectatorship, ritual behavior, or the casting of audiences into the work. On the contrary, there is a very intricate and intrinsic relationship between directorial concepts, *mise en scène* and performance enactment, which has historically evolved throughout the increasing dominance of the director's and stage designer's roles in the 20th century, exemplified by the significant influences we attribute to Stanislavski, Appia, Piscator, Brecht, Svoboda, Grotowski, Brook, Kantor, Wilson, LeCompte, Suzuki, Foreman, Lepage or Bogart. Performance architectures intimately conjoin the tasks of directing/conducting and designing for interactional performance. The "interactional" dimension has a technological side to it, historically understood within the scope of the theatrical avant-garde in Western and Eastern Europe, Russia, Japan, Canada and the Americas that formed my training and awareness. The design side, moreover, cuts across many fields of practice that fuse theatre with visual arts, music, architecture, fashion and engineering, among others. Composing or conducting atmospheres is always an interactional aesthetic; it only ever addresses the audience's conduct. We build performances under the moon light, and flow is a keyword for the conceptual spaces we shall now examine.

Let us for a moment look at the Japanese theatre traditions, especially the Kabuki theatre and its representations in the *ukiyo-e* woodprints and drawings, where one finds many depictions of actors in scenes recalling a play, a particular character, monolog, and a dance as well as the *joruri* music heard for this moment of dancing. Muto Junko, in her essay "Enjoying Actor Prints: Imagining the Voices of Actors and Music," speaks of the joy of "hearing" *ukiyo-e* actor prints:

> The majority of the buying public in the Edo period did not only enjoy the actor print by looking at it, but also by recollecting the actors' voices on stage, and the *joruri* music. This is possible because actor prints often depict particular highlights from a play where actors give long speeches, or scenes where they dance to *joruri* music. In other words, to the visual

world of *ukiyo-e*, which captures an actor's appearance and action, they add an element of voice and sound. One could enjoy an actor print much more if one could hear it as well as see it.

(Junko 2005: 10–11)

There is a word in Japanese for a particular moment when an actor enters the theatre on the *hanamichi* (runway), moving forward toward the main stage. Three-seventh into the way, the Kabuki actor who has seemed to float forward through the audience in his long flowing kimono suddenly stops on the *hanamichi*, and turns to the spectators and listeners. It is a greatly charged moment called *Shichi-san*, heightened by the percussive beats played by the musicians on the side of the stage, the percussionist now pointing his claves at the actor or the audience. It is a memorable moment I had experienced at a production of *Yukikeisei* (The Snow Courtesan) in Tokyo, where the courtesan entered in this floating style – in a role considered most difficult for *onnagata* female role specialists since they have to create the sense of the most delicate femininity (as male actors). The floating is both an effect of the particular way of making small steps, sliding forward, and of the garment that hides these steps and thus allows us to have the illusion of the *Shichi-san* being like a "landing," the actor having softly parachuted and glided into our spatial consciousness of the entire snow landscape we imagine to lie in front or us and around us. Upstage, above the snowy land, there was an image of the moon.

One is tempted to think such floating/dancing to be like moon light, charmingly evoked by the 18th-century courtesan Hanaogi in her calligraphy on transience and the fashionable life of beauty and display in the flashy world of the city. Most ephemeral and yet most sensual, dancing moves through space, becoming and always changing space of its appearances and dynamic potentials, moments to be perceived in space and through time. If we watch or witness this, our bodies float as well, in these heightened empathic moments. Our bodies were born to move, and we breathe and experience the world moving. This notion of the breath was highlighted beautifully in the documentary film on the dancer Anna Halprin's long career, *Breath Made Visible* (2010, directed by Ruedi Gerber). In this film we often see Halprin in outdoor locations, in the world outside the theatre, in nature. She practices bodyweather. The arrangement of theatre, where spectators are seated to be "stilled" in their sensing perception of movement, is an unnatural one, even as the histories of performance in many cultures show rather fascinating differences in the architectural visions that were brought to the construction of "containers" for action, choreography, singing, music, and the many forms of orchestration of sensory perception.[5]

From the point of view of the perceiving audience, a proscenium stage, an empty black box as much as a cluttered "set," an installation or a site-specific location – they can all indicate containers, and the architecture of a theatre or the material design of space (the set) function as structuring devices for the event, and the relationships between the place, the artwork and the audience.

Scenography is a somewhat underrated craft, compared to the attention given to directors, choreographers, composers, playwrights and actors/dancers or musicians. Yet scenographic practices – including set, costume, lighting, sound and visual projection designs – are fundamental for the construction of essential (dis)orientations without which in fact no performance would take place.

Designing a space and an atmosphere for performance sets up the crossing of thresholds, throws the engines of engagement into gear by precipitating the trust, the expectation and anticipation we bring to a space, the knowledge we have of a spatial environment, once we enter it, and the "tacit agreements" with which we behave in it (Newling 2007). Some of these tacit agreements become particularly challenged when the audience is invited not to sit still but to move into/across the stage. The scenographic manipulation of such agreements, and of physical materials and the visual and aural aspects of performance, may have had secondary (technical) status within the field of theatre, performance and dance studies. However, not only recent debates on the "postdramatic" (Lehmann 1999/2006; Worthen 2008), but the general impact of media on our cultures, along with the surge of digital practices, social networks and internet platforms, provokes fresh interest in participation, in the contingent "object" of performance, the *mise en scène* of remediation, re-mix and real-time performance drawing on the expanding techniques of digital interactivity.

In our new century considered to be an advanced technological era of pervasive computing, questions of space and movement hold crucial challenges for interaction designers and architects, and for those of us working in performance design with real-time interfaces and immersive digital projections (video, sound, animation, 3D graphics, VR, networked transmissions). The nature and quality of immersion has already come up repeatedly; I now add reflections on intimate and distanced immersion, as well as on the devices and immersive techniques. In theatrical contexts there rarely have been fully immersive virtual reality performances available to a broader audience; in new media art contexts, large-scale complex artworks (using stereoscopic VR head mounted displays for a single user who navigates inside a 3D computer graphics world) remain exceptional instances (such as Char Davies' *Osmose* in 1995), often limited to CAVE virtual reality environments and electronic visualization labs. We have rarely seen theatre audiences watching a play with 3D glasses (as they watched James Cameron's *Avatar*). Immersive performances consisting of a primarily auditory experience have been explored by a number of composers and groups like Blast Theory and Rimini Protokoll. In these instances spatial relationships are replaced by imagined ones physically residing in the headspace of listeners, as demonstrated also in Janet Cardiff's well known audio walks (*The Missing Voice*, 1999; *The Telephone Call*, 2001).

Has interactivity brought about new conventions of digital *mise en scène?* Another question is whether the commonplace use of video projections in contemporary stagings replaces the need for material sets and objects? It is a fair assumption that the gradual adoption of digital media across the field of performance arts generates alternative materialities that either disrupt existing scenographic practices or perpetuate established practices through new means. Without a doubt, we have entered a new era of stage design. The Prague Quadrennial, one of the most prestigious international exhibitions of scenography, announced a series of forum discussions centering on the theme of "Scenography Expanding," already in 2009:

> Throughout the past decade, scenographic practice and performance design have continuously moved beyond the black box of the theatre toward a hybrid terrain located at the intersections of theatre, architecture, exhibition, visual arts, and media. This terrain and its spaces are constructed from action and interaction. They are defined by individual and group behaviour, and are contrasted by distinct behavioural patterns. It is proposed here that spaces that are staged in such a way – spaces that are at the same time hybrid, mediated, narrative, and transformative – result from a trans-disciplinary understanding of space and a distinct awareness of social agency. These two factors of 'expansion' are seen as the central driving forces in contemporary scenographic practice and theory.
>
> (Prague Quadrennial announcement, 11 November 2009)

This remarkable statement laid the ground for the workshops leading up to the 2011 Quadrennial, with researchers (artists, curators, programmers, directors, dramaturges, critics, and theorists) invited to participate in international scenography symposia held in Riga, Belgrade and Évora, and to push a wider range of disciplines, genres, theoretical and artistic positions that comprise the relationships between spectator, artist/author and curator in contemporary scenographic/performance design practice. The 2011 Quadrennial, as well as the subsequent ones, clearly reflects a growing emphasis on scale, hybridity and body schema (action-perception) but also on the changing aspects of the venue for performance, on the intersections between the small (intimacy/proximity) and the large (spectacle/distance). Volume, proportion and tactile qualities crucially affect physical presence under these design considerations that amplify internal experience of an external/social environment, i.e. how we become immersed in a performance and take the existence of the world around us into our consciousness. This process is a circuit, a kind of criss-crossing between staging and internalization, and there are of course important intermediaries at work between bodies, perception processes, and the augmented 3D performance world.

Composing atmospheres, in my mind, implies design principles that reflect such circuits and an expanding dramaturgy which includes intermedial techniques of live cinematography/recording and processing interactive media, choreography, narration, projection and real-time synthesis. Taking our cues from the traditions of Japanese prints (*ukiyo-e*) mentioned above, I now describe the choreographic installation *UKIYO* (*Moveable World*) as my case study. First created by the DAP-Lab ensemble in 2009–10, the work's Japanese title literally means "Floating World" in English, and its composition is owed to the transcultural nature of the project involving artists from Europe and Japan.[6] "Floating" was the core conceptual metaphor for our production, after studying the Japanese theatre architecture and the *ukiyo-e* prints, especially the stunning drawings by Hokusai (1760–1849), a master of print making now considered a major influence on *manga*. In order to create atmospheres of a "moveable world," we worked with four primary dimensions: (1) movement environment (spatial design), (2) movement images (projections of digital objects and virtual spaces), (3) movement of sound (from macro to micro levels) and (4) sensual design (including tactile and olfactory stimuli). The compositional elements discussed below follow this four-fold scaffold.

Environmental atmosphere

Floating the audience implies dissolving any borders between stage and auditorium, and departing from the common convention of frontal staging used by the majority of dance- theatre productions that take place on a proscenium or thrust stage. If the performance includes screen-based elements (projections), then screens or projection surfaces have to be embedded in the physical space while allowing the audience to move freely and entertain multiple, changing perspectives. This is the first divergence from proscenium staging where in many cases the projection screen is positioned upstage or on the set behind the actors to allow everyone in the house to see the projected images. In some instances, especially in dance or opera productions, designers use a transparent front scrim allowing the projected images to fall onto the downstage area on occasion, while lighting actors behind in such a manner that they appear to move inside the screen images, which can produce powerful visual effects. This has been beautifully exploited by Paul Kaiser, Shelley Eshkar and Marc Downie's graphic compositions for Merce Cunningham, Bill T. Jones and Trisha Brown during their collaborations in the 1990s and the early 2000s, whereas historically the use of suspended screens on the dance stage can be traced as far back as Lucinda Childs' *Dance* (1979), a piece in which collaborator Sol LeWitt stretched a translucent scrim across the stage and projected the dancers onto it (the work was revived at The Joyce Theatre in October 2009, and again for the retrospective "Lucinda Childs: A Portrait 1963–2016" in November 2016).

The effect of the front projection is to create an illusion of cinematographic depth pushing through the flat screen, so to speak, whereas other designers have highlighted the flatness of the screenic by staging multimedia scenes on a very shallow stage, where actors and video projections are almost constrained into a deliberately flattened space. We see such a staging in The Builders Association production of *Alladeen* (2003), and we also reminded here of the influence of Robert Wilson's pictorial work with *tableaux vivants* in many of his earlier productions that emphasize the horizontal line and its surface modulations, while Wilson's use of vertical and diagonal lines gradually and inevitably introduce temporal dynamics, horizon, depth or the illusion of depth and multiplanar perspectives (enriched by his extraordinary use of light and color from the distant cyclorama to the front and to minutely isolated spots on bodies and object-sculptures).[7]

The embeddedness of screens can also occur in installations, for example, in galleries and museums or in public spaces that allow the use of facades or walls as receptive surfaces/membranes. The idea of the screen as a skin or membrane opens up obvious possibilities of using both front and rear projections, thus allowing the images to be seen from different angles. Physical space and screenic space are different from one another, and therefore complicate our work with 2D and 3D realities if we seek to combine them. Most importantly, in an environmental staging, there is always a *live presence* of performing bodies engaging audiences in real-time and acting in partnership with the embedded media (sound, visual projection, light, objects, etc.). In our experimentations with digital images in *UKIYO* we therefore speak of *screen presences*. Mixing or juxtaposing the body of a live performer with video representations of body can produce various combinations of spatial-temporal relationships between performers and digital images, and in some cases the screen presences can be called video actors. Of course the use of projections is not limited to realistic representations of actors: in many instances such images or graphics can convey architectural spaces, landscapes, objects, texts or abstract imagery, still and moving.

We include lighting as embedded media here, as the role of lighting design is of course crucial in environments, allowing changes or subtle redirections of focus, mood, tonality and intensity for actions – and their presencing – in particular areas of the total spatial aggregate. When you use filmic projection, you also need to pay attention to the lighting of the stage area surrounding the screens or projection surfaces to avoid interferences (when light falls onto the film projection) and increase contrast. Lighting designers who take inspiration from Wilson's exquisite work with *chiaroscuro* effects and color might try to experiment with the careful use of gels impacting the tonality of projected images on screens, for example, mixing a blue or red color onto a black and white film projection. The interplay of lighting and projection is a sensitive process of potential transmutations that requires careful testing of the various effects of illumination and chromatic modification, and of changing textures and intensities. If environmental staging transforms the space, then lighting and projection design, together with costume and sound design, need to collaborate closely in order to achieve the highest level of integrated visual expression.[8]

Real-time interaction with digital media poses the most advanced compositional challenge for the design, since *interactivity* generally indicates a computational, programmed environment or *choreographic system*, with controllable parameters in the software code, which enables live creation and modification of sound, image and lighting during the performance through performers and/or participants.[9] In such cases of live creation, the performers need to connect their actions or gestures to the screen presences in some manner, wherever these are projected behind, in front, above, below or aside the performing bodies. In reverse, the software system used for the overall "machining architecture" of the performance needs to connect its operational attention to the position, gestural action and movement of the performers. In technical terms, we speak of a feedback loop in such a system. In compositional terms, we might also think of whole interdependent organism as an intelligent system. We need to strive for balance in this system so that design principles for the environment can take into account the intelligence of the performers, body dynamics, proprioception, and interplay behavior which can be subject to uncontrollable occurrences (noise, imprecision, change, unpredictability, failure, etc.).

While the blurring of boundaries in an environmental staging is the result of abandoning the clear separation between a frontal stage (proscenium) and an audience seated in the auditorium, the transformations of the roles of performer, spectator and stage vary according to the degree to which actors perform the "intermedial stage" or spectators become participants that navigate the intermedial stage or even generate aural or visual responses from the system through their body movements/actions (as I implied in my reference to David Rokeby's pioneering work with *Very Nervous System*). Including the audience into the stage − or "casting the audience in the role of the performer, asking them to activate the work with their movements" (Rokeby 2019: 89) − invariably means to open yourself to the unexpected, to mix an open form into a planned choreographic concept. One might even think of such environmental staging as a dramatic atmosphere or *algorithmic atmosphere* in which it is mainly the audience that is choreographed or invited to participate in the transformations of the moveable world. Dramaturgically, such an approach takes into account the potential of uncontrollable situations. Even if interactive systems use "controllers," and algorithms, in their own surveillance manner (as camera vision or sensing system), identify and assess behaviors that occur, at the same time we might expect audiences to behave still within the parameters of *tacit agreements*, presuming that they will not physically prevent or disturb a performer, their own performance, or bring the event to a standstill. Some audience members, from my observations over the years, notice controller operations and deflect them or even "provoke" the sensing system (e.g. the Kinect camera), challenging its creative possibilities and constraints. The collective organism most likely will display certain self-regulatory mechanisms throughout the evolution of the event. For designers this is a most fascinating scenario, in which they can explore collective intelligence in a complex dynamic system of "emergent behaviors."[10] This collective sharing of intelligence is also not too dissimilar from the thoughts and associations that can emerge in the Social Dreaming Matrix that I mentioned above.

In our DAP-Lab productions, we use the concept of the environmental intermedial stage together with the notion of a "wearable space" − a space that becomes like a garment, so to speak, and allows the performer to fold and unfold space like an imaginary cloth or a tactile skin, a "shining skin" as Steven Connor calls the epidermal in his *Book of Skin* (2004: 53). In extension, wearability of the space affects the audience in almost equal measure, as the visitors become *Ohren-zeugen* and *Augenzeugen* (oral and visual witnesses) and also sense the tactility and elemental fabrics of the space. Our wearables, designed by Michèle Danjoux, and the suspended kinetic objects or fabrics hung in the space, are intelligent sensorial garments and artifacts, equipped with sensors, microphones or small speakers, which can transmit motion data into space (as well as record and amply sound). They are used as interactive agents, allowing the actor/dancer or visitor to affect the sonic or digital images becoming audible/visible in the environment. The garments proper, worn on the body, express a debt to Oskar Schlemmer's *Triadic Ballet*, his *Formentanz* (Form Dance) and *Raumtanz* (Space Dance) originating in the Bauhaus in the early 20th century and pointing to the prominence of sculptural and painterly design ideas for performance. In a previous production, *Suna no Onna*, we had worked on composing "wearable spaces" for the dancers on stage (Birringer and Danjoux 2008), highlighting the textural connection between the costumes worn by the dancers and the projected digital environment which behaved like the wrinkling and creasing of fabrics or like a luminous skin. The spatial design becomes the digital extension of the costume design for the characters. While such ideas are derived from fashion and painterly design, we also build a dramatic structure and focus on developing narrative material, characters and actions which can tell a story in a particular digitally projected environment programmed to behave in response to expressive input from the performers. The environment has an atmospheric character (or multiple characters). It is subject and object, a Floating World always in moving relationships with the "pictures" drawn by the actors or dancers.

Perhaps we can now anticipate the idea of the remix and the real-time modulation in performance, as it alludes to fashion, music and skin in a complex sense of interrelationship between material surfaces, on the one hand, and underlying (data) source code on the other. The increasingly seamless switching operations of the digital (encoded data and distributed data streaming) have been explored extensively in "scratching" (manipulating vinyl records in live performance) and digital audio processing (deformation and reprocessing of microsound), and such mashing up now also extends to digital video and the infinite range of image mutations or combinations enabled by software and networks. In some of my previous writings after the turn of the new century, I had commented specifically on telematic or networked performance (also called telepresence) – now ever more relevant again in the moment of the 2020 pandemic when performers are rediscovering ZOOM presences and the intimacies of "distant feelings."[11] In the theatre, such remixing and modulation, including the incorporation of "zoomed" telepresence-projections, need to take actual physical performers into account. Thus the idealized and abstracted mingle with the tangible and imperfected, the uncanny real of the human flesh and body.

Imagining a space that can be worn, in the sense in which a performer can model and transform a costume or become a mask, thus means taking the digital phenomena almost literally as actors, as living states of being and as being transformable data objects. As we understand digital performance, the visual and auditory dimensions of a work – and its entire sensorial impact marked by audio-visual tactility, plasticity, intensity, color, and multiple kinetic tonalities, rhythms, and pulsations of movements and image-forms – are continuously evolving relations between the physical and the virtual in a space that never stands still and has no (single) dominant perspective.

Scenography and projection

The design of the environment was the starting point for *UKIYO*. After studying the Japanese prints and noticing the frequent use of Kabuki actors as subject of the pictures, I became fascinated by the *hanamichi* (runways) of the traditional Kabuki theatre, and proposed to create an open space that would have five *hanamichi* forming an irregular pattern. As a first step, then, we created a model of the criss-crossing *hanamichi*, imagining the runways to be white dance floor strips on a black floor. Initially the runways would be flat, in later versions I planned to make them more dynamic by raising some of them slightly or creating irregularly ascending or descending trajectories. The runways are the primary movement areas for the dancers and the character they create. Throughout the duration of the installation, the performers are with the audience in the same large space, and on occasion "enter" the runways or occupy them performing, while the audience can follow them or observe them from any point in the space, in front, behind, from the sides. The digital scenography extends this space into virtual on- and off-stage spaces that are visual diegetic (narrative) space-projections, dialectically enhancing the imaginary world and kinetic atmosphere of the performance as a whole. The atmosphere thus creates a more demanding layering of simultaneous perspectives. All digital "objects" are mixed and synthesized in real time.

The reference to the *hanamichi*, on the one hand, places our spatial vision for the work into the context of Japanese Kabuki theatre and the preoccupation that the *ukiyo-e* artists of the 18th century had with so-called actor prints or audible prints depicting a scene voiced and danced to *joruri* music. The calligraphy on the prints depicting the actor *en-scène* would help the viewer to recollect the particular dance and stage speech. In Kabuki theatre, of course, the particular function of the *hanamichi* was to cut right through the audience, thus serving as stage and passageway, bringing actors and audience together in one space. In my design for movement/images, I felt that the physical passages could be echoed by the movement of images and intertitles on the screens. For the characters dressed by Michèle Danjoux's conceptual costumes, the *hanamichi* were also envisaged as runways (Figures 6.2 and 6.3).

From a bird's-eye view, the five *hanamichi*, on the other hand, are also a symbolic character in itself: I wanted to evoke an abstract labyrinth that draws the audience

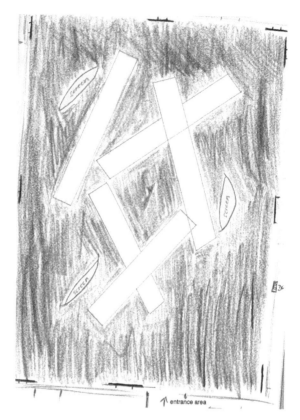

FIGURE 6.2 Drawing of *UKIYO* performance environment © 2009 DAP–Lab.

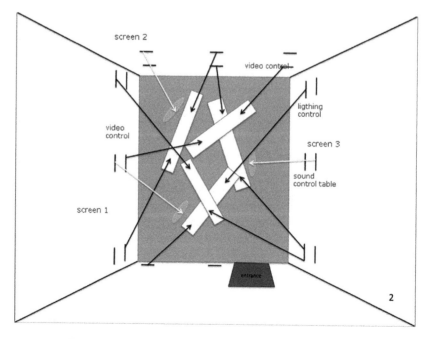

FIGURE 6.3 Computer rendering of *UKIYO* performance environment for lighting plot © 2009 DAP–Lab.

into a system of corridors to which we allude in the projected titles that appear on the three suspended screens during the prolog (one of the screens was later replaced by a weather balloon). The screens are raised 2.5 m from the floor and suspended from the grid, forming an irregular triptych for rear projection. On these screens the projected still or moving images unfold, in black and white and in color. Many of the images we shot are in fact "portraits" of our performers, composed in the studio or in outdoor locations. Some of the diegetic images are historical photographs – found objects recomposed into short animated films or layered into composite portraits/landscapes. The runways are white and thus allow special lighting to illuminate them or color them for the scenes of the dancers' performance, which we choreographed as a series of solos (following a particular code of costume colors – red, white, black/grey, silver, golden), and then as increasingly overlapping and simultaneous duets, trios and quartets that phase in and out of the fluid action as well being woven into the sonic environment created by the composer and the performers themselves. We did not choose a Kabuki play but created a dramaturgy based on choreographic motifs and the development of particular characters chosen by the dancers in response to some of the images I had brought to rehearsal.

For the scenographic process, these archival images were crucial. They were our "audible prints," so to speak, and the ones I chose refer to early 20th-century revolutionary communist history and the beginnings of the modernist avant-garde. Our source materials were the media-archeological research by Siegfried Zielinksi, especially his vivid writings on Russian engineering and motion study in *Deep Time of the Media: Toward an Archaeology of Hearing and Seeing by Technical Means,* and Christian Kracht's dystopian novel *Ich werde hier sein im Sonnenschein und im Schatten* – a surreal work depicting a century of (post)revolutionary process stretching geographically from Africa to Mongolia via the secret hiding places of the military engineers in the Swiss mountain *Réduit.* The archeological theme here connects Russian engineering feats with the transrational poetry (onomatopoeia) of Velimir Khlebnikov, Aleksej Kapitanovich Gastev, and Alexei Elseevich Kruchonykh, the audacious protective space suits of Cold War fashion, and the fictional virtual language investigated by the engineer Brazhinsky in Kracht's novel.

What I am describing here is a textual and visual montage method that we apply in our works (and that are common in postdramatic theatre). Since our ensemble had worked with dresses incorporating wireless sensors before, with garment and physical motion affecting the digital projection environment, I was drawn to the new language of communication Kracht's character Brazhinsky calls *Rauchsprache* ("smoke language"). Our dancer Yiorgos Bakalos silently performs the opening gestures and movements of *UKIYO,* welcoming the audience inside the *Réduit.* I project this ritual invitation also on the screens (in graphic text). Later, I quote the passage that we used as intertitles in the *entr'acte* that bridges Part 1 and Part 2, a silent movie shot in black and white with two actors, Mamen Rivera and Olu Taiwo, appearing on facing screens:

Language is a collection of symbolic sounds, it originates in a cosmos of unrecognizable forms – which are above all, never knowable.

But please explain to me – how does the smoke language function?

Well, we begin to speak what we think, we place it into the space. Then we can look at the spoken, we can walk around it, and finally we can move it. Since it exists, we can move it.

(Kracht 2008: 43–44, my translation)

With facing screens I mean to indicate that the audience can watch the dialog between the two actors on either of two suspended screens, choosing whether to watch Rivera or Taiwo – their dialog shot from two different camera angles. My conception of an atmosphere here becomes utopian: I project the notion of a gaseous smoke-like language into my understanding of the kind of moveable world created here. The plasticity of (design) languages lies at the heart of this performance installation; we deliberately relinquished any overt reliance on direct interactivity and controllers that require software programs registering data transmission (input) and generating reactions in the environment (output). Such a system of mapping is always limiting to the complex wholeness of visual, kinetic, auditory/vocal movement and somatic experience, since the dancer in interactive performance systems needs to pay undue attention to the quantitative motion or acceleration/deceleration.

UKIYO is a poetic universe full of *thinking images*, as filmmaker Raúl Ruiz might call such a composition which questions the very stability or recognizability of symbolic forms that software might operate upon within the overall performance system. Our "audible prints" are somatically experienceable since they are ineffable yet enactive, phenomenally perceptual and at the same time cognitively interdependent with the movement (e)motionally grasped and processed. When we thought of data or numbers circulating in the system, we wanted to think of them in plastic and musical terms. Plasticity and musicality were the main design inspirations brought to the scenographic, choreographic, filmic and musical process our ensemble engaged. Composer Oded Ben-Tal, for example, conducted early experiments with percussive gestures which we motion-captured to obtain data that could be used for the kinaesonic aspects of the real-time projection environment we were building.

How does motion capture technology relate to movement composition or movement projection? Generally speaking, optical motion-capture – nowadays more commonly used than the earlier magnetic systems which needed an exoskeleton placed on the performer with wires running to the computers – involves a system of multiple infrared cameras, computer hardware and software that enable digital 3-D representation of moving bodies. The capture involves the placement of reflectors in strategic positions on the performer's body; cameras surrounding the performer track these sensors in time and space, feeding the information to the computer for consolidation into a single data file. Mocap data subsequently drives the movement of simulated figures on the computer, where they can be mapped onto other anatomies in an animation program. With the animation tool one can draw out and reconfigure the abstracted motions and trajectories of the performed action. What we see are animated motion pictures; the captured movement phrases thus become the digital building blocks for virtual dance or interactive performances that explore possible, always newly manipulable relationships between live and synthetic presences.

In cinema, the representational and the animated are commingled, to the extent that we often do not know the difference anymore. In the theatre, the animated figures have to be projected onto screens and thus they assume a different kind of presence. The promise of motion tracking technology and real-time digital signal processing (also now available in motion-capturing) is the simultaneous exploration of a fluid environment in which movement or gesture can generate sound and animation: sound can affect video images, and animated images can in-form new movement possibilities or a new form of action painting (to which we shall return later). The re-programming of captured data with machine-learning technique based on artificial intelligence also opens up new pathways for thinking images that might behave according to their own logic. The experimental Chilean filmmaker Raúl Ruiz, in an interview with Kriss Ravetto, thinks of these images as "crazy poetry," the kind of shamanistic poetry that can insult the gods.[12]

In *UKIYO*, the shamanistic poetry was meant to light up some of the virtual scenes in Second Life, which join the performance in the second half, once the animated avatars arrive in the Floating World. No insults to any gods intended. In the first half, the kinetic data were combined with the physical sound gestures the dancers performed in the space, but we then decided not to use the motion capture data since the acoustic and live electronic performances created much stronger plastic and sensual rhythms than we could achieve by triggering digital data effects. Kinaesthetically, gestural interactivity with real-time environments (sonic or visual) can deflect both from the physical virtuosity or embodied expressiveness of the performer and from the unpredictable qualities and metaphoric richness of moving digital objects (films, layered animations, graphics, text). There was no reason – in the poetic context of our performance – to work with direct mapping and causal feedback and to ask the actors to move according to software parameters. Our audience was to experience "moveability" as a virtuality that was not overdetermined, in the sense in which digital programs determine, for example, the principal directions and speed of images: forwards, backwards, slow, fast, or the pitch and duration/dynamics of sound. We wanted indirect relationships to happen, to surprise us in a polyphonic manner, as if sound and voice were to transcribe, counterpoint, and also decenter the visual medium (cf. Birringer 2008a: 153–64).

One particular key idea for the costume design was Danjoux's plan to build small speakers of various sizes as sound transmitters into the garments, and to explore the visual aesthetic of audio technology along with a noise aesthetic that she felt inspired by. One of the dancers, Anne-Laure Misme, was given a set of inverted, dysfunctional speakers to wear on her breasts. Another dancer, Helenna

Ren, performs a choreography with two larger, sound-emitting speakers which she balances on a yoke placed on her shoulders. Caroline Wilkins performs live on an amplified bandoneon in Part 1, while in Part 2 her entire dress is designed as a bandoneon opening up and revealing, down her spine, a black leather strip with built-in speakers. Ren and Wilkins thus perform with wires, their amplified body/instruments connected to the sound system and into the sonic and image processing environments performed live by Ben-Tal, Sandy Finlayson, Paul Verity Smith and myself on the computers. Analog and digital processes – in which the sound-gestural choreography interacts with the visual and audible scenography, with sonic diffusion and the progression of image-movement in the projected environment as a whole – combine to create what we think of as the "curves" of *UKIYO*.

We studied the demonstrations of the chronocyclographic method in Gastev's Moscow Institute (C.I.T) by a female worker with an artificial arm: strike pressure of an artificial arm, a virtual arm, wielding a hammer. This also illuminates the notion of the "remix" as we understand its use in digital performance art and multimedia theatre: kinetic atmospheres can include archival histories in the sense in which techniques imply gestures, motion and duration. The remix is a function of sampling, as we know it from contemporary VJ and DJ cultures, hip hop and electronic music, and as we have always known it from the history of the arts. While playwrights often rewrite their sources or develop their original plot lines mixing in various references to the history of forms and dramaturgical conventions, choreographers and sonic artists can also work with specific improvisation technologies that afford them fluid layering of materials, motifs, voices and phrases. DJs working on turntables are creating the live mix in real time, just as dancers do who improvise together in contact improvisation.

Mentioning choreographers/dancers and sonic artists also reminds us of the distinctions between media. Physical contact improvisation (analog) implies an intermixing of corporeal terms (weight, effort, lightness, strength, subtlety, force, etc.) and a fluid alchemy of expressions, quick reactions to the environment, obstacles, the contingencies of space, dynamic changes in full awareness of temporal rhythms and the significance of breath (time). Sonic artists layer and collage audio tracks, recorded sound materials and beats, and also intervene into the materialities (vinyl, CD, phonograph, digital signals) of the mediums themselves, thus generating an often surprising wealth of "malformation" and recombination. Delicate alterations of the "source code" and deliberate uses of glitches and clicks have become featured aspects of electronic music, and it is somewhat surprising that the cracked media movement has not yet had a greater impact on multimedia theatre. In theatrical terms, we would like to make the concept of the remix more productive for an understanding of compositional possibilities and architectures in performance. Mixing tracks for us means the combination of different cultural ideas and objects, the cut-in and the cross-cut between sounds and images, between the analog and the digital, between action and projection. The mixing of live action and projection is one of the hardest challenges in the theatre, and I will take a closer look at this in the next chapter.

In *UKIYO* we are also mixing tracks from the East and the West, looking for opportunities to generate a new syntax with variable rhythms and their effects on the atmosphere. The underlying inspiration for the real and virtual scenography can be traced back to a single "audible" print, the image-diagram of a woman worker with a prosthetic arm whose execution of a percussive motion (wielding a hammer) was measured by Russian engineers to observe the force and the *movement curves* of the motion. Not only were such studies done for industrial purposes, using measurement to improve worker productivity, but they also figured prominently in the early history of chronophotography, cinematography and motion science (Étienne-Jules Marey, Georges Méliès, Eadweard Muybridge). The historical connection to contemporary motion-capture technologies used in performance and 3D animation, sports science or game design imposed itself on us forcefully. Our work deliberately reflects the prosthetics of capturing systems, but *UKIYO* offers transformations of the perverse "biogram" of a body becoming instrument (virtual hammering), slowly shifting meaning from the biomechanical to a different range of surreal and abstract connotations, including the recrafting of bodies and dislocated hands in our Japanese partners' experimentations with avataric behavior in Second Life.[13]

We had already planned, at the early stage of our designing process, to collaborate with partners from Tokyo, having exchanged ideas about how motion capture data might be usable in *UKIYO* to link the real space thematically to the virtual environment, or to generate choreographic processes that might flow between RL and Second Life, the material world and the immaterial world. We began collaboration with several telematic sessions during which we streamed camera-based film images of the "WorkerWoman" character between London and Tokyo, conjoining the hands and arms of Helenna Ren (London studio) and Ruby Rumiko Bessho (Tokyo studio). WorkerWoman was our first character, setting in motion some of the other scenic materials we developed during 2008–09. The development then focused on the remix of musical and movement ideas, along with the design concepts Danjoux proposed for the wearable garments capable of generating sound or incorporating audio-phonic instruments. The small wearable speakers, some dysfunctional and some fully functional, also offered a spectrum of visual references to sound art and the history of cracked media.[14] The moveable virtual world of *UKIYO* is "sounded" by the gestures and clothes of our dancers, and these sounds are then continuously remixed until the behavior of avatars in the virtual world, in return, slowly begins to change the musical chemistry. The virtual avatar behaviors also act upon the movement vocabulary of the dancers in real space, in fact their kinaesthetic transmission influences the atmospherics of the installation as a whole, since audience and dancers learn from the repetitive body techniques enacted by the avatars (Figure 6.4).

The challenge we faced was the "modeling" of the virtual world upon the material space. Our Japanese partners proposed a simulated model of the scenographic

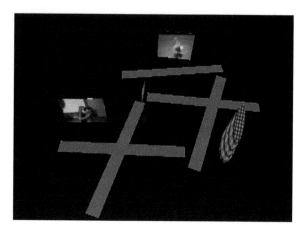

FIGURE 6.4 Model of *UKIYO* virtual space in Second Life © 2009 Courtesy of Kabayan/Inetdance.

drawings I had made, mapping the space in Second Life according to our physical environment, and using the screen spaces inside Second Life *UKIYO* as portals to the networked dance sent via live video streams. In this manner, the Japanese dancer in Tokyo could in fact perform in real-time with our physical performance in London, projected via the Second Life virtual environment into our digital *mise en scène* on the ground (or rather, above the ground, suspended and floating). This is a basic definition of what I mean by digital scenography: the live performance architecture incorporates analog, digital and networked dimensions. Its atmosphere is constituted as mixture of the organic-real and the virtual, the corporeal and the projected: there are sensorial oscillations that shape the being in and sharing this space. Performers and audiences are inside and outside the digital worlds simultaneously, and the screen canvases co-animate the localized movement narratives, as much as the movement characters and costume designs animate the images from the past and present, and even forecast the meanings that might be read into the dancing language of the avatars, their bodies, genders/ sexualities, identities.

Atmosphere is conditional – it is living, ecological and engineered, we are surrounded by artifacts and involved in experiencing the agency and importance of the kinetic objects, the acting and the observing/perceiving. This suggests a going beyond of any dichotomous oppositions such as active-passive – touching upon the dilemma of participation or interaction. *UKIYO* is a moveable world in which everything is possible in terms of participation (and withholding it, too). At the end, if we ever come to one, we could ask whether we are ever able not to participate and enact in an atmosphere? Perceptual experience inside a moveable world, is it not always inevitably environmentally specific and concretely apprehended? What does it even mean to think of a passive audience?

Rather, such hybrid spaces can evoke forceful and beautiful combinatories, impactful in terms of the known and unknown, assemblages of the expressive and intuitive with the logical and cool computational circuitry, the subtle bodily rhythms with the brittle digital graphics, mundane flesh with extravagant avatar torsos, the familiar with the alien or uncanny. Much could be said here about the nature of "digital space" or virtual environments, if one were to follow Deleuze's writings on the time-image (in cinema) further into the hybrid chronological and non-linear temporal coordinates of digital aesthetic production with multiple vectors of distributed space (Munster 2006: 172–73). If we think of stage design as grounding visual information (setting, place, context, etc.), then our scenography behaves more like the Flash web interface design Munster examines in her writings on digital materiality and the *deformations/compressions* of image information. Rather than delimiting spatial location, or providing spatial configuration and "scenery," the digital imagery behaves as a *smoke language* in the manner we reimagined Kracht's poetic evocation: we can walk around it and see it evaporating and re-emerging, layered pictures of photounrealism mutating into visually flattened anime or jumpy avatar motion in Second Life, unlifelike physics chaotically out of world and too slow or too fast. All strangely perceptible, and imperceptible.

The scenographic behavior is live, emerging, unfolding, and therefore a crucial aspect of the atmospherics. The digital *mise en scène* does not represent stable historical or iconic references points but behaves like the hieroglyphic language of cinema imagined by Eisenstein in the 1920s, layers upon layers, glimpses of revolutionary modernity appearing at high velocity (as if Buster Keaton had cranked the camera at different speeds), intercut with Japanese landscape stills, industrial übermarionettes and machine-women (reminiscent of Fritz Lang's *Metropolis*), animated manga killer vamps and haunted-looking butoh dancers that articulate contorted anatomies. The remix is full of shifting disparities, graphic intimations of failed biomechanism that humorously distract from the physiological and psychological dimensions of the labor performed by the dancers on the *hanamichi*.

The choreographic installation was first performed on June 1, 2009, in the Artaud Performance Centre (London). The positioning of the screen spaces posed questions. The Artaud space (21 × 15 × 10 m) was perhaps too large for three hanging projection screens (2.80 × 2.30 m), but on the other hand, the projected images did not overpower the physical arrangement. There was much space to breathe and to move. Spatial proportions and such issues of scale relations always need to be considered. No attempt was made to use projection as backdrop or as "extension" of the stage space, which is a common design feature with contemporary theatre. The "virtual scenography" for *UKIYO* has a sculptural character and needed to be three-dimensional. It also can be interpreted as a necessary compromise, perhaps even a limitation: the digital visual dimensions rely on points/positions in space, planes that vertically cut into the larger volume of the hall. The audience is not *immersed* in a virtual reality, but can well distinguish between the material space and the projected digital worlds. The suspended screens are like windows into these worlds of animate forms and image streams. If you work with projections and use screens, even in a panorama staging, the technical and spatial manner of the projection will always cause a certain dissonance, and it is the same dissonance between mediums that I described earlier when comparing the physical and the digital.

This dissonance *is* the nervous system of kinetic atmospheres. It draws us into an environment, and into its "anti-environment" as well, if we think of such atmospheres as a habitat that makes the visitor forget that this is a habitat, in the manner in which McLuhan and others had imagined that the ocean and the existence of water are not perceptible by fish or animals that live inside the ocean – the water is the medium of existence (Durham Peters 2015: 55). It exists in all its permanence and impermanence, its conductivity and varying depths, temperatures, currents, etc.

Sound-theatre in real space

The scenography for the real environment is straightforward. The five *hanamichi* are the material space for the movement characters; all garments and objects used in performance carry a strong sensual presence and plasticity, especially as the performers increasingly draw the audience closer to their actions, with members of the audience walking around and across the space, often coming into close, intimate proximity with the dancers, yet sometimes retreating, as if shying back from contact. The creation of sound performance is a leitmotif for *UKIYO*, in the manner in which I referred earlier to the physiological and psychological dimensions of labor processes. The characters – we named them InstrumentWoman, SpeakerWoman, HammerWoman, WorkerWoman, BirdWoman and MutantWoman – carry out motion sequences that produce sound through the garment textures, objects, and small or large speakers manipulated by them. A special emphasis is given to the gait and the manner of walking or acting with wired costumes that restrict motion sometimes to a few parameters. I like to think of such kinetic atmospheres as including high-wire acts: like the criss-crossing *hanamichi*, the electrical wires turn dancers and perceivers into acrobats, sometimes only imagining the connective tissue that holds this habitat together, powering it, enabling transmission of voltages – low voltage levels and charges. Our installation, if one were to pursue this metaphor, functions like a substation, sending out signals that are perceived by the audience (Figure 6.5).

Such play with degrees of freedom heightens attention to the movement's microperceptual qualities, the small gestures and the relations opened up between different sound frequencies coming directly from the body or originating in the amplified ambient sound that envelops everyone in the theatre space. The musicians and dancers in the ensemble work particularly on the inside/

FIGURE 6.5   Katsura Isobe as MutantWoman with audience, Kibla Art Center, Maribor, Slovenia © 2010 DAP-Lab.

outside auditory spaces and editing of sound, building a special series of sensory haikus that attract audience attention to the intimate sounds created organically. Michel Chion calls this "internal logic," when the sound-image develops, grows and is born out by the narrative atmosphere itself and the feelings it inspires (visualized sound), whereas the "external" intervenes and ruptures the flow, often creating discontinuity or changes in tempo (Chion 1994: 46). In the hybrid sound design for *UKIYO*, Ben-Tal's digital sound also floods the entire space. It is *acousmatic* in the sense in which sound artists (after Pierre Schaeffer) understood such sound to be heard without seeing their originating cause, the emitters or instruments. Often such sound was made only for the recording medium; for us the ambiguity of acousmatics was interesting since we explored the visible and the invisible, for example, when Ben-Tal sampled the bandoneon. The tones played by Wilkins on her instrument were recorded, but also the "breathing," when the instrument is expanded to take air in before pressing it out. Often the dancers are clearly visible as the subjective source for acoustic sound. Katsura Isobe wears a red sleeve design built out of small metallic plates that rustle when she moves to change shape. Her *hanamichi* becomes a rustling corridor (*rieselnde Fläche*). In some scenes, the spatial points of audition are mixed, for example, when Ren's SpeakerWoman manipulates the speakers that emit electronic stereo sound and complement her own physical phrasing. Ren dances the image of a rice field worker, seeding sound in the earth while also dropping actual tiny rice grains to the floor. In contrast, Bakalos and Misme perform to muted or aggressive industrial factory sound, inventing a series of variations on the percussive gesture (Figure 6.6).

At the end of Part 1, Wilkins performs the bandoneon lying horizontally on the floor, her instrument on top of her body as if it were growing from her, exfoliating, expanding and curving with her breath. The contact mic attached to the

FIGURE 6.6 Caroline Wilkins as InstrumentWoman wearing BandoneonDress in
UKIYO © 2009 DAP-Lab.

bandoneon picks up microscopic sounds of air and breath, her fingernails hitting
keys and wood, and transform them electronically into eerily high frequencies
that hover in the space, then reverberate and fade out. Wilkins' instrumental
music-theatre technique draws attention to objects/instruments and the relations
between gesture, object, image and sound and how the sounding processes become perceptible. This exteriorization was beautifully translated in Danjoux's
costume for Wilkins, worn in Part 2 after the silent Entr'acte.

In Part 2 of *UKIYO*, the sound appears to originate in the Second Life virtual environment and is diffused around the entire space, as we hear the sound
of claves and the high frequency voice of a Japanese singer intoning a haiku (see
below). Then the relationship between digital scenography, choreography and
music shifts, and attention is refocused on the directional sound coming out
of the small speakers in Wilkins' BandoneonDress which is constructed out of
fabrics echoing the construction of the instrument.[15] Wilkins begins to use extended vocal techniques to instigate a scene of onomatopoeic sound-poetry based
on principles of call and response (gospel, jazz). This is the concluding scene
of *UKIYO*, chaotically pointing to a sense of improvised collectivity emerging
from the world of the avatars into the real space, with audience members now
dispersed all across the theatre, literally inside the *hanamichi* and perhaps brushing
shoulders with the dancers, tuning into them or changing places with other audience members whose presence and articulated behavior have made themselves

felt as well. The masks of collectivity works: immersed in a mass, individuals are relieved from a phobia of *contact*.

The pluri-dimensionality of such a spatial-sonic atmosphere, and the fragmentation of multi-directional perspectives (of moving-seeing, hearing, sensing) into the criss-crossing "floating islands," proffer an experience quite similar to one I witnessed happening in *Shift...Centre,* a dance installation by the African choreographer Opiyo Okach (artistic director of Gaara Dance Foundation) presented at The Place Theatre, London, during the 2009 Dance Umbrella. The Kenyan company unravels the convention of spectating by creating a collective intensity of presentness that is, paradoxically, quite unfocused. An incalculable number of events are generated simultaneously, small and big, noticed and unnoticed, each person in the room sees what speaks to them, "stumbles" across something, picks something up, follows a changing relation, not able not to move and participate in the immanence of the whole. The political dimension of such work might be subdued, barely foregrounded, but inevitably such physical-musical performance provokes shifts in how we experience commonality and diversity (in his program notes Okach refers to lived multiple realities – "traditional African, Islam, Christianity, MTV...").[16] I take it that Okach deliberately uses the reference to "decentering" in his title – a reference that is now all the more unavoidable during the 2020 anti-racist protests (Black Lives Matter) and demonstrations that have had innumerable repercussions also in universities, museums and arts organizations, venues that are now reflecting on how their curation needs to shift the center. In order to allow for such affective instances of "social choreography" (Kozel 2008), where you become so aware of everything, the multiple viewpoints, proximities and distances, the spatial-rhythm (African bodies articulating), the ineffable shiftings of tonalities and sound-movement relations, the soft careful behavior of audience members moving around not wanting to disturb another's concentration or enjoyment and in fact looking for that in the other people's eyes, the performance design may need to decrease the amount of graphic information and relent, let go a bit, trust the rhythms that make the room reverberate.

The floating islands in *UKIYO* were both the white performance corridors and the black spaces in between: maybe we could even speak of a sea, with different areas of moving driftwood, or different aquatic techniques required for dancers and audience to propel themselves through the water. Affective intensity can be manifested through quiet subtlety rather than overpowering auditory or visual excess. Floating. Using the sky to imagine an ocean. Above all, the participatory atmosphere of the work revolves around gentle invitations to the audience, the elements beckoning them to come inside, to listen inside, to follow inside, to hear spatially and enjoy the increasing proximity to the performers or, if they prefer, to keep their distance. Social choreography, I propose, is largely about this sense of proximity and intimacy that can be established and *negotiated* in the performance environment. The intimacy (and the de-distancing I expect in the future) is a concrete relational form of *interface,* without front or back; the audience is inside the action and can practically feel its delicate touch, its breath and sound, on the skin. They will choose what they want to feel and how to respond.

Sensual space and virtual space: avatar choreography

Part 2 of *UKIYO* visually folds the virtual realm of Second Life into Real Life insofar as the computer-generated model in the virtual world and the animated choreography of the avatars now provide the framework for the unusual movement happening in the physical space. Part 2 opens up in the projected world which then folds into the physical action. The virtual scenes create a largely burlesque environment for the exploration of the "smoke language" mentioned above, setting up a rhythmic structure of short sung haikus that seem to propel the dance of the Kabuki avatars, computer-generated by our Japanese collaborator Gekitora. In the humorous (r)evolutionary logic underlying the scene, the human workers in the bustling crowd of the collective on earth (or in the sea) learn the new movement management from the avatars, some of which step up to the edge of the screen to look out from their virtual world and monitor the adaptations by their disciples. We composed these scenes with a tongue in cheek reference to the machine aesthetic of European modernism, the streamlined gestures of industrial workers in the age of automation, but also to Khlebnikov's exhortation, in the prolog to *Victory over the Sun* (1913): "You, the people who have been born and are not yet dead /...The theatre is a mouth! / Spectators, be an organ of hearing (be all ears) / And be observers" (Figure 6.7).

Toward the slow fade out of our performance, we imagined the whole space to have become a mouth, so to speak, with Wilkins (InstrumentWoman) intoning the birdlike trills and chants of her crazed Tutti Frutti interpretation of *zaum*, the

FIGURE 6.7 Ginyuu avatar looking out at Ruby Rumiko Bessho in rehearsal for *UKIYO* © 2009 DAP-Lab.

transrational one word lines of poetry invented by Kruchonykh and Khlebnikov for the ecstatic cubo-futurist tone of their collaborative opera *Victory over the Sun* (which the DAP-Lab interpreted in its own stage version, titled *for the time being*, at Sadler's Wells in 2014). The virtual environment in Second Life, meanwhile, had been opened to the public as well, and as I streamed the live sim of our virtual *UKIYO* onto the screens in our performance space, more and more visitors (rather, their avatars) were arriving and milling around. The performance, then, has two kinds of (a)synchronous audiences, a real one and a virtual one.

Naturally, the avatars of visitors in Second Life could not experience our real spatial-rhythm atmosphere which had become a mouth. What they did experience was the rather more abstract modeled environment of the *hanamichi* and Ruby Rumiko Bessho's adopted choreography streamed up live onto the miniature screens in the virtual world. She interpreted the human adaptation of the Ginyuu avatar choreography and developed her own variation on it, responding as well to the feeling of the mouth she gained from watching our performance via networked camera from my laptop (which also served as a communication/chat line to our collaborators in Japan during the premiere; Figure 6.8).

At one point, I sent my own avatar into the SL environment to walk around a bit, while with my other hand I was activating a last animation sequence to appear on the screens – a cosmonaut-avatar stepping into space and slowly tumbling into zero-gravity, looking down at our mouth and the human audience that looked up, in a strange moment of collectively shared perplexity, when certainties about what exactly constitutes reality, or "mixed reality," might be on hold.

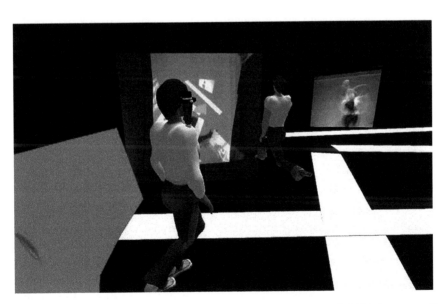

FIGURE 6.8 Visiting avatars walking around the virtual *UKIYO* © 2009 Courtesy of Kabayan/Inetdance.

Mixed realities, and undesigned atmospheres

Mutual mimicry – the beauty of creatures behaving in their habitats, merging and responding, teaming up or departing, all forms or media, so to speak, gesturing to each other and thus compounding the atmosphere. Swimming in gravity? Drawn by physical forces? Or by amphibious dreams, adapting to live in water and on land? Shifting between mutable space, under the moon that floats from field to field, as we learn from Hanaogi IV (in our epigraph)? What if the stage, our habitat, goes dark, and we imagine being without light, direction, location? What if the temperature drops, or increases in a way that we find it harder to breathe, our skin feeling hot, flaring up? Atmospheres in performance are unstable, just as I intimated in the beginning when I mentioned the white *hanamichi* in connection to floating islands. What if these corridors were like vapor trails in the sky? How can we ever predict how the temporary community of audience and actors behaves, what they experience and imagine, and how atmospheric conditions and the material/immaterial media act?

Michael Takeo Magruder, a scientist/artist attending one of our rehearsal workshops, talked about his conception of data environments, manifested in numerous installations that have received wide recognition, most recently *The Vitruvian World* and *Data_plex (economy)*. He mentioned a 2007 commission (by Turbulence/New Radio and Performing Arts, Inc.) to a handful of creative teams to produce artworks that would engage users across three distinct environments: the online virtual world of Second Life, a traditional gallery space, and the internet via the Turbulence website. Artworks operating across vast distances, time zones, cosmic space. Magruder believes that art events of this nature will soon become relatively commonplace within New Media Art practice. He also suggests that

> *Second Life's* accessibility and the user-centric nature of the environment opened new avenues to update and extend practice within the field of networked Virtual Art. Practitioners viewed the arrival of *Second Life* as a means to overcome the technological inhibitors of previous generations of Web and games platforms; their work would no longer necessarily consist of isolated virtual microcosms with limited or non-existent interaction between users.
>
> (Magruder 2009)

I am not sure now that Second Life and Flash survived. They may be gone. In the old theatre world, inhibitors also still exist: the notion of the "user" does not quite translate to performing artists working within dramaturgical and compositional frameworks of creation and public dissemination, nor does it translate to audiences either asked to spectate (say, a multimedia performance by the Wooster Group, or a West End show directed by Katie Mitchell or Thomas Kail) or to participate in environments that involve actor-audience interaction as in Opiyo

Okach's *Shift...Centre* or in Joumana Mourad's *IN_FINITE*. Direct interaction would change tacit agreements considerably, and I have not addressed situations where audience members are "cast" to do things for the performers, where they are delegated.

Referring to Second Life which housed the sim for *The Vitruvian World*, Magruder argues that this "fluid metaverse is the domain of its avatar inhabitants – the projected extensions of a disparate collection of individuals channeling their consciousnesses into the shared environment" (Magruder 2009). This idea of channeling of course is wonderful. A shared environment as community, and one that Eugenio Barba quite rightly considered a more secular, empty ritual (thereby implying a more horizontal idea of an event, not one having spiritual or religious efficacy, linking up "vertically" to spirits and gods), presupposes equal access and most likely a sustained involvement in its creative or other activities over a duration of time. Ritual suggests reperformance as well as some kind of transformative affect. We are back to the question of whether one can construct such transformation, provide for what is yet to come and use all available kinetics of storytelling as one might use prophecies or predictions of the weather, the oncoming storms and tides. We now live in an *interpandemic mode*, as Alan Read has suggested (Read 2020), and thus the kinetics I have described implicate an ethical understanding of hybrid living presences.[17] Presence and distance in a potentially quarantined performance space require attention to the fabrication of atmospherics, how breathing will be possible in such worlding. Zooming out, community building might be instilled into the character of online installations or networked co-authoring; yet not necessarily of performances taking place on a particular day at a particular hour in a particular location, featuring trained performers interpreting a score, however hybridized, and inviting the channeling from encroaching visitors. Attending is attending to one another.

Zooming out: the merging of *UKIYO* in London with a sims in Second Life was an experiment that we have only repeated a few times, so there is still a learning curve ahead of us to make practical use of a metaverse in theatre production or to incorporate other virtual and VR platforms (as these change and evolve and cannot remain). The *UKIYO* installation was performed over a few years (2009–11), then gave way to new explorations; it could not migrate to a networked site, as it needs the real-life performers in an actual, physically shared atmosphere.[18] When I look back now, after a few years, I do not even fully remember how to log on to Second Life. Around the corner now wait the VR or AR headsets, like the Microsoft HoloLens (with holograms that can display information, blend with the real world, or even simulate a virtual world). You can foresee calibrating your HoloLens to the space, calculating your social distance (Figure 6.9).

What can I suggest to practitioners who want to direct or perform in such mixed realities? For a digital scenographic practice that is not web-based but embedded in a physical theatre or performance venue, the hybridization of

FIGURE 6.9 Katsura Isobe [right] in organic Gingko leaves dress dancing the "creation scene" actuating 3D virtual landscape projected onto weather ball. *UKIYO*, Sadler's Wells © 2010 Michèle Danjoux/DAP-Lab.

material design, embodied performance and 3D graphics or virtual world design requires specific compromises. These concern primarily the physical integration of projected 3D virtual environments – or any artificial life and AI system – into the time and spatial human-scale relations of an event, of a gathering and its contingencies. The atmosphere is composed in the gathering, and if we think of the white and black *hanamichi* as floating islands, then all there is – participant actors, audience, media – is moving and sensing. The *channeling* goes both ways. Intelligent environments are always interface environments, resonant contexts that provide feedback to the performer. Audiences always provide feedback too. What I find most exciting is the manner of drifting that I observed in the audience, their spatial drifting but acute sense of attention to the qualities, the affective associations of the soundings, rhythms, motions and relations happening there, in front, behind, beside, below and above them.

Importantly, for a recognition of the *tekhne* of atmosphere making, performer techniques need to co-evolve with interactive digital systems and audience feedback so that analog processes – the vast range of human expressions and cultural perceptions – can thrive in computational climates. Such atmospheric climates are our future theatre. In curatorial terms, much debate is needed and is taking place, for the creation of adequate conditions for the display of new media art and performance installations (cf. Graham and Cook 2010; Obrist 2014; Foster 2020). Such curatorial concerns, now clearly at the forefront of all strategic planning in museums and cultural heritage sites, are less often addressed in performing arts contexts (academies that train the next generation of performers likely to work with interactive, real-time technologies), but they are needed to support research into mixed reality design.

Further illustrations from the various stagings of *UKIYO* could be reproduced here to draw attention to the floating patterns of performers and audiences. Some of these would indicate the closeness that some spectators chose in being with, and breathing with, the atmosphere and the dancers. They would indicate the casual, indefinite nature of atmospheres that include many design elements but also appear highly contingent, suggesting that tonal atmospheres cannot be completely engineered: the weather of a performance event cannot be predicted. Our audiences in Slovenia behaved differently from the ones in Japan. To take an example, audience members reacted very differently to the presence of an air pump – placed at a spot near the corridors, with its tube linked to the weather balloon that hung suspended above the audience. That pump offered a discrete invitation to use it, to keep the balloon afloat with air. A few audience members did use it, pumping air into the inflatable balloon, watching how the images caressed the smooth curves of the latex material.

The community of practitioners and interactors is widening: ritual techniques are constantly evolving too, and the crossovers that Magruder mentions concern a much wider cultural base than the performing arts. The scenographic imagination for atmospheric making in the theatre can only benefit from this.[19] In the following chapter, I shall look at a range of different works that illuminate compositional methods and how they draw from this wider base.

Notes

1 They were published as "Art Disincorporated: John Newling's Writings 1995–2005," Preface to *John Newling 1995–2005*, 2 volumes, Warwick: SWPA Limited, 2005, pp.11–17.
2 The son of a very old and close friend, Cédric Weidmann edits and writes for *Delirium Magazine*, see, for example: https://delirium-magazin.ch/section/category/home/how-likely-is-it-to-be-a-bat. I am grateful to Heiner Weidmann for sending me the bat story and photos of his new *Fledermaus-nistkasten*.
3 Monday, June 8, 2020, Ngozi Anyanwu and Jonathan McCrory (National Black Theatre), Daily Live Online Conversations with US and Global Theatre Artists, Segal Talks, curated by Frank Hentschker, The Graduate Center, City University of New York, with HowlRound Theatre Commons (a free and open platform for theatre makers worldwide).
4 As I write this, I remember that some time ago, when Polish theatre director Jerzy Grotowski began to embark upon his influential later paratheatre, theatre of sources, and art as vehicle phases, his collaborator Eugenio Barba (who had founded Odin Teatret in the mid-1960s and later the International School of Theatre Anthropology) published a book titled *The Floating Islands* (Holstebro 1979), later followed by *Beyond the Floating Islands* (New York, 1986). In *Floating Islands,* Barba did not address the building of theatrical spaces but the condition of exilic wandering, of dwelling in theatre as nomads would dwell in tents.
5 McCormick's latest book on *Atmospheric Things* speaks of such containers as "alluring elemental envelopments" and focuses on the release of balloons as a primary example of crafting forms of life in the air and of a tense relationship between envelopment

and the "elemental," here also acknowledging the inspiring influence of Durham Peters' book on elemental infrastructures and media that I discovered just before revising this chapter. I owe thanks to both of these authors (Durham Peters 2015; McCormack 2018).

6 I wish to thank all members of the DAP-Lab ensemble for their contributions to the project, and our Japanese partners in Tokyo for their role in the collaboration. The first version of *UKIYO (Moveable World)* premiered at Antonin Artaud Centre, Brunel University, June 1, 2009; the second laboratory took place at Keio University later that year. The expanded work premiered at Sadler's Wells and toured Europe in 2010–11, and inspired the later creations of the *metakimospheres* (2015–19). Project website: www.people.brunel.ac.uk/dap/ukiyo.html. *UKIYO* was supported by the DAP-Lab at Brunel University's Center for Contemporary and Digital Performance, a PMi2/connect British Council research cooperation award, and a grant by the Japan Foundation.

7 One of the most beautiful poetic analyses of theatrical lighting design appears in the chapter "The deaf gaze" in a catalog dedicated to Wilson's stagecraft (Morey and Pardo 2002: 45–91).

8 Indicating a significant expansion of traditional design training, the Yale Drama School announced the start up (in 2010) of a new M.F.A. in "projection design," which acknowledges the need for design artists of 21st-century theatre to understand and practice the inclusion of image projection media, cinematic techniques and digital imaging technologies into their stage craft (http://drama.yale.edu).

9 Kaisu Koski 2007 offers a meticulous set of definitions for some of the terms she adopts for her analysis of augmented theatre, technologically extended bodies, media content and participation in "intermedial performance," and although she bases her descriptions on her own work, as I do too, some of the vocabulary has been gradually established and shared over the past decade in various studies of digital art, design, and performance, including Manovich 2001, Bolter and Gromola 2003, Carver and Beardon 2004, Chapple and Kattenbelt 2006, and Birringer 2008a. In her own examples, which differ somewhat from my basic distinction between frontal and environmental staging, Koski succinctly observes that augmented theatre can use the traditional separation of performer and spectator as well as transform the relations through various combinations of environmental staging and the intermediality of the stage:

> The first combination appears on staging which allows a spectator to navigate physical space, but where she is connected to other stages by mobile communications. The second option is staging where a spectator is stationary, but the performance surrounds her. Part of the performance is screen-based material. The third staging allows a spectator to navigate between multiple stages. Some of the stages include screen-based material. The fourth staging I discuss is one in which a participant causes technologically performed aural and/or visual reactions through her body movements.
>
> (Koski 2007: 34)

David Rokeby was an early pioneer of interactive spatial composition (especially with sound), and he wrote the VNS (Very Nervous System) software particularly for his early installations (1983–91) designed to surround and immerse the audience in an environment. Rokeby has recalled the system as a type of "performance" – in public space of art museums or galleries – where the audiences were the performers themselves: their movement translated into sound drawing the bodies into further movement: through the "algorithmic processing that is coded into this rapid real-time feedback loop, behaviors can be discouraged or reinforced, altering your relationship with your own body" (Rokeby 2019: 90).

10 Environmental staging therefore is a practical example, in the theatre, of what engineers and computer scientists elsewhere examine in terms of their development of methodologies for building artificial systems (robots or virtual agents in particular).

One of the best introductions to this field of systems engineering is Pfeifer and Bongard's *How the Body Shapes the Way We Think* (2007). See also Kleber and Trojanowska 2019. Immersive performance, in which movement with body sensors creates a constant flow and feedback loop of sensory and perceptual data, which can control image/sound projections of Virtual Environments and permutations of the time-images, heightens proprioceptive awareness of the physical body moving in space, and stimulates a process of re-experiencing what constitutes self and identity. The acceleration and deceleration of perception of image and sensorimotor logic has made performance with wearables (e.g. sensors measuring neurophysiological functions) significant for artistic experiments with the physiological, the machinic, and the virtual (Birringer and Danjoux 2020).

11 See, for example, the series of telepresence performances conducted by Annie Abrahams and her collaborators Daniel Pinheiro, Lisa Parra and others over the past few years: https://bram.org/distantF/. I took part in several of them between 2016 and 2019 and I was particularly struck by the atmosphere of the (nearly) silent "telematic embraces" we enacted, for example, during *Distant Feeling # 3* (November 24, 2016), when we were invited to experiment with an online "séance" where the main goal was to experience the others' presence with eyes closed and no talking allowed. Together participants tried to get a grip on what, why and how energy flows when bodies are absent (and yet present) in online performance situations. In my notebook entry, I wrote that

> the idea of being silently wrapped in the telematic embrace is an experience that is effective in direct proportion to one's ability to suspend disbelief. Like Abrahams' work *The Big Kiss*, or Paul Sermon's telematic pieces, the sensation of intimacy is never quite 'real,' it is based on the willingness to believe and to allow closeness to become 'real' despite separation. For those who participated in this experiment, it was exactly that: the willingness to trust in the knowledge of the virtual proximity and connectiveness of the others. It is that knowledge that can be convincing enough to suspend disbelief and thus be silently wrapped in the telematic embrace. This work is a great model for how we might conduct ourselves on the Internet.

12 See https://vimeo.com/433722. This "Director's Cut" interview between Raul Ruiz and Kriss Ravetto-Biagioli (recorded November 27, 2007) was available online when I worked on an earlier draft of this chapter.

13 See Erin Manning's *Relationscapes* for an insightful discussion of early chronophotography and Deleuze's concept of the movement-image, linked to eurhythmics, Leni Riefenstahl's cinematic aesthetic, and the notion of the biogram as becoming-body (Manning 2009: 124–26).

14 From the mid-20th century into the 21st, artists and musicians have manipulated, cracked, and broken audio media technologies to create novel sounds and performances. Caleb Kelly 2009 describes how the deliberate utilization of the normally undesirable (a crack, a break) has become the site of productive creation, citing many artists, including John Cage, Nam June Paik, Yasunao Tone and Oval, who broke apart both playback devices (phonographs and compact disc players) and the recorded media (vinyl records and compact discs) to generate an extended sound palette.

15 See Danjoux 2017, where all the intricacies of her wearable designs are explained and analyzed in great detail. In hybrid performance architectures that include interactive systems and real-time synthesis, it is problematic to use notions of preformed "choreography"; in such contexts the choreographic is partly improvised and partly structured yet open to instant transformations and adaptations. See "After Choreography" (Birringer 2008b).

16 Michèle Danjoux and I attended this performance event and we were thrilled to listen to the voice too, especially of the women dancers who intoned some of their scenes in a powerful manner. For further information on *Shift…Centre*, see: http://www.gaaraprojects.com/shiftinggaarae.htm (accessed July 23, 2019) Diametrically

opposite to Okach's physical performance are examples of contemporary design which relinquish the physical altogether, as in Penelope Wehrli's *camera orfeo* installation (2008) which followed her pattern of video design for earlier works such as *Bluebeard's Castle* by removing all actors from the scene and replacing them with kinetic video-projective sculptures. *Camera orfeo,* subtitled "an auto-choreographic and media composition," shows endoscopic images of the vocal cords of a singer while singing the aria "Possento spirito/Orfeo son lo" (Monteverdi) and, among other things, video images of dancers which are fed into a circular system controlled through the random movements of the visitors. The musical, choreographic and visual source material is continuously recombined and transformed into a kaleidoscope of images and sounds through the use of cameras that register what goes on in the exhibition and performance space. Benoît Lachambre and Louise Lecavalier's *Is You Me* (Coda Festival 2009, Norway) is a dance duet staged on an all-white raked platform which literally sutures the movement into Laurent Goldring's continuous and highly kinetic video drawings and graffiti projected live onto the dancers. Projecting visual graphics down onto the performers is a trend noticeable over the past few years; a prominent example, discussed widely in the international dance tech community after its worldwide touring, is *Glow* (2007), created by the Australian Chunky Move company in collaboration with software engineer Frieder Weiss. For further commentary on these staging methods, see Chapter 7.

17 In a personal email, Alan Read told me that he just completed his new book and worked on the index during the outbreak of the corona pandemic when he decided to add "Pre-Pandemic/Post-Pandemic" as a marker of what he thought was the book's inevitable *inter-pandemic* status. *The Dark Theatre,* he added, was about bankruptcy and loss, and it was unlikely that theatre would ever reach a post pandemic resurrection.

18 Some more expansive writings on *UKIYO* have already published; therefore I did not think it was necessary to go into all the details here. Concerning the expansive design aesthetics for the wearables and the relationship between design and choreography, see Danjoux 2017. For some other aspects of atmospheric attunement in *UKIYO,* see Birringer and Danjoux 2013, and Birringer 2013b. Kathleen Stewart has written evocatively about such charged attunements, and how circulating forces and palpable rhythms, valences, sensations, moods, and tempi are generated as atmospheres (Stewart 2011).

19 On a grander scale, Tomás Saraceno has used inflatable balloon ideas in some of his more recent sculptural projects, the *Aerosolar Journeys* and the *Aerocene.* The latter, shown at Wilhelm-Hack-Museum, Ludwigshafen (2017), involves developing a range of flying sculptures held afloat by thermals, without engines, gas, fossil fuels or solar cells. Alongside the *Aerocene* project that now involves teams of artists, designers and scientists, the *Museo Aero Solar,* first launched in Medellin, Colombia in 2007, later exhibited globally, became a collaborative movement for fabricating a spherical sculpture made of reused plastic bags provided by participants, envisioned as a utopian thermal "flying object" where the updraft is provided solely by solar heat (https://museoaerosolar.wordpress.com/).

7
PROJECTION ENVIRONMENTS AND ANIMATED LIGHT

Light, haunted

Attention is created through focusing of perception, heightening our sensorial awareness of a person or object in space, of our own position in space, of movement, proximity and distance, of color, size, shape, weight, texture, scent or other properties of phenomena with which we come in contact. Light and darkness are two major contexts we are familiar with, and each condition influences the perceptive mechanisms, which are ingrained in our nervous system. In complete darkness, our eyes must adjust in order for us to be able to move at all or to imagine where we are in space. When an artificial light source illuminates an area in the dark, our eyes are drawn to it. I am not sure whether I would follow Arnold Aronson, the renowned historian and theorist of theatre set design, in calling the illuminated stage an "abyss" that stares back at us, but he wants to draw attention to the seductive lure of the theatre as a primarily visual art (Aronson 2005). For him the abyss is not a mouth but an eye that swallows. If the actor, under the spotlight, is looking at us without seeing us – spectators in the dark who are preying on the actor – then we might discover a curious paradox. Does the actor focus on the light even as it might blind her? What do the actor's eyes need to see? How is the actor focused *by* the light?

The matter of preying is perhaps not the track to pursue, though it might illuminate my recent obsession with underground sensings, caves and dark underworlds. Theatre director Herbert Blau, mentioned earlier, once dealt with the seductive lure of the theatre in terms of the actors not fully realizing how much they are stared at, by the "eyes of prey" (Blau 1987), and yet how occluded the guilty thing in the abyss might be – to which like Furies one is drawn irresistibly, trying to catch something in the unconscious, as Hamlet desperately wishes to do. The spectators preying are then also guilty of listening and peeking into the

DOI: 10.4324/9781003114710-7

sacrifices (as Pentheus did when spying on the bacchants' secret rites, his limbs then ripped apart by his mother Agave and the furious women), the groaning of the ghosts, the occulted or elusive source of the whole dynamics, that slippery edge of spectatorial (immersing) desire at the vanishing point that Blau addresses in his plaint about haunting appearances.

Hauntology is an immersive speculative genre that deals with such specters and ghost work, and our gothic imagination was fed by early cinema's black and white expressionist sets, with their steep angles and deep shadows. The vampire around the corner. The notion of *chiaroscuro* was often used to describe the high-contrast arrangements of darkness and light, and in the German language the critic Lotte Eisner used the adjective *helldunkel* to refer to her own sense of the abyss in expressionist cinema (*The Cabinet of Dr Caligari, Nosferatu,* etc.), with its lighting atmosphere a kind of "twilight of the German soul" articulated through shadowy, enigmatic interiors, or in misty, insubstantial landscapes (Eisner 2008). Even more intriguingly, a recent study called *AUDINT: Unsound: Undead,* collects excavations of peripheral sonic perception ("unsound") and the various ways in which frequencies are utilized to modulate an understanding of presence/non-presence, anomalous zones of transmission, and ultimately hint at life/death (Goodman, Hays and Ikoniadou 2019). Such themes of hauntology, the editors argue, have also inflected the musical zeitgeist and growing concern with spatio-temporal anomalies, Electronic Voice Phenomena, psychoacoustics, auditory illusions and a reinvestment in traces of lost futures inhabiting the present.

This idea of the future inhabiting the present resonates with my musical memories of Sun Ra and George Clinton's bands Parliament and Funkadelic (*Mothership Connection),* the linkage of black cultural heritage and the space age intoned in the sounds and lyrics of Afrofuturism. The space age and Moon landing seem a long time ago. The present is unmitigated; during the current wave of anti-racism protests and the Black Lives Matter movement, we grasp that the present, like the past, has been violently anti-black. The future needs to be invoked, as this unmitigated present is an abyss, an overshadowed *aporia* for any assumptive logic of discussing a "narrative of social, political, or national redemption" (Wilderson 2020: 15), an entrapment that Othello, even more so than Hamlet, cannot escape. The Studio Museum in Harlem staged a major exhibit, *The Shadows Took Shape* (a phrase from a poem by Sun Ra) exploring Afrofuturistic aesthetics (November 14, 2013 to March 9, 2014), and I remember seeing more than sixty works of art that looked intensely at what Afropessimist Frank Wilderson is resigned to call the *other as black,* as black alien and irredeemable, and yet as clearly undismissable, conjuring up negotiable utopias and other angels of history in relation to technology, time and space within African-American communities. The *Shadows Took Shape* recurred, in a performative sense – as shadows recur. Its long shadows were transcultural and resonant with repeating historical struggles and diasporic convolutions. In one of the film screenings held in conjunction with the exhibit, John Akomfrah (co-founder of the Black Audio Film Collective) showed *The Stuart Hall Project,* a haunting record of the massive social and political perturbations of post-colonial Britain. Produced from archival footage (British television and radio, images from his trips to Jamaica, with the music of Miles Davis), the film portrays exile, racism, hybridity, violence, and radical struggle – an ever unfinished conversation, all of which has been the experience of New World black and South Asian émigrés for some time. Shadows, then, that grow into a very dark, dark light.

In a more current re-reading of *The Eye of Prey,* less bothered with deconstruction but touched and inspired by Wilderson's *Afropessimism* (Wilderson 2020) and Frank Moten's notion of the *undercommons* (Harney and Moten 2013), I tend to see the burrowing into the deep or the cellarage as a lighting out into an under-acknowledged shared space, a latent commons: the positively alien and self-insulated commons we are in fact forced to channel. We think and dream darkness, we fear falling into it. In my family, on my father's father side, numerous grand uncles worked *unter Tag* in the coal mines, and all my life I have imagined what it must be like to work down below, in the deep entrails of the earth, with a small flickering light in front of the helmet. Shadows flashing by on the walls, geological layers or life lines of the earth are intimated, with coal as a sedimentary rock that is made up of organic carbon, the remains of fossil plants. Those existed since the late Carboniferous, about 300 million years ago. The coal seams have preserved fossil plants and insects, sandwiched in between thick layers of dark grey deltaic mudstone or sandstone. Rocks, minerals, metals: the whole geophysics of long durational "medianatures" (Parikka 2015) and material agents invite us to imagine or detect a hidden last picture show exhibiting the psychogeography of extractions from the earth (in later human history of engineering and technological development), and the shifting plate tectonics of the mind.

We must channel into such a commons and earth below. If the ghosts don't appear, Blau would have said, then the performance is in trouble. But ghosts are invisible, aren't they, and this reminds me of a haunting moment, some time ago in Chicago, when an Indian dancer, Avanthi Meduri, asked me in rehearsal to point a down-light onto her that would shine "through" her and into the ground under her feet. Permeable actor, light vanishing her or transfixing her, thus her presence drifting off perhaps or rising, translucent, just as I take Blau's emphasis on re-presentation as presence to imply that *presence in itself is not immersive*, not manifestly there but always mediating, shifting in its appearance, visible and not visible, transfigured. With dim lighting, a noticeable choice of lighting designers in many performances over the past years, I must strain my eyes to see the performer. I must squint my eyes and lean forward. Perhaps lighting design in theatre now strives to explore what Eisner called *helldunkel* and French cinematographers, more ecosophically minded, call the dusky time *entre chien et loup*.

As I explained in earlier chapters on DAP-Lab's installations, especially *kimosphere no. 4* (2017), composing a ritual for slow time/slow space – where the audience can drift into an atmosphere and undergo a kind of incubation – means to focus attention on audible and visible, tactile, multisensorial processes. On how these atmospheric processes, if we understand them as formations and elemental media, permeate us, how they flow through us or touch us like the wind, how they shape visibility and audibility, how they saturate or diminish. Audience are the guests. Our ensemble will act as the host. We quietly observe the behavior of our guests, their questions and responses, their participation, how they let the wind touch them, how they express their attention or hesitate, remaining still. This is not necessarily a purposeful social choreography in the sense in which Shannon Jackson has articulated the art of social politics (Jackson 2011). Our invitation, on a metaphysical level, is more of an open-ended gesture toward a sanctuary, a meditative-playful space rather than an enclosure where immersion is experienced as agitating, provoking – as we learn in *Schlingensief: in das Schweigen hineinschreien*, a documentary on the late director's innumerable theatrical provocations. Provocations can be insinuated more subliminally, with low heat. Participatory theatre is not necessarily more democratic, open and convivial than other forms of theatre that rely on a separation of stage and audience, of scene and observation – whether brightly lit or in dim light. The barely perceptible is often less suffocating than the spectacular. Navigation is poetry.

In his book, *Theatre in the Expanded Field: Seven Approaches to Performance*, Alan Read reflects on biopolitics and wishful thinking about "coming communities" (Read 2013): he considers *immunity* in his final approach to performance. Someone tells him of feeling shamed by the forceful invitation to members of the audience to participate in contemporary theatre works (of several known companies, such as Punchdrunk, Third Rail Projects, Shunt, Rimini Protokol, Nerve Tank, Nature Theatre of Oklahoma, and Toneelgroep). Their approaches to theatre are distinct, yet they favor a kind of emancipated participation, which, in turn, arouses Read's suspicion. In a provocative maneuver, he adapts the term *immunity* from Roberto Esposito's political philosophy where it refers to the policing of the somatic borders of the body (Esposito 2008). What threatens the borders (and the auto-immune system) is not a matter of what they include but of what they exclude. Immunity is granted to the individual against the common (the community) which – in immunisatory terms – is alien and toxic, since immunity in fact means not having anything *in common* with the alienating, the toxic.

This complex and contradictory problem invites further investigation, but first I want to mention sublime occurrences where community unravels positively. On a weekend not too long ago in time, an old musician friend, Phill Niblock, came to give a concert at Tate Modern in London. I went on a Sunday, and spent six hours at the museum as I wanted to explore the Switch House, the new addition to the museum built on top of the rediscovered old oil tanks of the Boiler House that became Tate Modern. The oil tanks are huge round coliseum-like structures. Niblock and musicians performed 150 minutes of heavy deep drone music, to four film projections of nature filmed in 1970; the concert celebrated early works from 1970 to 1971, when he lived in upstate New York and filmed the outdoors in a radius of 100 miles from his house. Niblock is the founder of Experimental Intermedia and has been performing for 50 years. Now 83, he brought along younger guest musicians, while audiences gathered from all over the place, young and old, near and far. The mood was wonderful, relaxed, energetic and spirited, I entered not really knowing anyone, and left meeting many, chatting afterwards and hanging out.

This is half the story. Earlier, before I made my planned trip to the Southbank, I got an urgent email from Chilean dancer Macarena Ortúzar, who had studied butoh in Japan on Min Tanaka's farm. She urged me to come in the twilight on Saturday, where Tanaka was to perform on the south terrace, in the fog. I got there a day too late, but I looked up the fog on the south terrace and stood inside it. Fujiko Nakaya's installation was in action. I got wet; children were hopping around laughing into the mists, screaming with pleasure as children do. Parents diligently took photos of the screamers with smart phones. A thin young Japanese man fidgeted with his long selfie-stick until he managed to photograph himself in the fog. I smiled at him and he smiled back. A pioneer of installation/video art in Japan, Nakaya had come to create this amazing fog sculpture, generated out of compressed water mist. The fog acts as a barometer, reading shifts in atmospheric conditions, reacting to the environment and rendering it visible and palpable to viewers. The fog was further animated by a light-and-sound-scape by Ryuichi Sakamoto and Shiro Takatani. And then, from nowhere, suddenly Min Tanaka, the butoh master, appeared and danced into and out of the fog, like a ghost dog (Figure 7.1).

FIGURE 7.1   Min Tanaka, dancing in Fujiko Nakaya's fog sculpture, Tate Modern, *Ten Days Six Nights* 2017. Photo courtesy Claudia Robles-Ángel.

Already in 1974, I read in the program, Tanaka had developed a unique style known as "hyper-dance" which emphasizes the psycho-physical unity of the body. He now teaches the *body weather* approach, enacting improvisational performances that abandon the stage in favor of parks, streets, seashores and fields in Japan or abroad; in the 1980s, he secretly infiltrated the former Soviet Union countries to perform as an act of rebellion. Ortúzar was there and documented the quiet act, as did Claudia Robles-Ángel, a sound and media artist who knew Niblock's work well and had explored audio-visual installation art in her own practice. When I write to them and mention the notion of the *ghost*, Macarena tells me:

> yes, like an appearing and disappearing ghost indeed; so beautiful, so beautiful; his dance tells me also this kind of complete surrounding to the world that pushes us, sometimes I felt his own disappointment for the world created by humans, with hopeless love. He is extraordinary really, just a huge amount of energy and the inspiration to keep working regardless of everything, age, time, struggle, cold, hot, everything and anything.

She then added that she traveled home crying: "I danced a little solo on Sunday here in Oxford, my way to say thank you to him in the silent distant."

I wanted to share this, since the sentiment expressed by the guest (Ortúzar) seems to confirm Read's notion of the immunisatory principle of theatre, which shames us, but not necessarily because it is not, and cannot be, an act of rebellion. It allows us to resist as false the offer of inclusivity. I feel energized too, all the same, and start to dance again a day later, from the distance, developing a small improvisation, called "Lemur," attaching a sensor on my right hand wrist called MiniBee (Sensestage). I move sounding through my body, from the wrist.

My lemurian dream is to dance in the twilight of the Red Island, our imaginary Madagascar which we tried to invoke in *kimosphere no. 4*. The blur of the fog sculpture, I would argue, is the sensation I have had for a while during the Brexit debate in the United Kingdom. There is no point now to be drawn into that here, I have already left England during the pandemic and the pandemic was not foreseen during the endless confusions over Brexit. The change there is only one change among other changes and other continuities, of a political and of a cultural kind. Briefly evoking the dance of Min Tanaka in the dim light, *entre chien et loup,* the blurry threshold time, between hope and fear, between the familiar and comfortable, the unknown and dangerous, I now only remember the glimpse of tremendous grace, love and affection generated in the witnesses of what was glimpsed and heard.

The *auditorium*, originally named as such to define a place of hearing, is a space of darkness, of blindness, at least in the modern era after the introduction of electrical lighting in stage architecture. In fact, when we listen intently we often close our eyes, and this also explains the difference between hearing and listening. The former is passive, while the latter is active. When we listen, we actively invite the sound to fill our mind, enrapture our thoughts and emotions and let our body vibrate. Sound is a spatial acoustic medium: the sound environment can envelop us or draw us toward its source, engaging us to decipher the meaning of the information. Opening the eyes, we project forward our anticipation of seeing something, our desire to see. Or we turn our bodies toward the sound we have heard. We focus the eyes which have receptors (in the retina) that send the information to our brain. The receptors begin to act on our imagination.

The physics of this relationship are important, as we shall find the vocabulary useful for thinking through lighting. In regard to light and color perception, the eyes are the receptor organs, synchronous with the sense of hearing (although this synchrony can be disrupted) and the function of the ears as amplifiers of vibrating particles. The rods and cones in the retina at the back of the eye are sensitive to dimness and brightness and detect the narrowly defined frequencies of light (green, red, blue). Responses from the three primary colors are processed in the brain to give us the perception of color. Interestingly, I re-learned these basic ideas of the physics of light, color and how the human eye reacts to color from working with images in Photoshop, and I am sure many of the readers recognize the same ideas instantly. Color, of course, is a property of light. The eyes see only a small part of the electromagnetic spectrum: visible light is made up of the wavelengths of light between infrared and ultraviolet radiation (between 400 and 700 nanometers). Taken together, these frequencies make up white light. White light can be divided into its component parts by passing it through a prism. The light is separated by wavelength and a spectrum is formed. Newton first discovered this phenomenon in the 17th century, and he named the colors of the spectrum.

If the ends of the spectrum are bent around and joined, a color circle (the color wheel as we know it from Photoshop) is formed with purple at the meeting place. For lighting designers, the use of color is crucial, and of course its use offers many challenges, particularly in conjunction with the more recent projective media (film, video, LED) that have come into play in multimedia performance. The basic challenge is: how much do you want the audience to see, and how do you want it be seen? How do you light an actor as compared to a dancer or a singer? Do you remember Robert Plant singing *Stairway to Heaven* in 1972, just his long blonde hair lit with a single spot from behind, the rest of Led Zeppelin and the stage in dim blue (black) light – creating the haunting, iridescent effect of halo around his head as well as the back of his finger tips (https://www.youtube.com/watch?v=xbhCPt6PZIU)?

Do you light the entire stage environment or only parts of it, and what mood do you want to create to support the dramaturgy of a scene or the characterization? Have we taken it for granted that dance productions use side lights (shins, for example) to draw attention to the movement of feet and legs and to enhance the kinetics of motion? Does color affect the mood of the actor as well as that of the audience listening to/watching the actor? What do we want to make happen with the shadows or silhouette effects on the face/body with down and back light? How do we use front light to create specific contours, or isolations, or filling the stage gradually and with gradations of fill light? Color has three distinct properties: hue, value and saturation or intensity. Design composition suggests that in order to understand color you must know how these three properties relate to each other. If we use film or video projection, the relations between lighting instruments and film projectors/LCD projectors are equally important. The stage or performance area can be lit with a variety of instruments (front light, down light, back light, side light, moving light, etc.), and just as with the sound environment, we need to see light as a dynamic medium: it can continuously change, it can move, and its intensities and temperatures be modified. If you fade out the light, it grows dark again on the modern stage. The abyss beckons.

The *dispositif* of projection space

Theatrical performances seldom take place under natural light conditions anymore (except when they happen outdoors), and thus the development of design in the 20th century was hugely affected by the introduction of electricity and special lighting instruments that are able to create spots or defined areas, as well as a range of moods and psychological effects on our visual and kinaesthetic perception. From the point of view of the performer, the idea of "projecting" voice, gesture, movement and the charged presence of body and mind on stage is a well-known concept that lies at the heart of practically all theatrical techniques of performance. Projecting, in this sense, means enlarging or emphasizing, creating focus on the act that needs to be seen and heard by those who are present. In the earlier chapters, attention was already drawn to this, but here I want to examine the role of "projection" from the design perspective, and thus focus more on the environment of expanded mediation (projected film, video, computer graphics, sound, etc.), and the *dispositif* created for the performer and mediating technologies.

Unlike theatre theory, which after Brecht speaks of the "apparatus" when it looks critically at the operations of the stage where actors enter and exit and things mechanically go up and down (using the fly space) or rotate on a turntable or are moved about by stage hands, film theory since the 1970s, following Jean-Louis Baudry, has preferred the notion of a *dispositif*, the French word meaning "disposition" or "arrangement."

> Le dispositif cinématographique aurait la particularité *de proposer au sujet des perceptions «d'une réalité» dont le statut approcherait de celui des représentations se donnant comme perceptions.*

> (Baudry 1978: 45)

Philosophers of media and social/political theory such a Foucault, Deleuze, Guattari, Flusser and Zielinski became hugely interested in the notion of the *dispositif* in the 1970s and 1980s, utilizing it as a conceptual category for examining environments (material, technological, medial) or regulating, strategic frameworks that are configured in certain ways making it possible for certain types of phenomena to occur. While Foucault tended to emphasize the regulatory and panoptic formations that produce power, knowledge and subjectivity, Deleuze and Guattari became more interested in the drifting and disjunctures between heterogeneous elements in a multilinear collective assemblage or *dispositif*. In other words, they acknowledge that arrangements are precarious and cannot always control outcomes. The lines that compose a multilinear ensemble – referred to as *agencements collectifs d'énonciation* or "collective enunciations" – can change direction or become unbalanced and forked. When human and technological processes are intimately intertwined or cannot be easily differentiated, the component materials, forces, energies, rules, conventions, and lines of communication are not stable, their contours are not fixed but subject to a series of variables (Deleuze 1992b: 159). Or as Durham Peters expresses it so eloquently in his chapter on "Cetaceans and Ships," referring to earth and marine environmental craft/technique and the mimicry of nature and technology:

> A medium reveals the nature it rests upon as the ground of practice. The concept of media is thus amphibious between organism and artifice. We cannot help but explore the astonishing and sometimes comically diverse morphological and functional range of living organisms as a historically sedimented set of solutions to problems of existence.
>
> (2015: 112)

Durham Peters stunningly analyzes the ship (seacraft) as one archetype of "artifice-become-nature" and "craft-become-environment" (102). The problems of existence relate to perception, and the vehicle-medium or artificial environment: in our case not the ship but the theatre. However, his list of the elements of seacraft is highly suggestive at the very least, when he describes it being a "veritable seedbed for innumerable arts such a navigation, steering, leverage, reading the sky and stars, mapping, timekeeping, documentation, carpentry, waterproofing, provisioning and preservation, containerization, division of labor, twenty-four hour surveillance, defense, fire control, ballast, alarm calls, and political hierarchy" (105).

Imagining the theatre or the cinema as a vehicle-medium, in Marshall McLuhan's sense, and thinking of reading stars in the sky, an installation by Hajra Waheed comes to mind which beautifully manifests a particular (inverted) perceptual experience. Shown in her *Hold Everything Dear* exhibition at the Power Plant Contemporary Art Gallery in Toronto (September 21, 2019–January 5, 2020), this work turned out to be the most audience pleasing of all. Placed in one of the rooms of the exhibition, *You Are Everywhere (a variation)* is an immersive installation inside a rectangular black box. It invites audiences to lie on a wooden floor, which is lit to resemble a starry sky. You look up by looking down beneath you. Merging earth and sky, *You Are Everywhere* allows visitors to experience an almost psychedelic twist of reality or star gazing. Having a starry night sky collapsed so that we do not have to stare up – this may provoke a sense of hope, or turn out to be a disconcerting disruption, letting us sink into a chaos of earthliness or the abysses below that peek through the cracks. Waheed's own comments suggest that it is an

> immersive floor installation…meant to shift audience perception by inviting them to lay across the floor and anchor themselves against the night sky: there is this quiet sense of being adrift where for a moment, earth and sky collapse.
>
> (Waheed 2019)[1]

Artists who design such immersive spaces with light, sound, silence and an atmospheric environ of course expect certain reactions to happen; when I experienced the installation, I did not feel anchored but rather completely at sea, so to speak, having fallen off the ship, perhaps into the open waters as I imagined. If we now take the obvious historical fact that there have been various kinds of arrangements of film-viewing and film-experiences, stabilizing or anchoring no doubt, but then include more recent developments in televisual media, VJing, club event, rock concert and computer culture, the heterogeneous assemblage of vehicle-medium components can manifest itself in many different kinds of dispositions. For example, there is the particular *dispositif* of spectatorship attributed to the classical or mainstream narrative cinema: its basic building block is the dark room of the movie house where spectators focus on the images on the screen rather than being made aware of the projector, the cinema's architectural space or the industrial apparatus of film production and distribution (not to mention the piano or Wurlitzer organ that have long disappeared).

The thin screen is the component Vilém Flusser has in mind when he writes that images are "significant surfaces." In most cases, he argues, they signify something "out there," and are meant to render that thing imaginable for us, by abstracting it, by reducing its four dimensions of space-plus-time to the two dimensions of a plane. The specific capacity to abstract planes form the space-time "out there," and to re-project this abstraction back "out there," Flusser calls "imagination," and he might be thinking of both the producers and the spectators who have the capacity to project and decipher images, the capacity to codify phenomena in two-dimensional symbols, and then to decode such symbols. But then he adds that the significance – the meaning of images – rests on their surfaces from which some dimensions have been suppressed (Flusser 1999).

The cinematographic *dispositif* therefore produces moving images by removing those other dimensions from the spectators' gaze. In the dark room, we are in front of the screen, and there is nothing else. The mediating principle of cinema requires that the depth of projection is denied so that the depth of field may exist. Again one might say here that this is the magic of the cinematic, producing an obvious immersion effect, but the sleight-of-hand also reproduces the medium itself. As it was the case with its ancestral forms – the magic lantern and shadow play – the cinema is an ensemble of techniques to make light fall on a surface. The seemingly empty space between the projector lamp and the screen is where cinema really happens. In an online forum on "The Thickness of the Screen," which examined the material properties of audiovisual media, Gabriel Menotti scrutinized this immersion effect, observing that just like the depth of projection, some other dimensions must be there but cannot be negotiated, in order for cinema to exist as such. Since they house the principles of the medium, those fundamental distances do not seem to be available for creative operation. When they are effectively occupied, cinema shows itself expanded – as sculpture, installation or performance.[2]

Without going into a deeper analysis of the experience of cinematic medium, we can posit that the cinematographic *dispositif* is different from the viewing and projection arrangement of a multi-screen video installation, and different again from a theatrical or site-specific multimedia performance or a participatory on-line virtual environment such as Second Life. Online environments come alive through the (inter)active graphic interface, pushing the earlier textual interfaces in MUDS (and MOOs) further by propelling a certain visual fixation that extends the manual operation of keys or mouse. In Second Life, the graphic space takes up much of the attention of the users and designers who concern themselves with the appearance of the avatars, virtual objects and landscapes. Even if the Second Life worlds are permeable to the extent that other digitized information can be streamed in (video, audio, etc.), they remain screen-based of course. The theatre remains different from such virtual environments as it exists in three-dimensional real space, which may or may not include projective media in the scenography for its kinetic atmospheres.

Theatre spaces can be large or small, contained in various kinds of architectures ranging from the proscenium to the black box. Larger stage productions generally use set design and props, and large ballet or dance productions often use the white back scrim to create a particular color field or texture as backdrop. The black box theatres tend to be smaller studio theatres or warehouse-type spaces used for experimental productions, stripped down and simplified arrangements often deploying no set design but only performers and light. The light in theatrical production is quite different from the projector light in the cinema, since the instruments are generally hung from the grid above (and just in front) of the stage, and together with side lights often used in dance, the lighting enables a three-dimensional sculpting of the performance space and the performers who act in it. However, there is a particular irony in the theatre architecture of large proscenium and opera stages, since the proscenium frame tends to create a gap between stage and auditorium, turning the frame into a "picture frame" that highlights the stage as a "stage image," especially when curtains go up at the beginning of a play to reveal an elaborate set or a naturalistic *trompe-l'oeil* scenery.

The stage design, in this case, tends to create a picture effect, reducing the actual depth of field to a framed picture that the audience looks into, peaking into the mouth – and it is the very effect that Wagner had wanted to achieve when he worked on his concept of the *Gesamtkunstwerk* and overlooked the construction of his Bayreuth Festspielhaus which was custom built for the control of aural and visual perception he desired for his music drama. Inevitably, such design bears a certain resemblance to the projective art of the movies over the last 100 years. The spatial arrangement and the ideological *dispositif* are similar, since both in the cinema and the proscenium theatre, spectatorship is constructed through the production of an illusionary reality – a projected world. This illusionism, which can be traced back to the baroque era, was of course contested in various ways by the modernist and postmodernist avant-gardes during the 20th century. As I mentioned in Chapter 1, already in the very beginning of the 20th century the Swiss theatre designer Adolphe Appia drew attention to radically new scenographic concepts, focusing on music and abstracted rhythmic forms interacting with light. In his collaborations with Dalcroze at the newly built Hellerau Festspielhaus, he worked with ideas of open environments in which performers and spectators shared the same spatial volume which was designed like a musical space, a rhythmicized *mise en scène* – often displaying only simple geometric modular elements such as raked stairs or platforms – coming alive through the movement of the dancers and the sound. Appia was preoccupied with the plasticity of light and motion, and in his writings (*La musique et la mise en scène*, 1899) he formulated this vision enthusiastically:

> In itself, lighting is an element that can produce unlimited effects; restored to its freedom, it becomes for us what the palette is for the painter... through projections that can be simple or complex, stationary or shifting, through partial obstruction, through varying gradations of transparency, we can obtain an infinite number of modulations.
>
> (Quoted in Bablet and Bablet 1982: 43)

How radical this conception must have been can only be imagined today, since spaces such as Hellerau, with is vast empty hall measuring 50 m × 16 m × 15 m, did not become a prototype for the mainstream theatres of the modern age. But Appia's experimentation with (indirect) lighting and rhythmic spaces have been influential, even today among the digital artists and virtual reality designers who are sensitive to the animated, transformable spaces Appia proposed. Michael takeo Magruder, as mentioned in the previous chapter, is one of the contemporary artists/scientists who have worked directly from the Appian vision. He created a virtual 3D reconstruction of Hellerau in the multi-user environment of Second Life for *Theatron,* an explorable model of theatre architecture in virtual space, and *Rhythmic Space(s)/Meeting Places,* two joint performances created in Second Life and in Hellerau's Great Hall for the CYNETart_07 festival. For *Meeting Places,* he asked Hellerau's technical team to build a white, life-size stair case configuration consisting of three symmetrical platforms onto which the virtual environment (built in Second Life) was projected via a large-format digital projection system. Using slowly transforming colors – red, green and blue – the modular staircase was at once the real and virtual world for seven performers who improvised a Dalcrozian choreography to live piano music. Magruder also explores the mingling of virtual and real architectures in site-specific projections, as in his project *Vanishing Point(s)*.[3]

Over the last half century, we also have witnessed various forms of video art, video installation and expanded cinema which proved influential for today's digital era of media arts and performance. The editors of *The Art of Projection* rightly suggest that today's digital video installations or sculptures are deeply indebted to the earlier video art experiments of the 1960s and 1970s as well as the traditions of experimental film and "expanded cinema."

> Video installation is a contemporary art method that combines video technology with installation art. It is an art form that utilizes all aspects of its surrounding environment as a vehicle of affecting the audience. Its origins tracing back to the birth of video art in the 1970s, it has increased in popularity as the means of digital video production have become more readily accessible. Today, video installation is ubiquitous, visible in a range of environments – from galleries and museums to an expanded field that includes site-specific work in urban or industrial landscapes. Popular formats include monitor work, projection, and performance. The only requirements are electricity and darkness.
>
> (Noble 2010: np)

Margaret Noble slightly underestimates the requirement for a spatial architecture of course, without which the darkness would not mean much. But the continuities between generations of video artists was superbly captured by the 2007 exhibition *Beyond Cinema: The Art of Projection,* curated by Joachim Jäger at Hamburger Bahnhof (Museum for Contemporary Art in Berlin, originally a train station built in 1846). What was also apparent in this exhibition was the significance of physical expression and performance in many of the works displayed, not only of the subjects "inside" but also of the viewers "outside." When referring to the performative dimension of the projected art, the curator speaks of *embodiment*, especially in regard to the inclusion of the physical presence of the viewer in the projection interfaces, thus pointing to the current trends for interactive art, physical computing, augmented reality, locative media and pervasive computing, which draw both on the experience of interactive performance and the compositional and time-based techniques of other media forms.

Commenting on a multi-screen projection by Doug Aitken, Jäger says:

> Viewing a projected multi-screen film work is therefore closer related to the experience of theatre and performance. There is no singular view, there are many, depending on your position, your relation to time and space. Of course, you could describe the filmed tour of Doug Aitken through the

island of 'Montserrat' as it is presented in the seven films of the installation *Eraser*, but then you would only experience one aspect of this work. The narration would not reflect the full connotations of the installation. In order to grasp the full experience, you have to take into account where you stand inside the installation, and what you are looking at in this or that moment.

(Jäger 2007)

This reference to the total installation space for projective art is important, as the empty black box or darkened gallery allows the visitors to choose their spatial relationship to the projected images, and thus the design for the projection is the first task, deciding where to place the screens or surfaces for the projection, and determining the size and scale. When I design projection for a performance in a theatre, I first imagine the space empty and explore various positions for the screens, testing the look of the projection (and type of screen, white or anthracite) and studying the "cinematic" effect and the color composition of the image material. The projected images (on larger screens, using LCD projectors with various types of brightness/lumen) always look different in real space, compared to computer screens on which we do the editing of the digital video or film. The first task, therefore, is to play with the manifestation of projected 2D images in a real 3D space, looking at the light quality: the hues, values and saturation, at the aspect ratio to be used (4:3 or HD 16:9 or other), and at the darker space around the edges of the projection. In this darker space, one can then slowly begin to imagine lighting an actor or dancer or a whole group. Proceeding in this way, the "writing" or composing for performance needs to have the filmic script and the textual or physical/choreographic score prepared ahead of time. If you have already developed the projective script for a performance, then you can test the image qualities and develop your design strategies for the whole multimedia work right from the beginning.

It is equally possible to start with the choreographic plan for the performance and work with the actors and dancers first, spelling out the action and delaying the decision-making process for the inclusion of the projection. However, if the close interaction between performance and projection (video or real time graphic visuals) is a fundamental aspect of the dramaturgy, then it is necessary to begin with the "cinematic," "cine-scenographic" or "videographic" space – the space of projection. In integrated digital performances, the digital architecture of the audio-visual environment needs to be spelled out first, so that actors or dancers can begin to work within the augmented reality and also sense its magic and emerging atmosphere.

I deliberately mention the sensuous and magical nature of working with projective light and film as I want to argue that the materiality and the symbolism of projection have a profoundly animistic dimension that is often no longer recognized, not because the silver screen or the early connotations of cinema have become devalued, but because the animism of matter, of the non-human, of plants, stones, trees, mountains as well as machines, projectors and screens, is generally not taken seriously in Western techno-scientific and literate cultures. Projectors, the amount of lumen, the screen, black and white film or color video, silent or with sound, still or moving images – all these are aspects of a technical system which is intertwined with our sensorial experience of entering bodily into a projected world. It is possible, therefore, to imagine the technical objects as machining performances implicitly full of sensory properties and interconnected affects that – just like the incantations of shamans and sorcerers in oral, pre-technological cultures – can alter perception and reorganize the senses. Such machining performances also take place in public, so to speak, in theatres as communal public spaces that afford the sense of the symbolic and a shared frame of reference.

Everyone knows the story of early cinema, but it is worth repeating. The first large film projector was created in 1895, and the Lumière brothers used their first portable camera to make short films on the fly (the film width they used –35 mm – and speed – 16 frames per second – became the industry norm until the onset of sound), focusing on recording scenes from everyday life, without storyline or the craftsmanship later developed through editing. The Lumières famously projected a scene depicting a train rushing headfirst toward the camera, causing audiences to scream in shock. These short early motion pictures remained mini-documentaries until the involvement of Georges Méliès, a French inventor and magician who transformed moving pictures into moving stories.

Méliès' fascination with cinema arose from an accident. While shooting, his camera jammed; when the film was projected, an optical illusion was created and the object being photographed seemed to disappear instantaneously. Later Méliès incorporated such "trick shots" into the stories, thus beginning to experiment with the magic of sleight-of-hand or editing (cutting), including such cinematic devices as the fade in and fade out, the dissolve, and the use of stop-motion photography that was the origin of animation. Apart from innovative uses of the camera, Méliès shot his filmed narratives as one might watch a theatrical play. The camera stayed in a static position, recording a staged production, one full scene at a time. Méliès directed his masterwork, the psychedelic *A Trip to the Moon*, in 1902, just a little more than a century ago. The film is about 14 minutes long (the average length of contemporary video installations which are looped), and has around 30 scenes, depicting a rocket ship landing in the eye of the man in the moon, and also featuring chorus girls, strange space creatures that appear and disappear, and an undersea landscape. In the following, I shall examine what is meant by such spaces and analyze examples of projective art and performance to illuminate what has been discussed so far regarding light and projection and their animistic atmospheres.

Video installation and projection design

The three examples I want to look at are video installations by Andy Warhol, Peter Campus (both from the early phase of projection-installation art), and Diana Thater (from the late 1990s). They could be called historical examples, and may show the trajectories to the kind of immersive work mentioned earlier, by Hajra Waheed, or any number of younger artists working across media and installations art (it was noticeable that the 2018 Turner Prize exhibition at the Tate Britain featured large-scale projection works for all four finalists – Charlotte Prodger, Luke Willis Thompson, Naeem Mohaiemen and Forensic Architecture; in the same year Sondra Perry showed her impressive digital animation *Typhoon Coming On* at the Serpentine Gallery, London). My historical examples also show the different use of black and white projection/color projection and arrangements in space, indicative of a widely common projection *dispositif,* and especially revealing in terms of how images are displayed, how the space of a gallery or museum display is organized for the viewer-visitor (Figure 7.2).

The first example shows a medium close up of the two-channel 16-mm film projection of Warhol's *Outer and Inner Space* as it was displayed in *Beyond Cinema: The Art of Projection* at Hamburger Bahnhof (some of the single channel works were also shown at Hamburg's Kino Arsenal) and as it might be displayed in other museum or gallery black boxes. The "black box" (as compared to the white cube) has become the standard spatial organization for video projections, and today LCD projectors are mostly hidden from view and hung high up under the ceiling to throw the images against a black wall where the exact size/ratio of the projected images falls onto a prepared (painted or screen) white area on the wall. Most often the image seamlessly lies on the wall surface, and unlike the early days of projective art, the apparatus of the film projection is not materially present in the space: the viewer looks through the "window" at the projected world, so to speak, standing in the room, walking about or sitting down on benches or chairs sometimes provided by the gallery. Providing an empty black box with images projected against one wall is of course a curatorial decision (as is the transfer from film to video). It clearly diminishes the sculptural and theatrical dimension of such work, assimilating video to painting and the hanging of pictures.

FIGURE 7.2 Andy Warhol, *Outer and Inner Space*, 1965. Two-channel 16 mm film projection. Black and white, sound, 33:oo. © 2009 The Andy Warhol Museum, Pittsburgh, PA. A museum of Carnegie Institute. Installation view, *Beyond Cinema: The Art of Projection,* Hamburger Bahnhof, Berlin 2006.

Some artists tend to work against this kind of illusionism and arrange the 16mm, 8mm, or video projectors in the room so that one cannot ignore the projection devices and is perhaps even drawn to the particular staging conditions of the image, as well as the sound made by the (film) projectors – the mechanics of the apparatus. In the photos of early video art you can see a projector on its box on the floor (e.g. Peter Campus' *Prototype for Interface,* 1972), and cables that run along the floor of the gallery and the smaller monitors sitting at floor level at an angle from the larger screens, thus requiring a different awareness of the sculptural-technical articulation of the setting (e.g. Diana Thater, *The Best Space is the Deep Space,* 1998). The curator of *Beyond Cinema* points out that the distinct experiences of certain kinds of media are becoming less available today when many visitors to exhibitions cannot easily distinguish between film and video, or a video tape and DVD, even though these media have their specific qualities and – especially during the expanded cinema era of the 1960s and 1970s – were developed in a sculptural *dispositif* with experimental styles both conceptual and performance-based.

In Warhol's *Outer and Inner Space,* the images are seen side by side, and based on the historical evidence gathered from the 1960s – preserved in the Warhol archives – we can deduce that this particular work was shot on both film and video and created at a time when video art had barely come into existence. Yet Warhol was already experimenting with film, and in the fall of 1965 shot *Outer and Inner Space* with Edie Sedgwick in the context of his *Exploding Plastic Inevitable* collaborations with the Velvet Underground and his many ventures with the "stars" in his entourage. *Outer and Inner Space* is Warhol's first double-screen film, incorporating the video footage he shot of Sedgwick. It is considered an important transitional work, since the double-screen format was very important in his later cinema, for example, in *The Chelsea Girls* (1966) and the expanded film and intermedia collaborations on *Exploding Plastic Inevitable.* Warhol's use of video in the making of this projective installation is illuminating, as he places his glamorous superstar Sedgwick in a dialog with her own video-taped image, shooting two reels of film (each 33 minutes long, in close up and profile) with Sedgwick positioned before the video monitor, responding to and illuminated by her own video image. First shown in 1966 and then largely forgotten for many years thereafter, *Outer and Inner Space* seems to meditate on the distinction between film and tape, provocatively suggesting a complex double reality of (real-time) recording and simultaneous feedback loop – a technique that informs much video art from the 1970s on and now recurs regularly in digital and inter-active installations.

In keeping with our interest in video actors and media casting, it is apparent in the Warhol films that the 22-year-old Sedgwick, although not a stage actor, here acts on camera, smiles, smokes, talks, and yet often stares off camera as if trying to keep a conscious connection with the video of her image projected behind her or maintain her focus on her stream of consciousness talking. It is hard to make out what she is talking about, due to the audio distortions created by the four layers of her discourse. There is sound on all four channels; therefore we often hear all four heads speaking at the same time which makes this soundtrack nearly incomprehensible. Near the end, the electronic breakdown of the video image of the videotape is followed by the flare-out of the film image at the end of the film reel. Like the enormous serial work, *Screen Tests* (made between 1964 and 1966 at the Factory in New York City), Warhol's *Outer and Inner Space* experiments with combinations of the same or similar portrait, flattening space and destroying perspective by undermining the relations of surface and depth (the video Sedgwick appearing "larger than life" in the film Sedgwick). Warhol experimented as well with the positioning of projections in radically different *dispositifs*, namely, both the cinematic space of strict concentration (the world premiere was at the Filmmaker's Cinematheque in New York City) and the wildly distracted happening environment of the *Exploding Plastic Inevitable* where there would be three to five film projectors showing different reels of the same film simultaneously, a similar number of slide projectors, moving spots with an assortment of colored gels, several pistol lights, a mirror ball hung from the ceiling, numerous loudspeakers blaring different pop records at once, along with a couple of sets by the Velvet Underground and Nico, and dancing by Gerard Malanga, Mary Woronov or Ingrid Superstar complete with props and lights that projected their shadows on the wall (cf. Joseph 2009: 71). What a fine entangled assemblage!

Lastly, what is also significant in *Outer and Inner Space* is the early experiment with doubling and repetition (the simultaneous looping of the films), close to Warhol's technique of silkscreening multiple images of the same face onto the same canvas. The concept of serialization underlies the production of the nearly 500 portraits of *Screen Tests* (between 1964 and 1966), capturing the subjects in stark relief by a strong key light, and filmed with a stationary 16mm Bolex camera on silent, black and white, 100-foot rolls of film at 24 frames per second. The resulting two-and-a-half-minute film reels were then screened in "slow motion" at 16 frames per second, which means that any exhibition of the *Screen Tests* needs to place these rather short silent films into a loop structure over the duration of the display. The issue of looping, stretching, slowing down or multiplying film or video projection, so common in the work of contemporary artists such as Diana Thater, Bruce Nauman, Douglas Gordon, Tacita Dean, Pierre Huyghe, Pippilotti Rist, Steve McQueen, Paul Pfeiffer, and others, would need to be examined in terms of the specific techniques (how action is imaged or images act) used in the editing of content as well as the material strategies for the spatial arrangements of projection.

Perhaps all projective installations are at the same time prototypes for interaction, as the title of Peter Campus' 1972 video work, *Prototype for Interface,* promises: namely if the viewer is imagined to be integrated into the encounter and thus into the projection's kinetic atmosphere and quality of experience. The use of closed-circuit video (in which a camera captures and projects the visitor) clearly became an early vehicle of the examination of self and other, of doubling and self-reflection in and through the medium – thus also a meditation

on circulations and what cybernetics explored through feedback loops as processes of regulation (within all control systems). Campus was an early pioneer
of video projection, exploring the particular properties of the medium (e.g.
double-channel chroma-keying of images, closed-circuit loops) and focusing on
self-perception and other issues of self-expression through the moving image,
involving the visitor engaging with their own reflection (on a glass) as well as the
moving image (video self-projection). Furthermore, in *Prototype for Interface* we
notice how physical gesture and embodied presence of the audience member in
the exhibition space draws attention to the mediating technology itself.

In these explorations, Campus often treated the camera as an extension of
the room, while seeking to use video cameras and projection (often through
superimposition) as alternative models of binocular vision and the human (or
animal) perceptual system. Campus believed that video was the first "real-time"
visual monitoring system that had become available, real-time defined by a human sense of simultaneity between an event and the perception of it. Television
indeed was the first moving image technology allowing for live or virtually con-
temporaneous broadcast of an event, yet Campus' research in neurophysiology
and cybernetics at MIT led him to test the video camera as a tool to alter our
relationships to a preconceived reality by offering new juxtapositions. Bill Viola,
another video artist, considers Campus' early seminal works (including *Dynamic
Field Series* and *Double Vision*) as investigations into the phenomena of human
existence, questions of identity and the nature of the self:

> In the tapes, like a scientist conducting a controlled experiment, Campus
> methodically, almost clinically, dissects the nature of visual perception be-
> fore our eyes. But unlike a scientist, he uses himself as the subject and, most
> significantly, he extends this subjectivity to the camera itself. Unlike many
> of his contemporaries who used the surveillance camera as a detached,
> fixed observer documenting the performer's actions, Campus assigned an
> active, independent ontological status to the camera eye. It variously takes
> on the position of the artist himself, his reflection, an outside observer, a
> mental self-image, a double, an unknown protagonist, the room, an eye, a
> hand, an animal, an insect's visual system. However, like a mirror of many
> facets all converging inward, the works keep returning to Campus himself
> and ultimately become a portrait of the Self searching for the ground of
> Being, peeling back layer upon layer of reality in the process.
>
> (Viola 2010)

Warhol as well as Campus undertake their experiments with camera and projection at the same historical moment when spectacular new forms of electronic
information and mass media (analyzed by Marshall McLuhan's influential writings in *The Medium is the Massage* and *Understanding Media*) begin to take hold
of the popular imagination. While their preoccupations may seem to differ,
with Campus probing into quiet ghostly encounters of self with its double, and

Warhol producing dissonant, often aggressively threatening multimedia montages within the all-encompassing audiovisual environments of *Exploding Plastic Inevitable*, their works hold a similarly strong impact on our understanding of the atmospherics and theatrical potential of film/video projection. We know that the Velvet Underground's sets often consisted of long dissonant improvisations amidst the various moving light and film projections, and we can conceptualize this idea of dissonance and distraction as an effective strategy of "quasi-cinema" audiovisual performance working against any easy suturing of the audience into the spectacle – the harmonizing identificatory processes of realism in theatre. The deliberate disharmonizing of visual and auditory "tracks" in the live performance work of The Wooster Group has now become a familiar deconstructive technique, but it is a technique largely found on the experimental, not the mainstream side of theatre.

The visceral impact of Warhol's multiplication of visual and sonic stimuli must have been shattering and disintegrating – an effect that is not quite the same as the Brechtian *Verfremdung* but works toward dissolving the medium or playing with intermedia promiscuity to an extent that the singular focus of structuralist film or single-channel video is abandoned. As a main legacy of expanded cinema, such disfocusing moves toward a new paradigm (installation, environmental art) where there are potentially no spectators but only participants. In terms of the multimedia *dispositif* in the theatre, we can therefore always distinguish between two main strategies: the first aiming at focused spectacular attentiveness which aligns theatre reception with the cinematic experience, and the second aiming at polydirectional, synaesthetic and interactive immersion which tends to dislocate any linear viewing creating confusions of space and self or "spatial and subjective interpenetrations" (Joseph 2009: 84).

Campus' *Prototype for Interface* was only the beginning, but here "closed-circuit" already means a confusion of outside and inside, the participant experiencing herself from outside herself. In the photograph of Diana Thater's installation (Figure 7.3), we see the small TV monitors strewn around the gallery space, adjacent to the larger projection screens. Here the participant walks between boxes and membranes, and it is worth remembering that it was during the 1960s – when Campus and Warhol, along with Nam June Paik, Wolf Vostell and others, began to experiment with both multiple monitors and multiple projections – that this shift became possible at all. The transition from the cathode ray tube monitor to video projection allowed Campus to abandon the physicality of the box and generate images that consisted of animated light alone.

Thater combines both systems in *The Best Space Is the Deep Space,* and in comparison with Campus we see how the box and screen arrangement to some extent limits the properties of the image. Campus sought to define the image as independent of a physical screen when he used the glass surfaces or other surfaces, allowing him to create images of almost any size and shape. Viola suggests that projecting directly onto the wall also made the image part of the space and architecture, evoking memories of medieval fresco painting where the painted image

FIGURE 7.3   Diana Thater, *The Best Space is the Deep Space*, 1998. 2 Video Projectors, 2 Video Monitors, 4 Media Players, Free-standing wall and existing architecture. Installation View: *Beyond Cinema: The Art of Projection,* Hamburger Bahnhof, Berlin 2006. Photo by Fredrik Nilsen © Diana Thater. Courtesy the artist and David Zwirner.

was directly applied into the wet plaster of the wall, becoming indistinguishable from the wall surface.

On the other hand, Thater's spatial arrangement of various monitor boxes and screens creates a kind of labyrinthian set, a "forest environment" that necessitates movement and continuous relocation on part of the participant. Such video installations correspond more readily to the multi-directional scenographies of performances that work with multiplication or dislocation of perspective.

Polyscenic theatre/Sentimental machines

Diana Thater's *dispositif* for her video installation *The Best Space is the Deep Space* allows analogies with any number of contemporary theatrical productions which incorporate film and/or video into the *mise en scène*. As I suggested, Thater's particular constellation of screens and monitors requires that the visitor moves around the space to view the projections or monitor images, and the localized sound emissions, over-viewing them all, from a distance (with side-way glances), or stepping close up to peruse a single image. This dislocating or mislocating implies shifts in the techniques of perception and sensation as well as a sequence of changing viewpoints. In theatrical production, such sequencing can only be achieved for a moving audience in cases of site-specific promenade performances.

In all other cases, the continuity of different scenes is acted out for the seated audience and lighting plays a critical role in shifting attention, from one side of the stage to another, or to particular moods and emotions.

The distinct advantage that theatre has over cinema, video art and installation art is the range of simultaneities it can produce in the temporal flow of an intermedial performance, for example, through the multiple and varying interrelationships that might be created between the onstage actors and the screen actors/actions, or the onstage physical realities and the projected worlds – the screened if absent virtual locations. The "intermedial," as Greg Giesekam has offered, is perhaps a more precise term than multimedial, if we draw attention to performance *dispositifs* that do not merely exhibit film, video and computer-generated images to support the scenography, for example, by expanding a sense of location, but in fact script productive interactions between performance and projected materials in a digital environment which substantially affects all the conventional structural elements of theatre such as the role of actor (creation of character), acting style and choreography, dialog, text, action, scene design, costume design and lighting design (Giesekam 2007: 8).

A digital theatre *dispositif* – and here we think of the comprehensive environment for an interactive/real-time performance in which simultaneous, mixed realities can be presented, processed, layered, and recombined – offers an expanded notion of the stage, a tangible, kinetic atmospherics that includes different sensory surfaces as well as more hidden, sedimented or withdrawn elements, invisible frequencies, radio-activities. The scripting of interfaces, and of polyphonic, nuanced interplays between human and technical ensembles, affords a range of cross-cut presences that dissolve boundaries because they are processual, ethereal. Multiple scenes are continuously coming into being or appear superposed: the "projections" of performance, or "bodies and machines at speed" as Adrian Mackenzie has called such transductive processes (Mackenzie 2002: 25), are intensely dynamic and formative, not already formed. These projections have to be choreographed, so to speak, brought into dynamic temporal-spatial relations with actors or dancers. Due to the nature of the media and data flows involved (sound, video, graphics, animations, etc.), the performance range of "actors" – of the cast – extends to all elements and combinations of elements that are capturable, distributable and networkable. In terms of the dynamic networked environment created, along with live recording/capturing technologies in-play, the performance of live media is perhaps inherently unstable and unchoreographable. And yet, through the experience of practice, casting and compositional controller-tools join in the development of new techniques that can be rehearsed. The polyscenic theatre is animated through the forces of the cross-cut relations.

Over the duration of a theatrical performance, such interrelations of course can shift and become modulated, and a photograph from a stage production by the Wooster Group, the Builders Association, Complicité, La Fura dels Baus, Teatro de Ciertos Habitantes, Klaus Obermaier, Robert Lepage, Cynthia Hopkins, Wayne McGregor or William Forsythe can only be misleading as it captures just an isolated moment from a scene. I have collected many photographs from the works of these artists. Having witnessed the performances, I often remember the moments from the continuum but sometimes I am also not sure that I have actually seen (heard) them. The photographs, like the "audible prints" of the *ukiyo-e* tradition, may delude me into recognizing something that was not "rendered" explicitly; and certainly, here we also touch upon the limits of explaining atmosphere or making it explicit, re-rendering its movement and stillness, its moods and tensegrities. If we were to analyze several scenes from a production, we would perhaps recognize certain recurring patterns in the spatial configuration. Thus the production photograph also reveals fundamental elements of the theatrical *dispositif,* for example, capturing a gestural relationship in a given moment of theatrical time between an actor and a screen, or an ensemble and its onstage placement vis-à-vis monitors and screens positioned in the *mise en scène.* Such gestures can be revealing of particular intensities (visual, tactile, spatial) or tonalities, even rhythms, that we can glean from the ways in which a body might appear to be suspended in motion, in a backing off reaction or a forward lunge or thrust.

A single photograph also captures a particular moment of the image that is projected in that moment of onstage interaction, just as the still from Thater's *The Best Space is the Deep Space* reflects a view of that moment in the linear as well as looping projection time of the videos. Interestingly, some video artists represent their work (if an exhibition produces a catalog) in a sequential manner, letting the reader re-imagine the viewing situation of the medium as a machinery that animates frame after frame. Martin Arnold's *Deanimated* is an especially rich exemplar of this method of representing an exhibition of films: the catalog is designed to offer a temporal reflection of sequential frames in the scaling and placing of pictorial documents (Matt and Miessgang 2002). The contemporary performance literature has rarely accomplished such feats.[4] At most we see a few stills from a production, teasing us into guess work as to how dramaturgical decisions were made, and how the performers dealt with variously shifting real and virtual scenes produced live through image/sound technologies and the (re) processing of materials (and vice versa: how the system environment responds to multiple and overlapping presences and other times, past, present and future).

Similarly, how do we reflect on the lighting, the changes in the quality and color of light in a scene, the subtle relationships between light on the actor or set and light in the image space of a projected scene? While it is impossible here to offer sustained readings of lighting design in various isolated stage productions, we can point out some of the characteristic functions of light and projection in multimedia performance. These functions are directly related to the different materialities and ontologies of the theatrical and the digital:

1 Theatrical performance takes place in physical space (the here and now) and, however multidimensional the staging of its action, therefore always deals with limitations of the volume of the stage, the visual organization of

perspective (sight lines, arrangements of objects in space, the proscenium arch with its frontal staging or the panorama in environmental or pervasive staging), the visibility of the actor and the closed continuum of time as a structure within which actors distend the static space through which they move.

2 This real space is actionable by the actors, i.e. it can be impacted upon and affected directly through physical force and the physical manipulation of bodies and objects, and lighting directs attention to such action.

3 The digital is not actionable in the same manner but is controlled through interfaces and processing (data analysis).

4 Fluid moving projected image spaces in augmented theatre suggest foldings of the virtual into the physical space, bringing immaterial data into the material environment. The incursion of the digital (and non-linear time) also changes the linear temporality of theatre.

5 Lighting, and especially its contouring capabilities in the use of hue, value and intensity, also acts upon space-time, its plasticity and the depth of field (or flatness).

6 Lighting generally conflicts with visual data projection: images cannot be lit in addition to the projection like, unless color and hues are used to change the entire colorization of the film/video. A black and white image, for example, could be altered by using pink or violet color lighting in top of the projection.

7 Computer-generated graphics or video in digital performance are projected like light or can be considered another dimension of lighting.

8 The advantage of real-time technological media is obvious: they are flexible in time and space, and projected data (video, audio, computer-generated graphics, interactive media, animation, etc.) inscribe screen spaces and virtual spaces, with their various operational, spatio-temporal and display qualities, into the physical-analog space.

9 The presence of media instruments (e.g. cameras, microphones, sensors, laptops) within the stage environment draws attention to technological equipment and live processing (what is often referred to as "real-time"), making the audience aware of the dialectic between premeditated/controlled and unfolding/uncontrolled elements.

10 The challenge, already foreseen by Adolphe Appia in his definition of the plastic elements of the stage, is the disconnect between different plasticities (actor and image, stage space and image/sonic spatialities and temporalities).

This problem of a potential disconnect between physical action and visual velocity/sonic intensity is often underestimated or not acknowledged since there is no clear-cut distinction between the analog and the digital – they are inevitably a combinatory feature of the intermedial stage where design concepts generally develop the *coalescence of the physical and the virtual, the analog and the digital*. This coalescence, moreover, is already always assumed on the level of the digital where diverse media forms are equalized as data, and where sound sources and visual sources are equally analyzed and rendered as data streams.

However, we remember that Appia had already anticipated the essential dilemma of the screen and "significant surfaces" (Flusser) when he recognized that among the four plastic elements of the stage – perpendicular painted scenery, the horizontal floor, the moving actor, and the lighted space – the static painterly space was at odds with the movement of lighting and actor. Instead of "setting" the world into the picture within the theatre, Appia discarded vanishing points and experimented with scenic modules and the performativity of "rhythmic space" as an open, dynamic field, with lighting as the unifying form of the plastic elements (Oddey and White 2006: 95; Brandstetter and Wiens 2010: 7). This was revolutionary at the time.

The legacy of Appia's rhythmic spaces must therefore be considered influential for contemporary design, especially the modeling and scripting in 3D virtual environments and the movement/animation of scenic spaces that are projected or that juxtapose lighting with digitally manipulated video projection. Real-time interaction within a computational environment makes the transitions between theatrical and digital performance very fluid and complex as the interface processes facilitate a mashing of different realms, for example, the live, the recorded, and the live-generated recording. If we take "sampling" and live processing, as it is done in music/sound art, as the core model for real-time performance, then capture and immediate processing/synthesizing of data become the hip hop or return beat model of contemporary composition – a mode of production already anticipated in Jacques Attali's cultural philosophy (*Noise: The Political Economy of Music*) some time ago. But while this conceptualization is fundamental for an understanding of the extension of the theatrical toward virtualization processes, toward simultaneous digital non-linear time-spaces and distributed presences, the actual physical tasks of lighting and scaling the scenes are never easy. The architectural modeling of screen surfaces and their fabrication, and the placement/calibration of projectors (or positioning of speakers and interface devices in the environment) are always tediously complicated. Lighting instruments have to be hung and focused in such a manner – if back lights, down lights, front lights and side lights are used – that their light diffusion can be controlled very precisely in order not to interfere with the filmic projections.

The image projections, at the same time, need to be positioned in such a manner that actors or dancers can engage stage space without needing to worry about the necessarily close relationship or contrast between body and image, or the shadows their physical bodies might throw onto screen space. Screens can be opaque, semi-transparent or wholly transparent, and screen surfaces can also be fragmented and distributed. In the latter case, if digital projection is hitting many areas in the stage or if a production uses large-scale projection, a separation of filmic light and stage lighting is difficult, unless the designer decides to mesh color light with the animated light of film or digital video. In the online "Blue Room Technical Forum," conversations among designers reveal many basic but often fascinating insights into such challenges, for instance regarding the issue of brightness (https://www.blue-room.org.uk/). One designer refers to it as the first problem, noting that "theatre lighting is bright (1kW per unit and sometimes 50 or more units live at any one time): current projectors struggle to compete with that." Second, he suggests using wide angle lenses for video projection so that one does not have to be too far away from the screen (either for rear projection, where sufficient depth would be needed, or to avoid shadows if front projected). Third, it is important to use good keystone facilities to make the image look "square" if this is desired. Finally, when using more than one projector it is necessary to calibrate each image precisely in order to avoid that each projected image looks different.

Another designer comments on the advancements in projector technology citing the impact of new moving projectors from companies such as High End, Robe and Barco, which have brought the entire idea of using projection within a show environment to a new generation of lighting designers and show designers who previously did not have the ability (financial or technological) to bring such tools into smaller scale shows. But then he warns that like the introduction of any new technology, it has also created the usual overuse and misuse of projection, with LD's using the same stock content over and over, along with the simple positioning of one centered screen onstage. However, he concludes, the ability to quickly and accurately create wide screen and multi-projector projections must be seen as an advancement, if one remembers that not too long ago technicians would spend hours lining up projector grids. This is now becoming much easier and less time consuming with the advancement of onboard technology.[5]

Among other details related to the shape and design of the screens, the question of image aspect ratios (4:3 or 16:9) and image quality (DV or HD) is always a consideration. I discussed earlier how the light of projected images in video installations tends to capture the attention in a dark space. In the theatre, the digital scenography requires careful balancing acts between projected space and stage space; if the production uses live camera feed, then there is specific lighting needed for the capture of the actors. Cameras do not like darkness, unless they are infrared. Microsoft Kinect cameras which have infrared capabilities have been used for interactive scenarios (to generate data), for example, bringing up haunted night vision imagery in the manner of the extended search and destroy scene near the end of Kathryn Bigelow's *Zero Dark Thirty*. They are used far less often for theatrical or cinematic impact

In the scene depicted from Berlin Volksbühne director Frank Castorf's production of Dostoyevsky's *Erniedrigte und Beleidigte* (Figure 1.3), we see two actors in front of the enclosed apartment near center stage, and another scene – taking place inside the apartment and thus hidden from direct view – projected onto an elevated screen upstage left. This projected scene, as the production establishes early on, is filmed live so that we know the screen images are coming from a live feed. Actors in this performance enter and exit projected film-space as they enter and exit the stage action and vice versa, a constant flow exists between real and mediated spaces, but the lighting appears evenly distributed on the whole hybrid set. The scenes filmed live (but hidden from view) are lit equally well, otherwise the projected scene would not have the necessary contrast and clarity.

The lighting in Castorf's Dostoyevsky production is not as differentiated or particularized as we perceive it in the complex projections typical of Klaus Obermaier's stage works, for example, *D.A.V.E.*, *Vivisector,* or *Apparition*, where light and video projections are absolutely crucial kinetic techniques. Obermaier intensely integrates digital video projection into the choreographic stage action relying on a precise dramaturgy of placing pre-recorded film exactly onto and behind the performers, isolating some performers through down light while leaving others unlit, distributing the moving images onto bodies and screens in a manner that evokes the memory of Appia's visionary concept of the "creating light" – light that creates expressive rhythms and luminous events. In *D.A.V.E.*, the dancers move between back-screen projections and front projection onto their own bodies: the digital animations are mapped very precisely on top of the body, suturing the physical presence of a dancer into the projection. A similar approach is used for *Vivisector*, except that in this piece there is no back screen and no other light source except the video projector. The performance is visible only through the animated light of body-projection, and although the director choreographs actions of real bodies, the streaming light of the video very nearly *becomes* corporeal.

These performances have a very strong osmotic and anamorphic quality, and in this sense they play with theatrical and filmic illusionism (the two-dimensional surface oscillating with the three-dimensional bodies of the dancers). These stage works are spectacular, literally, and there are some extraordinary sequences of projections of (images of) stomachs on real stomachs, biological gender disturbances, virtual images on real bodies, which confuse our perception of space-time and identity, especially as Obermaier often manipulates the time code of video to accelerate the speed of images, reverse the flow, or insert quick animations and numerous jump cuts to disrupt the linearity of the image movement. Physical and image movements are in perpetual oscillating conflict, and film is literally grafted onto the skin creating tactile tele-presences that makes us sense the "wearing" of phantasmagoric pictures. Obermaier's performers interact with the video projection system in a highly precise manner; their choreography is rehearsed, following a tight cue structure, and the performers enact, so to speak, the projection script but cannot change or alter to it.[6] Unlike interactive real-time performances, these stage works are pre-scripted and their focus is on precise image superimpositions rather than physical improvisation of movement feeding into computational parameters to allow emergent possibilities generated in real-time. Obermaier's professional background as a theatre director is apparent here; the actors use precise blocking to be positioned in the exact places where the video projections meet them to create the total stage picture or composition (Figure 7.4).

FIGURE 7.4 *Apparition,* Klaus Obermaier and Ars Electronica Futurelab, featuring Robert Tannion and Desireé Kongerød, 2004. Photo Gabi Hauser and courtesy of the Klaus Obermaier.

It is tempting to contrast the approaches, used by theatre directors and choreographers, to precise mapping or explicit interactivity, on the one hand, and indirect, generative and emergent systems, on the other, but contrasting dramaturgies may more generally be owed to the narrative (figurative) or abstract (non-representational) tendencies of a particular work. Over the past decades, directors and choreographers have increasingly made use of both the "polyscenicness" of multimedia performance – exploring narrative as well as radically non-linear dramaturgies enabled through the interplay between live performance and digital projection – and the particular qualities of "painting with light" and complex de-linear time modules. The digital control operations in the theatre (using specialized software for visual projections, sound, light and MIDI or DMX protocols), and the architectural design software available for modeling and virtual rendering, have introduced a range of creative possibilities for design and programming of performance. Compared to the older traditions of directing and choreographing, the practice of programming live audio-visual media performances has surely entered its crucial formative phase. Technical and artistic issues inevitably become entwined; programming a computational environment may be quite at odds with directing the thickness of body interacting with other bodies in a communal context, rather than a machining architecture alone.

Furthermore, once we think of multiple projections and real-time signal processing in "responsive environments" (a term for hybrid spaces that Myron Krueger already used for his 1969 *Glowflow* installation),[7] which involve the audience as participants, it is difficult to stay locked within the traditional performing arts frame of theatre and dance. What I suggest, however, is that this frame has already been shifted early on, not just by Appia and other modernist artists, but by the dynamic repositioning, both phenomenologically and ontologically, of the audience in relation to the "virtual" – to the infinite fold of potential theatrical dis-organization. Within a longer purview of theatre history, we can draw connecting lines between Appia's modular staging and lighting designs, Loïe Fuller's *danses lumineuses* (notorious for their use of color slide projections onto the billowing costumes Fuller manipulated in her dance), Josef Svoboda's polyscenic film projections in his productions with Laterna Magika, and Robert Wilson's painterly use of lighting in his visual libretti for theatre and opera.[8] Performance companies in Japan (Dumb Type) and Latin America (Margarita Bali's huge urban projections of her *Pizzurno Pixelado* choreographies in Buenos Aires; Ivani Santana, Tania Fraga and Diana Domingues' work in Brazil) have created equally impressive experiments in kinetic light and projection design. Moving from drawings to the three-dimensional time-space design, especially in Wilson's case, reveals an intense preoccupation with mediums and surfaces, and the stratifications and sedimentations of the stage images.

South African multimedia artist William Kentridge, widely known for his celebrated animated drawings, originally worked in the theatre and used his techniques for ingenious interplay involving actors, puppets and film projections (in early work with the Handspring Puppet Company). More recently, in *Seven Fragments for Georges Méliès* and *Journey to the Moon* (2003), Kentridge plays himself as the central actor in his studio, drawing and erasing on paper with charcoal, ink, and paper cut-outs, performing curious additive and reductive processes, which look like magical manipulations of space and time but are simple cinematic tricks, appearing and disappearing acts, sometimes effected with a simple change in the projection speed and direction. Kentridge's epic work (with composer Philip Miller and musical director Thuthuka Sibisi), *The Head & The Load,* which came to Tate Modern in July 2018, amalgamated music, dance, film projections, mechanized sculptures and shadow play to tell the untold story of the millions of African porters and carriers involved in World War I. It touched me as powerfully as his other recent installations, such as *O Sentimental Machine* and *The Refusal of Time* (shown at Whitechapel, 2016), which always have a strong suggestive power to evoke historical predicaments, violence and guilt (this goes back to his early animated films, e.g. *Felix in Exile* and *History of the Main Complaint*). *The Refusal of Time*, especially, with its five different all-surround-projections, exudes a suffocating atmosphere through its relentless, pounding beats (clockwork-like, engine-driven) and the often shrill or dissonant brass music. The beats fill the space as if spat out from the strange engine in the center of the room (the "lungs" as Kentridge calls it) around which animated films swirl and stagger with painful references to the history of colonialism and the imposition of European time on the African colonies. It is the circular staggering that lingers in the memory for a long time.

Kentridge's collage and palimpsest technique is breathtaking (his drawing has the kinetic force of Goya's unflinching *Disasters of War* and *Caprichos*), as is his consciously repeated use of striking symbolic objects or devices in different constellations (such as the gramophone horn or old photography tripods). In a lecture on his work method, held at Princeton University in 2015, Kentridge emphasizes the "studio as process" and his "thinking through material," which I consider crucial for an understanding of his complex "archival" projection-performances and redistributions of sculpted or found mechanical objects.[9] The lecture illuminates his method of collecting such objects or rummaging through old film archives (working on *O Sentimental Machine* in Turkey and tracing Trotzky's exile, locating the beach front hotel on Büyükada Island where he stayed) to come up with surprising new image remashings, such as the extraordinary hybrid tripod-gramophones or the ghostly liquid gestures of Trotzky inside a tank that fills with water, as we listen to the recovered audio recording of the French-language speech Trotzky had prepared for Paris.

This speech, which is worth mentioning here, expresses Trotzky's belief about the limited programmability of human beings, suggesting that unfortunately they are "programmable but sentimental machines," unreliable if they fall in love. Trotzky's visions about the future of Russian communism are projected onto one screen of Kentridge's installation, which is like a stage set for an office or waiting room. There are four doors, two on either side, all showing further video projections. Some of these other screen projections parody the "sentimental machine" idea through a slapstick performance in which Kentridge, portraying Trotzky, shouts orders at his secretary with the horn of a gramophone. She becomes increasingly panicked, typing everything the megaphone pours into her, printing endless sheets of paper off her typewriter that swirl around in the air, engaging in a mad dance with the gramophone-megaphone-man. Turning into "sentimental machines," the two performers with megaphone are depicted in a surreal duet reminiscent of early Russian, German and French cinematic language (with a touch of the Marx Brothers) and its slightly jerky movements in the action filmed from a single constant camera angle. Kentridge comments on the atmosphere of "passionate absurd" that he strives for her, associating the "useless machines" he sometimes builds with the inconclusive, edgy utopia lying within revolutionary dreams and desires: on the edge of disaster.

On a scale hitherto unimaginable, Kentridge also worked on designing opera productions at the Met in New York City, the Dutch National Opera, Opera di Roma, and the English National Opera.[10] In late 2016 I saw his animated projections for Alban Berg's *Lulu* on the vast stage of the London Coliseum – a sensory onslaught that leaves one exhausted. But at the same time Kentridge's fragmented animations capture something vital about Lulu's desire and the way in which she can never remain a comprehensible or stable object of desire that the men around her try to construct her into. Kentridge also animates a stark 1920s Weimar expressionism with his black ink drawings, and the splintered libretto words splayed over the set appear to arise from within the main character's subconscious, drifting upward through Lulu's flickering desire and the manipulations of the male gaze that pursue her. In an interview with the stage designer, the collaboration between Kentridge's projections and her modular set is articulated eloquently; Sabine Theunissen notes that she conceived the stage as a single architectural interior, with two sliding panels that function as both stage frame and video projection screen, to partially or completely reveal a rear space, once in full width, but more often only on one side of the stage (Figure 7.5).

FIGURE 7.5   Alban Berg's *Lulu*, stage direction by William Kentridge, set design by Susanne Theunissen. 2016. Photo: Yasuko Kageyama-Teatro dell'Opera di Roma. Courtesy of Opera Roma.

> From the very beginning, the videos and the set have been composed together to achieve a total interaction and complementarity. I think fragmentation was an important keyword. I also like to regard the panels not only for what they reveal but also for the part that they cover and keep secret. They suggest an intriguing, unseen, dark space behind them that the spectator can eventually imagine, just like the unrevealed dark corner in Lulu's story and mind, referring to early psychoanalysis, questioning fear and desire. Two states that are probably the core of the story.
>
> (Picchi 2017: np)

Theunissen also mentions that with the angled walls, oversized room and crossed staircases she wanted to heighten the dynamics of the interior, having studied Pabst's early film *Loulou* (1929), to find a corresponding intensity to the close-ups in cinema, "not only counting on the projected big scale drawings but also dealing with the stage itself, the realism of details of the interiors, the intimate scale, along with Alban Berg's music and William Kentridge's artistic language." Regarding Berg's composition, Theunissen suggests that Lulu's vocal music sounds

> like a long breathing… requiring smooth space transformations rather than clear visual breaks like in the case of a classical curtain. Therefore, all transitions have been conceived to fold and unfold a scene magically, using the sliding panels simultaneously. These work like big erasers, turning and

rolling furniture, opening doors and blowing video images away. This has led us to mechanize a lot of the objects on stage requiring endless sessions of technical training to get the rhythm of the movements right.

(Picchi 2017: np).

Lastly, regarding the role of the projected videos, Theunissen argues that

the interaction between the video and the set is always important in our collaboration; the set can be regarded as a very elaborate background, having its own identity, its color, its materiality, its relief, its fractures, transforming itself, moving, sliding, offering infinite combinations in the merging with the black and white video. It is astonishing how the brain responds to those combinations. We do not know what is in the movie and what is a real object on the stage.

(Picchi 2017: np)

What struck my attention, when watching these "sliding" dynamics and the flow of the many projections, was the phenomenon of Kentridge's method of drawing and redrawing, what I earlier called his palimpsest method. As it was articulated in the earlier animated films, the drawing-filming technique here also has a troubling side to it: there is an over-writing (erasing, smudging, recreating), accompanied by extraordinarily subtle tonalities of music/sound, breath voices, crickets, operatic music, acousmatics, which disturbs the "line" as something that can hold attention (I mean in regard to truth or historical fact or psychological affect). The lines disappear, even as they are continuous drawing and space making. But something always disappears, a kind of blooded thought that vanishes, as if it could not be repeated. Something is sucked out of the funnel of the cone-shaped horn and vanishes.

Projections and animation techniques are manipulations of the visual medium, and of the different spaces, surfaces and volumes the medium plays. And they are burrowings too, just as Blau's theory of ghosting imagined, after his reading of Kafka's short story *The Burrow*. Animated projections convey a sense of the artifactual, perhaps unearthly in its receding allure and its withdrawals. Simple animation tricks with mundane objects can evoke hallucinations, gravity-defying space ships and other uncanny visible/invisible juxtapositions, vanishings, or absurdist tricks of the imagination, as Kentridge calls them when he refers to Trotzky's Groucho Marx-like gestures or the drowning megaphone and other mechanical accidents. Animation techniques are also effective in conjuring up fairy-tale and manga atmospheres, where myths, cartoons and science-fiction can intermingle. Contemporary companies have taken advantage of these potentials. The Builders Association, Big Art Group, Hotel Pro Forma, Societas Raffaello Sanzio, Nature Theatre of Oklahoma, or Rimini Protokoll, for example – they draw from the plastic languages developed by earlier generations of modern artists and pioneers of expanded cinema, and the inspiring convergences we have witnessed between visual art, photography, film, theatre, dance and opera.

A peripatetic season outside

I conclude with a brief analysis of the staging methods in productions by The Builders Association, a company that exemplifies how screenic projection can clearly become central as a design concept which to a large extent subordinates all other physical staging options. I noticed this for the first time in *Alladeen* (2003), a magnificent collaborative production (with the London-based arts group moti-roti[11] directed by Marianne Weems, which submerges the stage actors almost literally into the large-scale flat/lateral screen projections stretching the width and height of the stage. A few scenes are played in front of the screen, but most often we see the actors through the downstage scrim in a staging arrangement that has very little depth. Rather, the actors, who perform the roles of call operators in an imaginary call center in Bangalore, India, are placed along a horizontal axis close to the front screen and in front of a back screen on a shallow staging area. The front scrim can be raised and lowered, and here the lighting (down and back light) is crucial for the creation of opaque or translucent scenes and the constant mixing of heterogeneous images – a kind of nested conglomeration of graphics and projected text, pre-recorded video and browser-style windows that open up live webcam images of the call operators who speak to their customers in the United States (Figure 7.6).

The actors literally act out Indian "call operators" whose telecommunicational role playing in an American global economy makes them live-edit their voices and their images, adjusting their images to what they imagine the callers want to

FIGURE 7.6 The Builders Association, *Alladeen*, 2003. Photo courtesy of The Builders Association.

imagine, as well as how they like to imagine themselves. As audience members, we watch how the actors (and technicians) manipulate the webcam-generated images of their wishful selves on the projected surfaces of a fully technologized proscenium stage (faces merged with faces of actors from a popular TV series). Live synthesis, enveloping all. The most remarkable aspect of the *mise en scène* is the apparently seamless blending of live actors and live images in this shallow space, generating an image-scape, so to speak, that leaves little room, if any, for action and spatial relocations or interactions in a theatrical sense. No burrowing here. Most of the acting is completely static, and shifts or developments of character are created entirely through video editing, image manipulation and voice. Temporal and spatial shifts in the performance are effects of image projection and kinetic video scenery, which distinguishes the aesthetics of The Builders Association from those of Robert Lepage (and Ex Machina), for example, where we see a much more evocative transformational play with objects and intricate use of sets/stage architectures (e.g. the monolith in *Elsinore* or the long tile-roofed Japanese house in *The Seven Streams of the River Ota*).

In *Continuous City* (2008), the lateral staging is maintained but instead of the large projection screens, the Builders Association devised a more disorienting and puzzling fragmentation of image space, using a multitude of differently sized smaller motorized screen panels that created a mosaic-like overall tapestry. This quilt of image-windows of pixelated web streams and Skype transmissions presents a massive, almost Time Square-like urban media façade of tele-presences, social networking messages, talking heads, telephone calls, web chats, blogs, data clouds, blurred information. All this looms over the little girl downstage who is presumably "at home" with a nanny while her peripatetic father travels around the globe and calls in, forever distant but somehow connected through the network. Although her presence seems dwarfed by the overwhelming data-scape of social networking, the performance powerfully grasps the inseparability of her role – her fundamental loneliness – from the constitutive distancing effect of her father's communications via "social media." Like some pathetic latter day Marco Polo (and perhaps the Builders Association sought inspiration from Italo Calvino's *Invisible Cities*), he sends his reports from distant cities that he visits as a traveling salesman, never present and yet tele-present – an alienating, distracted ghost-like figure that might as well exist, as in the picture below (Figure 7.7), only in a dream. Or in the wrong time zone. Sam, the little girl, has a virtual relationship with the absent father, and the actress Moe Angelos, who plays Sam's nanny, Deb, uses each stop on the *Continuous City* tour to arrive at the location several days before the rest of the company, exploring the area and beginning to write blog posts about it. During the performances, her character continues to write these blog posts while she is seated at her onstage computer.

The shallow staging of the office/home is similar to that in *Alladeen* and in *Super Vision*, another work that shows us mostly static actors in profile or from behind, their faces enlarged on the screens. Presumably, patrons might relate

FIGURE 7.7   The Builders Association, *Continuous City*, 2008. Photo courtesy of The
Builders Association.

to Deb (and the show in general) because she is blogging about their city and
embodies the current cultural obsession with social networking technologies.
While this adds a peculiar site-specific touch for the local audiences, *Continuous
City* dramatizes a rather more disconcerting tale of the placenessness, disconnect-
edness and nearly autistic form of social networking.

Margaret Weems, the director, appears intent on revealing the brittle nature of these network connections (and fading cell phone signals), exploring also
the precariously blurring lines between fact and fiction when she has the actor
Harry Sinclair, who plays Sam's father, call actual relatives to create a sense of
immediate real-life connectivity which, however, amounts to very little. His
boss, played by Rizwan Mirza, runs Xubu.cc, an actual Web site (run by The
Builders Association) where patrons can record videos of themselves that may
be incorporated into the show to provide some kind of contemporary version of
"closed circuit" video (as I described it above in reference to Peter Campus' *Prototype for Interface*). The audience, in other words, is potentially incorporated into
the supervisory structure of the projections. And yet, this immersion tactic also
appears shallow and forced – it pretends a manner of topicality and place-fulness
that the performance dramaturgy cannot quite map, there is no cohesive glue to
a place via Facebook messenger. And that may well be intended by the creators
of *Continuous City*.

Remarkably, the complex interactional network technology in this Builders
Association production certainly succeeds in creating the kind of polyscenicness
and non-linear temporal structure characteristic of postdramatic performance.
The strongest feature of the work is its projective architecture – displaying many
aspects of the kind of modularity, serialization and unframeable excess of visual
data information (the fragmenting streams of webcasts and blogs) which I have
discussed in regard to the dynamic distribution of spatial perspectives and the

simulation of hypermediated locales. On the other hand, theatre critics have pointed out the overall flatness of the acting in *Continuous City,* criticizing the falling away of acting techniques that could hold up to the spectacular projective visuals in the staging (Marranca 2008: 203), and this comes hardly as a surprise. *Continuous City* is clearly dominated by the excessive animated character of the projected data-atmosphere, with its mechanically operated panels slightly moving like the wings of a flock of birds.

Its virtual choreographies of webcasts and tele-presences intimidate and, largely, overwhelm the physical presence of actors on stage, reducing their ability to "embody" their virtual reality on the physical stage. Or rather, they present their reductive protocolized embodiment in a very literal way. These simulacral effects seem fully intended, and thus the work approaches an almost documentary-style technological network realism, which seamlessly integrates live presence with the virtualization apparatus (web browsers, filmed material, live relay, computer monitors, CGI animation, lighting, sound, video projection, electronic music, editing and rendering) in order to build undifferentiated "dramatic" encounters that appear like real-life stories. The atmosphere that is engineered here has a strong inorganic density to it, perhaps close to a feeling of claustrophobia that, in 2020, has become all to shockingly familiar to many people who communicated through ZOOM during the lockdowns of the COVID-19 pandemic.

Continuous City is thus also a strangely flattened, incestuous circular system, projected onto a theatre stage in a consistent method of re-mediating the actors' actions through live feeds. The staging is completely intermedial, rarely do we see any cracks or failures of the *dispositif,* even though Weems seems to suggest that the Builders want to expose the compositing techniques rather than project the technologies as a massive ornament: "We definitely are refocusing the stage picture by bringing the media into the foreground," she tells Bonnie Marranca in an interview in *Performance Histories* (Marranca 2008: 193). This is not a statement that can be easily unpacked, as some observers feel that the actors function entirely through the technology and the performance resembles a "live movie" on stage. Marranca asks to consider the development of the performer in such stagings, wondering whether the technological tools call into question any notion of a sustainable "presence" in the theatre and therefore imply a new concept of character, of dramaturgy.

> What is the new dramaturgy for this theatre? How do you look at objects in your work? What about light? How do people move in these worlds? I wonder if you have to rethink a lot of these things, or simply rely on the performers?

Weems answers:

> The work is developed very, very minutely in the rehearsal process – in fact, there are so many layers that they are almost undetectable to the human eye because every moment in the performance is tightly designed

and scored. For instance, in rehearsal the performers have to spend a lot of time standing in one place on stage so we can get the light and the video in exactly the right balance with them, physically. They know that every single second of the show is built around where they are on stage, and they work very closely with the technical artists to align with video projected behind or in front of them, or with a camera picking up their image. The sense of extending into the network that then extends out to the audience is very visceral, and very closely created in alignment, in every moment of the show. They know they need to work in concert with the media – it's there to extend their reach but it's a force to be grappled with

(Marranca 2008: 192–93).

Weems' commentary on blocking – and working "in concert" – in her multi-media productions is illuminating, echoing my observations of the precise image choreography in Obermaier's direction of *Apparition.* This precision is necessary when "grafting" the actors into the animated flow of the projections. But is also shows the conceptual contrasts to live interactive media performances that work with aleatory and chance modes of production or generative processes not controllable in the tight manner Weems prefers. The Builders' digital stage is an intelligent environment of highly *controlled protocol,* a congealment of superpositions which, to an extent, forces the live performers to inhabit a strictly "scripted" (in the coding sense) flat illusionary three-dimensional world of image projections – an unnervingly paradoxical, constraining matrix.

In the next chapter exploring such technographic atmospheres, I shall look at a different type of intelligent, sentient environment for performance, giving attention to live audiovisual media performance and a divergent approach that favors the generative process, the "making of" (to a certain extent), though not necessarily in a radical Brechtian manner of exposing the "lungs" and labor of the machineries. But in conclusion I want to refer here to the *amplification* needed for the animated surfaces of the Builders Association's projection design: since the productions use electronic sound and amplified video/web streams, all actors' voices are miked and amplified, too, in order to accommodate the grafting of real performers into the imagescape. A vastly underexplored area in such intermedial productions is the dense relationship between sound and projection/light. It is addressed in some film studies and analyses of sound tracks but rarely ever in performance analysis. I end this section on the polyscenic qualities of intermedial performance with a few questions derived from Michel Chion's *Audio-Vision,* which you can further adopt for a whole range of material and immaterial aspects of the theatrical *dispositif* illuminating the presentational and representational dimensions of a production.

Pointing to the technical and figurative comparisons between image projection and sound projection, Chion suggests that we can notice technical differences in the kind of framing used for images (camera movement, panning or crane shots, etc.), or the kind of scale relations between screens and stage setting, and then ask how the sound behaves in regard to variations in scale and depth.

Does the sound, or the lighting, ignore such changes, exaggerate or accompany them? Chion then moves to the figurative comparisons, compressing them into two complementary questions that may appear simple but harbor important revelations:

1 What do I see of what I hear? I hear voices, a machine, a street, an airplane flying by, a dog barking, a door being shut. Are their sources visible? Offscreen? Suggested visually? implied by narrative logic of anticipation? As echo or as a vibrational mood evoked by the fading of the light, the passing of summer?

2 What do I hear of what I see? I hear the footsteps of the actor walking across the stage, the dropping of an object onto the floor, the rustling of a raincoat, and the pouring of water into a glass. But I don't hear the wind that moves the tree branches shown in the image projection, nor the diminuendo of the horse's galloping into the distance in the next shot or the excited breath of the woman who appears to lean against the tree now, with an agitated look on her face. The face draws near (the shot zooming in), a dark shadow now falling onto her eyes, her amplified voice quietly reciting a few words, from a dream perhaps, without that her lips are moving at all. The man on stage drinks from the water glass, impassively staring at us, then laughing as if he responded to the voice.

In such scenes in intermedial performance, onstage/offstage and onscreen/offscreen realities intermingle, and symmetrical questions, as Chion suggests, are often difficult to answer since the potential sources of sounds in a scene are more numerous than we might ordinarily imagine (Chion 1994: 191–92). Although there are similarities in visual and sonic design or dramaturgy, the poetry of theatre and film arises from the ambiguities, the intangibles in the atmosphere, and our interpretations of the relationships between sound, moving image, light/shadow and emotive intent. Chion proposes that by attending to these kinds of questions we come to realize also the asynchronous tensions and potentials (of anticipation or retrospective illusions) of audio-visual and tactile relations. We realize both negative sounds in the image – when the projected scene calls for them but the images do not produce them for us to hear – and negative images in the sound, their presence merely suggested by sound track. "The sounds that are there, the images that are there often have no other function than artfully outlining the form of these 'absent presences,' these sounds and images which, in their very negativity, are often the more important" (192).

In intermedial productions, as it was demonstrated in The Wooster Group's recent *La Didone* – an audio-postdramatic staging of a 17th-century baroque opera by Francesco Cavalli and Giovanni Francesco Busenello meshed up with re-enacted scenes from the futuristic 1965 Italian horror movie *Terrore nello Spazio* ("Planet of the Vampires") – specific emphasis on the voice and on auditory perception can contribute to a more comprehensive understanding of directing audio-visual performance and facing the various challenges posed by projection and the "animation" of sonic and visual content. The Wooster Group's style of mashing materials and using video in theatre, as we shall also see in the next chapter, is exemplary for one strand of contemporary intermedial performance, closely connected again to the contexts of video art described above, which integrated a televisual/video aesthetic into the stage dramaturgy, whereas The Builders Association's projection design features the wide-screen filmic approach that is more generally used in dance and visual theatre – in scenographies that rely more explicitly on the kinetics of movement. The power of voice (sound), however, remains an intricate imaginative challenge, demonstrated nowhere more beautifully than in Ivana Müller's choreography for *While We Were Holding It Together* (2006), a stunning performance in which, during the entire length of the piece, the five performers remain in a motionless pose, transfixed in an enduring *tableau vivant*, slowly and carefully evoking all kinds of surreal scenes, dreams, memories and fantasies purely through their (amplified) voicing. "I imagine we all have big black moustaches. We are in a combat operation…" "I imagine looking through a keyhole, to see who's coming up the stairs. We've been hiding in this apartment for the last three weeks." "I imagine us still here in three hundred years, being discovered by an archaeologist."[12]

Notes

1 Hajra Waheed's comments on her installation *You Are Everywhere (a variation, 2012–19)* in the exhibition *Hold Everything Dear* at the Power Plant Contemporary Art Gallery, Toronto (2019–20) are taken from a video where she introduces the works assembled in the show and discusses the transitions between various visual, sculptural and sonic (video) works, suggesting that these different objects and forms in her work allow transferences from one state of being to another, offering "a meditation on undefeated despair." See: https://www.youtube.com/watch?v=M3oko3oHVn4.

2 This was a discussion on the empyre soft-skinned space that took place during the month of September 2009. Gabriel Menotti's comment was posted on September 2: http://lists.artdesign.unsw.edu.au/pipermail/empyre/.

3 See: http://www.takeo.org/nspace/sl005/. This particular project conjoins Takeo Magruder's long-standing use of computational processes and virtual environment composition with his collaborator Hugh Denard's studies of the playfully illusionistic and fantastical worlds of Roman fresco art.

4 In the visual/performing arts, perhaps the most stunning exhibition catalog/monograph ever published, using serial, sequential film/photograph strips as performative documentation of durational, time-based art, is *Out of Now: The Lifeworks of Tehching Hsieh* (2008). Between September 1978 and July 1986, Hsieh realized five separate one-year-long performance pieces in which he conformed to simple but very restrictive rules throughout each entire year (a year of solitary confinement in a sealed cell; a year in which he punched a worker's time clock in his studio every hour on the hour; a year spent living without shelter in Manhattan; a year in which he was tied by an eight-foot rope to the artist Linda Montano, etc.). I saw the retrospective exhibition in London and vividly remember, as one of the most striking documents, the film of *One Year Performance 1980–81* (a.k.a. *Time Clock Piece*), a work for which Hsieh set out to punch a time clock, every hour on the hour, for one year and document each punch with one still, taken with a 16mm film camera. The result that was exhibited is a 13-minute time-lapse film that features Hsieh, as a jittery figure with sprouting hair, alongside clock hands that spin wildly and a punch card that fills and

depletes with the hours that pass. These stills are also reproduced in the monograph (pp. 128–158), thus enabling us to consider the small but noticeable internal variations at a reduced pace. There are also reproductions of the maps that document the artist's seemingly homeless existence, the meandering daily travels around Manhattan in *One Year Performance 1981–82* (a.k.a. *Outdoor Piece*).

5 See: http://www.blue-room.org.uk/lofiversion/index.php/t24610.html. These discussions took place since the end of the 2000s and have increased ever since. More recent discussions also look at the relations of camera to lighting when performances are being recorded or a live stream is planned, for example, https://www.blue-room.org.uk/index.php?showtopic=74758.

6 See: http://www.exile.at. The choreographer's website displays a vast collection of very impressive images from the performances that reflect the projection mapping. It also includes a host of publications: http://www.exile.at/ko/publications.html.

7 Krueger's responsive environments were created in the same era and historical context in which Bob Rauschenberg and Billy Klüver provoked attention by the path-breaking *9 Evenings: Theatre and Engineering* (1966) which brought artists and engineers together (subsequently followed by the founding of the organization Experiments in Art and Technology [EAT]) for a large-scale festival for electronic and interactive performances and demonstrations at the New York Armory. I tried to analyze the various interactional scenarios created there in two chapters of a previous book (Birringer 2008a). These interactive environments were later extended by David Rokeby's influential *Very Nervous System* installations, and are today reinvented in large-scale immersive compositions by companies such as teamLab, Random International, Luzinterruptus.

8 Having been invited to contribute to a book-length critical study of Robert Wilson's singularly important new opera, *Einstein on the Beach* (1976), I tried to discuss Wilson's painterly approach to lighting as well as his proto-digital performance concepts (he used film projections early on in his theatre stagings). See Birringer 2019.

9 For Kentridge's vivid lecture on "thinking through material," as Belknap Visitor in the Humanities (Princeton University), delivered on October 14, 2015, see: https://mediacentral.princeton.edu/media/William+Kentridge%2C+Belknap+Visitor+in+the+Humanities/1_4ssfyuxq. An exhibition at the Liebighaus in Frankfurt (March–August 2018) also featured *O Sentimental Machine,* and on that occasion Kentridge talked about the work in a lecture which previews *The Head & The Load*: https://www.youtube.com/watch?v=rQnpBYpXYE8.

10 Kentridge's drawings and sketches for the *Lulu* projections were also exhibited (in November 2015) at Marian Goodman Gallery in New York City during the run at Metropolitan Opera. A limited-edition artist book, *The Lulu Plays* by Frank Wedekind, featured illustrations by Kentridge (Arion Press). For his drawings, he reached back to German Expressionist woodcuts (e.g. Kirchner) for inspiration to design the fractured projections, often applying black ink onto actual dictionary or libretto pages, to accompany the tale of a femme fatale's descent into prison and eventually death. For set designer's Sabine Theunissen's comments on her collaboration with Kentridge's direction and visual projections, see her 2017 interview in *Scenography Today*: https://www.scenographytoday.com/sabine-theunissen-lulu/. See also Kentridge's beautifully evocative drawings for an opera project about the Chinese Cultural Revolution: *Notes towards a Model Opera* (Bejing: UCCA Center for Contemporary Art, 2015), and the catalog *More Sweetly Play the Dance,* ed. Marente Bloemheuvel and Jaap Guldemond, Amsterdam: EYE Filmmuseum/nai010 publishers, 2015.

11 motiroti was a London based organization which used the arts to achieve intercultural innovation, producing collaborative work in theatre, performance art, installation and short films. Since the mid-1990s, the company made its name internationally, with several acclaimed and award-winning art works that transformed relationships between people, communities and spaces. They ceased operation in 2014. The Builders

Association are based in New York City; founded in 1994 by Marianne Weems, the company has created numerous multi-media productions and toured internationally.

12 Cf. http://www.ivanamuller.com/works/while-we-were-holding-it-together/. For a recording of Ivana Müller's highly regarded choreography, see: https://vimeo.com/235764472. The length of the performance is 66 minutes. The power of imagination is evoked through seemingly autobiographical voices or voiced fantasies, and in many places of the choreography (only in a few instances supported or replaced by sonic sound effects), it is the character (gendered rhetorical flair, intonation) of voice alone that carries the audience and moves it along as if in a highly elaborate ritual of social dreaming. I have given this last section of the chapter an oblique title, referring to Amar Kanwar's film installation *A Season Outside,* which I witnessed at the 2015 exhibition *Experiments with Truth: Gandhi and Images of Nonviolence* (Menil Collection, Houston). It was fascinating to me how the filmmaker used his voice over (as autobiographic narration) to question issues of "decisions" (and memories) on violence and truth. I felt he offered his voice/narration to us to prompt us to examine our own decisions about historical truth, and to "arm" ourselves to be enabled to face it. I mention this here in relation to Kentridge and the other audio-visual works discussed that include or interpolate historical markers, images or subtexts. As Kanwar uses the film as an occasion for his own introspection, in the process he uncompromisingly brings into public view the full extent to which violence is rooted inside us, in all of us (perhaps in Gandhi as well).

8
THE STAGE AND ITS SCREEN DOUBLE

Screening the stage/Re-scripting the stage

In the last chapter, we looked at intermedial performance from a design perspective, introducing some principles and characteristics of "media scenography" in theatre and distinguishing staged performance from screen-based media art and installation art. If projection, lighting and amplification are key elements of mediation in theatre, they are also intrinsically connected to the performers in the stage presentation format of performance. The performer occupies the stage and interacts with the environment which includes the audience. In the following, the challenges of directing the performer interactions will be taken up and I shall ask how performers interface with media materials and real-time media generation.

As soon as one thinks of interfaces, new definitional problems arise, as patterns in interactivity in contemporary culture clearly show a much wider spectrum of *dispositifs* beyond the stage format of performance. Thus choreographic installations, interactive exhibitions, telepresence and networked performances, games, locative and other sound art and audio-visual environments, all of which might involve audiences as participants, need to be mentioned. The main operational distinction remains the one between presentational theatre, on one hand, where the form of staging involves actors and the aesthetic form exists independently from audience activity, and participatory intermedial artworks, on the other. The latter has no need for actors but implies that the audience engages directly with the work, navigates its environment and affects or influences the emerging atmosphere, form and content of the work. An intermedial performance environment can also be hybrid, involving actors and the audience inter-actors at the same time; it can also include the option for the visitor of not-to-interact, which you might argue defeats the purpose

DOI: 10.4324/9781003114710-8

of an interactive artwork, if we take gaming cultures seriously and accept the claim that unless you *play* the game, the game does not take place. There might be occasional cases of art games, 3D interactive installations or online multi-player environments where you might choose *not to play* and just observe the avatar idling, and listen to the soundtrack, waiting for something/nothing to happen. This is your Bartleby choice, of course. Rather than dwelling on the numerous variations of such constellations, I want to emphasize that the creation of an intermedial stage – the form of the staging and the particular qualities of content – always involves important directorial decisions. Although it requires quite distinct decisions and applications (of the technical apparatus), it must still be seen within the modern tradition of *Regietheater*.[1] These decisions guide the "role" of the inter-actor, whether it is a professional performer who rehearses with the system, or whether it is an audience member invited to play and be both spectator and immersant. The role of immersant in interactional performance environments is demanding, as we have seen in some of the earlier chapters. At the end of the book I shall return to it, but first I want to address directing choices and dramaturgies for actors rehearsing with moving images, sound and amplification.

Exhausting interaction and graphic phantoms

Around the beginning of the 21st century, observers of the performing arts noticed that the fusion of dance and technology had produced a few significant stage works that startled audiences and drew attention to *digital interactivity*. Projections of virtual dancers appeared on screens in Merce Cunningham's *Hand-drawn Spaces* and *BIPED*, emanations or graphic phantoms that fluttered in space while the real dancers performed the choreography on stage. In *Ghost-catching*, Bill T. Jones' animated figure danced a virtual solo, at times alone and then with multiple copies of "Jones" spawned from the data extracted (motion-captured) from the performer's body. The virtual Jones was heard talking, grumbling and humming, which gave the animation an eerie sense of the surreal. A voice in duet with animated motion data, drawn body-figures, moving nerves, colored lines. In Trisha Brown's *how long does the subject linger at the edge of the volume…*, projected graphic creatures interacted with the dancers on stage as if attracted to the human bodies and their movement. The jagged geometric creatures as such (irregular triangles, squares, rectangles, lines) remained indeterminate images hovering between abstraction and figuration, unaware that they were acting. Yet the graphic phantoms, sometimes referred to as digital doubles, have become supplements, algorithmic emergences and somatic ghosts allowing us to reflect upon the mediations between bodies and technical beings (Figure 8.1).

Digital performance art poses its particular challenges, if we think of directing algorithms or devising such interplays between graphic phantoms and actors. It runs into limitations that concern both compositional practice (e.g. the

FIGURE 8.1 Merce Cunningham Dance Company, with Paul Kaiser/Shelley Eshkar.
BIPED, 1999. Photo: Stephanie Berger.

dramaturgical placement of interfaces for trained performers in a stage work) and
the participatory agendas of interactive design for audiences (who are not trained
in the interfaces or cannot necessarily navigate the programmed parameters intuitively). In artificial intelligence research, engineers are working hard toward in-
stilling learning capabilities into their creatures: intelligent technical organisms
might learn from the behavior of the audience.[2] And processual systems (artificial life, multi-agent populations) develop their dynamic (self)reconfigurations –
their emergence, just as we have known for some time (at least since Xenakis)
that computational parameters in software-generated real-time processing allow
for random or stochastic variations, and other methods to accelerate, slow down
and alter operations on sound synthesis.[3]

The question whether participatory design is actually achievable or desirable in stage-centered performances was addressed by the *Pixelspaces* symposium
"Re-Scripting the Stage" at the 2011 ars electronica:

> Interactivity and participation have been core elements of media art since
> its very inception. In performances and installations produced in recent
> years, more or less successful attempts have been made to put this im-
> manent interactive element in the hands of the audience attending the
> performance – for example, through the use of various tracking technol-
> ogies. In addition to the attendant problems associated with people's in-
> ability to grasp the connection between cause and effect, the process of

enabling audience members themselves to generate sounds or visuals often quickly results in the exhaustion of the performance's aesthetic, emotional or intellectual quality. In the spirit of our contemporary Age of Participation in which social media and a digital lifestyle set the tone, we will conduct a transdisciplinary discussion on innovative participatory scenarios for the multimedial stage–audience context …in the future.

(Pixelspaces program)

Even if it were euphemistic to speak of the "Age of Participation," it is pertinent to inquire about the exhaustion mentioned here, and about the aesthetic, emotional or intellectual debt of performances that deploy technological interfaces to generate new methods for bringing interaction (and audience participation) into the theatre.

Re-scripting intimacy: wearing black and white or color

The title of a platform of digital and live art events, workshops and symposia in 2007 – "Intimacy: Across Visceral and Digital Performance" (Goldsmiths, London) – aroused curiosity in the context of a technologized world of after-effects and the forwarding/sharing and meshing up of social media content. If one reads and experiences the world through code and dispersive network channels, one wants to know what kind of *intimacy* could possibly be involved, how it moves "across." Implied in the movement there is also an assumed difference between the visceral and the digital, and this assumption concerns us here, too. I want to reflect briefly on the practical workshop I was invited to conduct during that platform (with design collaborator Michèle Danjoux), then look at the re-scripting of the stage, adding some critical thoughts on its screen doubles. These comments do sway along the curved path, across the visceral and the digital, which many of us are describing as we learn to perform in computational ecologies. And as we collaborate with (or train) dancers, actors and musicians to act out certain behaviors in such environments of after-effects, adopting interface operations and bio-information. I propose to define intimacy, in this context, as a heightened compounding sensorial opening into which we surge. Some call this experience immersive. We call it sensortized.

Dancers and actors alike rely on a very specialized physical training regime, a deep knowledge and intimacy of their bodies and voices, their vocal and bodies' structures and relations to/in movement, space, and change in time, movement through change and through effort. Specific techniques, if you think of William Forsythe's choreographic vocabulary, require the execution of complex isolations and isometric patterns, inversions and fragmentations at lightning speed. The Tadashi Suzuki method emphasizes physical ("animal") energies and a focused relationship of the feet to the ground, the gravitational attraction for the earth which the lower half of the body feels. Many Asian and Southeast Asian acting and dance techniques, and also Peking Opera acting and vocal skills, are highly coded and require years of training. Other techniques, for example, in Western contact improvisation, spark intensified perception of the movement continuum, in touch with others, sharing distributed weight, strength, lightness, a measured giving and taking, initiating and reacting, a kind of deep listening to others, and a sounding/breathing with combined energies, between ground and air, space-time of uncertainty and expectation.

Interfaces imply the "between" – the sense of connection and convergence, grasping and letting go, a facing of one another, a touch or conversation that also implies proximity, a closeness as in an embrace when I allow the body to touch another body, sensing the other through the clothes. The same sensing applies if I touch another body with my shoulders and spine, or if I focus attention on peripheral vision-touch and what happens on the sides of my motion or position in space. A kinaesthetic continuity is experienced in such moments, perhaps it is a kind of intimacy that we do not immediately know. Now we generally do not think of being intimate with machines, or being physically close to someone at a remote distance, even as our senses obviously extend into space and connect us to what we cannot see or feel.

José Gil has described the space of the body as "the skin extending itself into space; it is the skin becoming space" (2006: 22). The physical, we can infer, is not the digital. The computational space-time differs from, yet also repeats coordinates of, human corporeal experience (by filtering data input), and thus a growing number of theorists now speak of digital embodiment and the "folding of digital code into the biological" (Munster 2006: 56). Indeed, the fold is a critical issue, a crease in perception sensibility; in fashion design it is a common concern, not a baroque metaphor. Interaction design could learn from fashion. Clothes are folded and unfolded all the time, pleating creates shapes, surfaces rub against each other, our skin is a sensor – or if you like, in keeping with a more computational language, a "filtering actor." Textures and colors of clothes transmit signals, communicating our choice of how we feel or want to express intention and attention, exuding our vitality, emotions, preferences and idiosyncracies: our affective states and how they change, from one day to the next, from one season to another, constant in their ephemeral idealism, functional and excessive (against the logic of function), psycho-somatic. Sometimes we wear clothes that are not comfortable, but we wear them because they excite us.

Before I continue to speak about clothes, projection and wearables, can we ask how we carry ourselves in this ecology now presumed to be "second nature" – a world of pervasive and ubiquitous computing in everyday life, and a world of innumerable artificial performance scenarios constructed for audiences (now also called users) to engage in? Have not games and gaming worlds become huge attractions? Do virtual worlds require a new cartography of the body? It was anticipated that we would see the rise of a new interactive art that might replace the theatre – and its fundamental grounding in the performance of the body – with arenas in which the "languages of new media" (Manovich 2001) and the political and material formations of digital culture are played out. While the theatre of old was believed to be based on an ideal symbiosis between the state (*polis*) and the stage (*theatron*), today's creative industries or social networks operate in a globalized network-world of interconnections which are of course largely reflective of an incongruously heterogeneous, dispersed, multicultural, multilingual and socially polarized universe. In such a universe, the question of *intimate interfaciality* is perhaps quite pressing, it goes to the roots of our sociability, to the roots of assumptions about a (same) reality we do not share; and even more so, it provokes ethical questions about our right to (inter)personal privacy in post-panoptic societies where almost everything is under surveillance and trackable. With YouTube, no more Here Be Dragons. Erik Kluitenberg therefore speaks of "delusive spaces" and imaginary/fantasmatic media, when he addresses the "disillusion of the subjective" (2008: 12).

I begin by looking at the imaginary, asking whether we have learned new interface conventions that enable artistic visions to comment upon the perceptions of the smooth translation between the physical and the virtual.

> Are you dressed as a range of scalar values submerging the screen, there's something dear julu that must be beyond or in the midst of the other side of the tree, surely the use of values better written point to newer sources? Is a range of scalar values submerging the screen, there's something dear julu that must be beyond or in the midst of the other side of the tree, surely the use of values better written point to newer sources dressed as you? Are you in your thing, are you in your flesh, ah don't answer... Is Julu wearing your..., are you wearing your thing?
>
> (Sondheim 2009: np)

And perhaps it is not smooth, indeed. Avatars often dress badly, just wearing their thing, but they can fly. My first example (and it was the topic for the workshop taught by Michèle and myself) refers to the idea of dressing up with new "wearables" – their skins becoming space. Fashion, in this context, is an important marker of transitional times of "self-fashioning" and image making (demonstrated exuberantly in the exhibition *SHOWstudio: Fashion Revolution* at Somerset House, London, September 7–December 23, 2009), as it continuously reorients our senses to clothing as an expressive medium that functions like a screen for our social nature as human beings. It projects. Our clothes also facilitate social interaction and are a form of self-organization – managing our personal appearance in different contexts and situations and provoking responses from those who interface with us and "read" the projections. They are an expressive medium communicating pleasure and signaling how we feel about ourselves, how we excite ourselves or want to fly into someone's face.

In our workshop "Bodies of Color" we began with a brief trackback to the compositions of Brazilian artist Hélio Oiticica whose early work developed from abstraction and 2D painting-collages to increasingly 3D works and sculptures, then boxes (*Bólides*) composed with found materials, installations, architectural models and social environmental projects. His work of the 60s and 70s culminated in the *Penetraveis* and *Parangolés* series. In the late 70s, just prior to his premature death while in exile in New York, he created several installations called "Quasi-Cinema" (audio-visual installations for audience-participants, based on his utopian and metaphysical principles of *vivencia* and the supra-sensorial). The *Parangolés* are provocative propositions for a sensory experience in motion: they are "wearables" (inhabitable fabrics, colors-in-action) to be felt and experienced, not just seen, and they disperse layered fabric structures into luminescent colors refracting light. "Performing" them, i.e. wearing them, folding and turning them inside out, thus creates changes of the form, the interaction of interior and exterior surfaces, color and light, through tactility as a vehicle of perception. Thinking of them as media skins, I consider them significant forerunners of our contemporary experiments with wearables (Birringer 2007). But more importantly, they also push across the threshold of the palpably erotic, and perhaps even the hallucinogenic.

The idea of "Quasi-Cinema" also reminded me of the era of black and white silent movies, and early abstract, dada and surrealist films of the 1920s and 1930s (cf. Hans Richter's *Rhythm 23, Rhythm 25, Filmstudy;* René Clair and Francis Picabia's *Entr'acte,* or Fernand Léger's *Ballet Mécanique*), especially with regard to the spatializations of time and the fascinating if peculiar cutting and mixing of slow motion, fast motion, jump cuts, distortions and parallel editing. A few years ago I was working on two short films that were inspired by early cinema. *a thousand machines* (2017) was a reworking of Fritz Lang's *Metropolis* turned into a study of dancers as workers in underground controlled by the heart machine (clock). The heart machine is a symbol for the electric pace-making of capitalism's underground factories; in my remix of *Metropolis* I added the arms and legs of dancers to associate the instrumentalizations that historians like McCarren have illuminated (cf. her important study of *Dancing Machines*). The other film, *Sisyphus of the Ear* (2016), portrays an older man's protracted and dangerous climb up a steep gravel hillside in a quarry. The climb fails and the protagonist keeps sliding down to the bottom, to repeat the climb over and over again. The images compose a study of futility (and an existential scenario indebted to Camus' 1955 story about Sisyphus where he speaks of it being set against the "unreasonable silence of the world" here compounded by the not visible condition of an "inner wind" (vestibular disorder, tinnitus and hyperacusis) troubling the protagonist. The audible conditions of the inner wind are kinetically evoked and dynamically and dramatically interpreted through the percussive instruments (music for percussion and electronics composed for both films by Paulo C. Chagas). Thus the aural and the tactile-visual engender each other's restraint. What interested me most when working on the performances and the film shoot was my relationship to "wearing" not only specific clothes (my father's coat for Sisyphus' ascent) but imaginary sound vibrations in my interface with the filmic and real objects and material environments. The environments in both films were "weather" for my imagination; they were also envelopes for sound waves, Sputnik radio signals and "earbodies" (cf. Birringer 2017a, 2018b; Figure 8.2).[4]

FIGURE 8.2 *a thousand machines*, film by Johannes Birringer, with music by Paulo C. Chagas. 2017. Photo: DAP-Lab.

All summer long (in 2016) I was preoccupied with the rehearsals, film shoot and editing, without having received any word yet from the composer. My silent film choreography stimulated historical research, following some visual flashes taking me back to the very early 20th century. I imagined Loïe Fuller's billowing costume when she danced her serpentine light and colors. A spectral being, ghost of fluid motion. How did she step up? From the historical evidence we know that Fuller practiced her serpentine dance – an evolving *genre* of the skirt dance that had become big entertainment in vaudeville theatres and music hall revues in the 1890s – as a multisensory experience in which her whirling fabrics interacted with colored light, magic lantern projections and other optical stage devices (we don't seem to know what music she danced with).

> Conceived practically in parallel with the birth of cinema, the serpentine dance has a unique legacy as a phenomenon which is at once proto-cinematic and cinematic, and, more radically, one which foreshadows expanded cinema and multi-media shows. To early filmmakers, the organically whirling silk fabric offered itself as an ideal *medium* through which to assert motion and time as cinema's two vital properties.
>
> (Uhlirova 2015: 21)

I looked for early Fuller dances on film, and the shorts I found are silent, so I can only imagine what was implied by audience responses at the time, for example, when a New York Herald reporter who went to the Folies-Bergère wrote about the theatre "in complete darkness, the audience very still," a violet light then shining upon Fuller entering in her outstretched wings of silk, "the music […] dramatic, weird, sensuous, and dreamy by turns" (Hindson 2015: 77). Dance philosopher Laurence Louppe claims that "very early in the history of dance modernity the traditional association between dance and music became intolerable – at least in terms of the received norms as, for example, the idea of 'bending' dance reductively to specialised musical forms." She goes on to suggest that "we owe it to Isadora Duncan to have dissociated dance from so-called ballet music," and that

> the great periods of radicality in dance and modernity (1910–30, 1960–70) have thus given rise to works danced in silence, a practice of Wigman as well as Jooss and Leeder in Germany in Laban's wake or by Doris Humphrey and José Limón in the United States.
>
> (Louppe 2010: 220–21)

Comparing processual or generative interactive performance to their multimedia forerunners a hundred years ago, I wonder why I have not come across a history of silent dance, or a history of *danse concrète*. Early sound art had a fascinating kinetic energy, if you only recall the *intonarumori* of the Futurists, but also the implied concrete sound effects of the motion designs of Schlemmer's Bauhaus dances or the constructivist sets of the Russian theatre avant-garde (e.g. Lyubov Popova's architectonic designs for the Meyerhold productions). Louppe's examples are somewhat paradoxical but pertinent for our reflections on the extended skin of expanded cinema – on the *sound stage* of theatre and dance. She mentions Humphrey's *Water Study* (1928) pointing to the dancers' *breathlessness* or use of audible breathing, and also to Laban's and Wigman's effort to break the dominance of music by replacing it, not just with silence, but with sounds from other sources, including groans, voice, language. Louppe implies that it was necessary to *un-mute* dance (Figure 8.3):

> The ritual and sacred character attached to silent movement had to be profaned in favour of an open expressivity […] This confers on the presence of language in current contemporary dance works a particularly unsettling role – that of *an elsewhere* to received codes and traditional definitions.
>
> (2010: 224)

It is clear today that language, voice, breath and sound effects can play many different roles on the choreographic stage, not only that of reparation and retrieval of something ruptured from the body or of something unbearable or embarrassing. In this context Michel Chion's comment on sound in cinema is fascinating: he argues that sound was used in the beginning to cover up the unpleasant sound of the projector (Chion 1994), whereas Louppe implies that dancers' noise on stage was considered ugly or indiscreetly visceral. Early physical transformations of the modern dancing body through the use of technologies are discussed in Rhonda Garelick's book *Electric Salome: Loïe Fuller's Performance of Modernism*,

FIGURE 8.3 Loïe Fuller imitator in *Serpentine* dance costume, 1902. Photo: Frederick W. Glasier.

pointing to Fuller as one of the pioneers of early modern dance and stage technologies. Her *Serpentine Dance* and *danses lumineuses* – such as *Fire Bird* – presented innovative movements of body and lighting technologies so powerful that she left her audiences at the Folies-Bergère breathless.

Fuller's captivating effect is attributed to a specific way of moving with her tools and materials, the combination of her body, costume and lighting instruments, and the disembodied rising and falling of silken shapes (Garelick 2007: 4–5). This rising and falling also associates elemental conditions and the distinction between ground and air or upward/downward motion, just as I alluded earlier to Sisyphus being *worn down* – or *outworn* – and deflated by the continuous lapses and slides downhill, roughing up the skin of the body as it rubs against the gravel stones. Extension and in-tension in the Sisyphus choreography conveys an atmospheric sensation of friction or downward tension, whereas the design features of Fuller's *danse concrète* are developed in an upward relation of movement to resultant *bunraku*-like floating shapes of the dance – Fuller as floating image-apparition and hidden manipulator of the animated costume. What the film exhibition *Birds of Paradise: Costume as Cinematic Spectacle* (London and New York, 2010–11) foregrounded was the extraordinary manner in which Fuller overturned the relationship between dance, space/place, and sound by making *her own body a screen for the image*, and thus for film and early 20th-century moving image/capture technologies. She danced her inaudible choreography receiving the light projections, animating them with the "billowing folds of cloth whose undulating secrets her arms…" (Louppe 2010: 226).

What *Birds of Paradise* does not reveal is early film's attitude vis-à-vis inaudible choreography or silent movement, and the tension between forms. Of course one could imagine silent film perfectly tuned to the deferral of sound toward its inaudible boundary, to the delays and returns – the phenomenon of how we are performed and subjected by sound's *ungraspability*. As Manco suggests in his writings on "ear bodies" – we need to become aware of how much *sound, bodies and their movements* are intermingled and mutually generating. From lost places, an auditory layered work of listening through veils of raucous splendid silence, sounding folds of space are slowly enhanced perhaps, bridging the impossible. In a synaesthetic sense, we may in fact hear Fuller's ghost. We may see the color of different membranes that flutter like airy architecture.

When we conducted our "Bodies of Color" workshop, participants were invited to explore the contemporary wearable sensorial interface (technologically augmented and supported) as folds for performance – playing with silken fabrics, cameras, and light projections, wearing cloth and discovering garments as sensors, touching upon the erotics of materials and feedbacks, small sounds and tiny noises emanating in distance/intimacy, along with glossolalia and static and breathing, interacting in a tactile-sensorial manner within the real-time synthesis. The synthesis is software processing in and for an environment of images, sounds, colors, and tonal stages. The tonal stages (again one could think also of pitch entanglements, frequencies between ground and air, low and high, half tones and overtones) increase and decrease according to the performer's location within the environmental space. Interestingly, there was some resistance among our group to work with textiles: sensual technology (viscerally tried on) came as a surprise to some who may have signed up to delve into hard core programming or behind-the-camera work. Gradually, as the space filled with color, this resistance broke down, and the focus shifted to the skin, full-body movement, the touch of light.

Screens, flickering

> unter der haut knistern die wellen:
> *we begin*
> *in the dark;*
> es gibt einen neuen ton
> der vibriert in dir wie die alte zeit
> > (Eva Maria Leuenberger, *dekarnation*)

If I were to say "alles weitere kennen Sie aus dem Kino" (*the rest will be familiar to you from the movies*), I might promise too much, or too little, depending on your preference for live theatre or live art, the latter priding itself for the longest time on an intrinsic resistance to mediatization. The cinema's relationship to theatre has of course evolved through several stages. It may have been one of "creative destruction" ever since the younger medium's inception at the end of the 19th century, as the editors of *Against Theatre* argue, but early modern cinema developed its formal means as an aesthetically distinct medium not only by denying a dependency on theatricality. It also borrowed freely from theatrical aesthetics, acting styles and conventions of both realism and anti-illusionism before refining its crucial resources of shot scale, mobile framing, camera angles and editing, its wide range of rhythmic cutting patterns, focus, resolution and close-ups.[5] The paradoxical boundary relations between cinema and stage continue to be intimately dynamic throughout the later 20th century, and the mutual adaptations or conversions of techniques, framing devices, spatio-temporal properties, etc., would require very careful delineation if one were eager to detect anything like an anti-cinematic prejudice in mainstream Western theatre. There are also numerous playwrights who wrote for the cinema or adapted their own plays into film scripts. The above mentioned Christoph Schlingensief, like Rainer Werner Fassbinder or Sergei Eisenstein, was a theatre artist and a filmmaker – not an uncommon combination.

An insistence on live presence and the uniquely ephemeral was provocatively argued, however, by Peggy Phelan's well-known exposé on the "ontology of performance." The counter-argument regarding mediatization/reproduction obviously gained ground as the cultural economy of the digital gradually established the overriding presence of media technologies in all areas: the digital incorporates performance, and performance already knows – and is received by audiences in this manner – that its remix and reactivation is inevitable.[6] You can see its trailers and excerpts on YouTube beforehand and afterwards. Friends send you their personal clips via social media, Facebook already alerts you to other excerpts shared and forwarded by others. Less concerned with the issue of recording and documentation as such, I wonder how the presence of filming-on-stage produces (or not) familiar effects of the movies or the clip and trailer culture. How are contemporary intermedial stage atmospheres composed, what arrangements circulate or are still tested, developed or discarded? How are we becoming attuned to these compositional arrangements and milieus; how can we describe our affective experience of the flickering screens and "vibrating tones" in the theatre (as the poet Eva Maria Leuenberger evokes such an atmosphere in *dekarnation*)?

Actually, the word *dekarnation* opens out into complex territories, implying de-corporealization, and in an archeological and ethnological sense referring to processes by which a human corpse or an animal carcass is freed from all soft parts, so that only the bones and, in the case of animals, the antlers or horns remain. If the theatre held out a promise of the not yet all too familiar remains, how do we value its visceral matter, plasticity and spatial poetry, its vital parts, bodies on stage, its expenditure of energies and the "wholly materialized gravity" Artaud defended ecstatically when he described the Balinese dance (rather than Western masterpieces) in *The Theatre and its Double*?[7] How do we value its embodiment when we watch its artificial construction (and decomposition) on stage? It is still fascinating to read Artaud today and ponder his comments on "spatial density" and his dismissal of the "stammering" he associates with spoken theatrical language and dialog.[8] In his "Theatre and Cruelty," bemoaning the loss of the idea of theatre, Artaud adds: "Cinema, in its turn, murders us with reflected, filtered, and projected images that no longer *connect* with our sensibility, and for ten years has maintained us and our faculties in an intellectual stupor" (Artaud 2010: 60).

At an early stage of the history of the movie industry, this critique still comes as a surprise given Artaud's involvement in film projects from the mid-1920s to the mid-1930s. The portrayal of his role in Carl Theodor Dreyer's *The Passion of Joan of Arc* (1928) is mesmerizing and sensual. My own interest in the fusions of the digital, cinematic, and theatrical is to some extent inspired by Artaud's original conception of filmic work in immediate contact with the human body – what he called "raw cinema" – as I understand him of course to have been aware that mediation (filtering, projection) is inescapable in film.[9] At the same time, my work as a choreographer on the contemporary digital stage – a realm of real-time interactive processing, computation and the virtual – has made me curious about the strategies of theatre directors who may not primarily compose with movement and the kinetics of captured/processed motion (as pioneered in Lucinda Childs' *Dance,* 1979, and later demonstrated in dance productions by Merce Cunningham, Bill T. Jones, Dumb Type, Troika Ranch, Klaus Obermaier, Chunky Move, Wayne McGregor and many others) but with narrative, language and actors trained in naturalism and psychological realism.

Looking back over the past twenty years of working in the international dance and technology scene, I have rarely seen an interactive digital dance piece that did *not* use projected/screenic images on stage. Screens, and what I think of as screenic arrangements, are a fundamental presence in the design. Media output is part and parcel of the aesthetic of composing movement that affects and modulates graphic images and the doubling of movement-figures on screens. Visual interactivity requires an explicit, affective and *reciprocal* relation between dancer and modulated image. But the architecture of sensing (camera vision and motion capture systems installed on stage or used prior to the live/projected dance) is different from theatre productions I have observed where *cameras become actors,* so to speak, gaining an unmistakable role in the dramaturgy and participating in the *mise en scène,* moving around the performance space. This is one of the key differences, say, between Cunningham's *BIPED,* which features a large open and *empty stage* (using back and front scrim projection of virtual dancers), and Katie Mitchell's live film productions with her densely cluttered stage *dispositifs* and box sets.

What I want to explore in the following case study – Mitchell's adaptation of Strindberg's *Fräulein Julie* – is the role of camera-as-actor on the theatre stage, and therefore the particular role of such a camera-agent and its consequences for the projection of kinetic atmospheres and for an adaptation, a re-writing of a play or the devising of dramatic content. The presence of film within the theatre has tended to be a minor phenomenon associated with the experimental avant-garde, not with the realist conventions of modern drama on the actor dominated stage and the kind of psychological realism associated with Stanislavski's training methods. Having read about Piscator, Meyerhold, Eisenstein, Tretyakov and Svoboda, my first exposure to this "minor literature" was in New York in the 1980s, watching video/performances created by director-auteurs such as Elizabeth LeCompte (Wooster Group), John Jesurun, or Robert Lepage. And by the 1990s the use of video projection had also become an increasingly frequent feature of the growing dance and technology movement in North America, Japan, and Europe.[10] These directors, along with choreographers deploying kinetic image projection and digital manipulation of captured movement, worked with their own ensembles and were able to develop their own particularized aesthetics and staging techniques. Mitchell seems surprised when she is asked whether she considers herself an auteur, but comments eloquently in several interviews on directing and cinematography in multimedia productions, referring to her recent productions as "camera shows."[11]

When a theatre director, invited by a mainstream venue such as the National Theatre or the Berlin Schaubühne, decides to turn a classic modern play into a "camera show" or "live film" on stage, some rather interesting questions arise regarding the directing choices, cinematography, scenography and acting, as well as regarding the gradual impact of multimedia decompositions of realism in the wider context of postdramatic performance, effected in the United Kingdom, for example, by companies such as Station House Opera, Complicité, 1927, Imitating the Dog, or Punchdrunk. In spite of the work of such ensembles, it could be argued that there are no widely established conventions of live cinematography in Western theatre, and thus a given (and limited) range of acting styles or devising techniques may not be transferable to "live film" nor benefit from the promotional protocols of dominant popular-cultural media and their camera work (from television shows and advertising commercials to Reality TV, music video, and sports coverage with multiple angles and slow-motion replays). If such conventions were established, one could not explain why Jay Scheib's Chekhov adaptation, *Platonov, or The Disinherited*, produced at the Kitchen in January 2014, received such curious attention in the *New York Times*.[12]

Mitchell's directorial work, ever since her productions for the National Theatre in London (e.g. *The Waves*, 2006; *The City*, 2007; *Women of Troy*, 2007; *...some trace of her*, 2008; *Pains of Youth*, 2009; *Beauty and the Beast*, 2010), invites mainstream audiences to reflect on their spectatorial role and the level of distraction they might experience watching the maneuvering of technological tools such as cameras, tripods and cables on stage, and the busy work of Foley artists producing sound effects. Surely any fundamental ideological insistence on the primacy of unmediated performance is put to rest in Mitchell's precise blocking of actors on stage who take shots with cameras and rearrange the lighting for the shots. In the interviews mentioned above, Mitchell speaks of the "ugliness of the chaos of construction," which she seems to enjoy as it is in service of constructing beautiful images, "finding the right image, and then the behavior inside it," as she claims in reference to *The Waves* and *...some trace of her* (the adaptation of Dostoevsky's *The Idiot*). Among the productions I listed, her *Pains of Youth* was not a camera show but, as she mentions in the interviews, a "relief" and a return to "straight theatre," working with actors in psychological realism without the chaos.

In the context of postmodern and postdramatic performance, after the recent decades of hybrid experimentation across all media and live art, the defense of the live actor on stage (straight theatre) is not urgent business any longer. After the first video monitors appeared on stage, followed by filmic projections coupled with complex electronic sound effects, we probably should have come to expect that the media might as well be natural dimensions of contemporary stage craft providing a multiplicity of interactions and intermedial play on stage for discerning audiences who have grown up in a culture of remediation. The theatre, to paraphrase Brecht, can remediate anything it likes to; it can in fact foreground and expose its framing devices and technical infrastructure, it can ambush the repetition of hegemonic representation. The forms of intermedial play, however, are not self-evident, and I therefore suggest to take a closer look at Mitchell's strategies and provide insight into her foregrounding what we may not know from the movies.[13]

Mitchell's version of Strindberg's *Fräulein Julie* is a good example of an intermedial performance inviting us to watch a play that unfolds on a sound stage as a live filming of an action adapted from a dramatic source – "after Strindberg." Over the past six or seven years, after her early deconstructive *Seagull* production (at the National in 2006) which received a very divisive critical response, Mitchell has continuously elaborated her experiments with film and the effects of filmic technology on her stage choreography. Her directorial control over the action on stage appears ever more meticulous. Cameras, camera operators, editors, sound equipment and sound makers have joined actors as performers: the production features nine actors/operators and five cameras in the total synthesis of this composed theatre (just as Wagner had predicted already in the 19th century, however with a focus on music drama). It stops short of going as far as Heiner Goebbels' theatre installation, *Stifters Dinge* (2007), a "composition for five pianos with no pianists, a play with no actors, a performance without performers, a no-man show," which displaces actors and musicians to draw attention to the synthetic operational assemblage.[14] Brecht's Epic Theatre, based on interruption (and formulated against Wagner's totalization), is also side-stepped, as there is nothing that is ever interrupted in Mitchell's production. But her seamless synthesis bears unconventional features since on the mainstream theatre stage we rarely think of the cameraman or the camera as an actor. I cannot be sure whether the producers think so, either. The photos handed to the press oddly mention only the Schaubühne actors visible in the frame, not the camera operators (also visible), even though the actors (Jule Böwe, Luise Wolfram, Tilman Strauss) themselves would not appear at all on screen without Stefan Kessissoglou and Krzysztof Honowski (camera), and would not be heard without Laura Sundermann and Stefan Nagel (sound). The actors and film scenes would be silent.

The three Schaubühne actors, and in addition, the actually silent Kristin double (Cathlen Gawlich) do share most of the camera work during the performance, and we also see and hear Chloe Miller (cellist) on stage. Adding to the confusion, Kristin's hands on film are credited to sound artist Stefan Nagel and actress Luise Wolfram. Back of the house sound mixing and film editing are done by co-director Leo Warner with Gareth Fry and Adrienne Quartly; additional music is composed by Paul Clark, and lighting designed by Philip Gladwell. The program credits "further camera takes, sounds, and voices" by the ensemble.

The role of the sound makers/musicians is of particular interest in this constellation, if we recall that early cinematic technology – the camera as instrument of visual technology – created a "silent" genre which overdetermined the kind of acting that was integrated into the film apparatus. As Marko Kostanić argues in a short but brilliant essay on the "Choreographic Unconscious," early motion pictures

> irreversibly influenced theatrical gesturality and acting. Cinematic thinking first appeared at that time, meaning that films no longer functioned as a technologically facilitated way of documenting the theatrical *dispositif.* Apart from the theatre as an accessible method of representation, one of the reasons for the 'time-lag' in the evolution of cinematic thinking was the original fascination with the invention of the medium. The discovery of motion pictures resulted in an inevitable desire to show as much liveliness, movement, and intensity as possible.

This analysis of motion (and motion pictures) – as gestus or variations of gesturality – offers a very stimulating insight into what I have called kinetic atmospheres throughout my writing, especially as it links theatre and cinema as conditions, and *dispositifs,* of processes through which matter and medium unfold and are sensed, and through which action and the actionable are transmitted. There is perhaps also a sense of turbulence in the atmospherics of gestural dexterity, an excess of motion as Kostanić implies:

> That is the register in which the cinematic acting of the time evolved, which used a burlesque, accelerated, and caricatured variant of almost incessant theatrical gesturality in order to become equivalent in persuasiveness to the ultimate sort of newly-discovered persuasiveness – a faithful reproduction of reality. But then, primarily owing to Griffith and partly also to Kuleshov's experiment, there was a break. Using the potentials of montage and close-up made it possible to enter the hitherto inaccessible space of theatrical relations and made the previous type of gesturality and its corresponding persuasiveness obsolete. This led to a sort of repression of the actor's body and, accordingly, to the narrative relevance of immobility, neutrality, and the focussed body. The crucial

thing was that it was no longer the movement that was choreographed on film; it was the gaze, which automatically created cinematic psychology and suspense.

(Kostanić, program notes)[15]

In the following, I will use certain terms such as "mixed reality," "montage," or "modulation" in order to develop a critical vocabulary able to do justice to the complex filmic scenography of Mitchell's directing, but also to raise questions about the extent to which the (Hitchcockian) cinematic suspense and fastidious play on the psychology of the gaze – "repressing" or decarnating the actor's body – work effectively on stage. The use of the close-up and what Kostanić calls the "focussed body" is certainly one of the more remarkable and also problematic features of Mitchell's *Fräulein Julie*. She does achieve a tranquil poetic sense of realism in the subtle foregrounding of the very creation of such close-ups, often done downstage right at a table that serves as a miniature film set for special camera takes of "mirroring" scenes enacted with and by the "Kristin double" – the actress that silently substitutes for the character of the cook in the household in Strindberg's play.

The slow close-up poetic realism I detect reflects a similarity to the long shot in Tarkovsky's films, and thus the narrative relevance of immobility and neutrality for the acting in theatre poses itself as one of the main questions. Motion stillness would also be a considerable paradox for any choreographic thinking linked to live films and camera shows. Contrary to what Mitchell seems to believe when she asks her actors to step inside the shot to "behave accurately," it is the camera that acts accurately to frame the behavior and compose the emotional affect. Another question might be whether the camera-driven live film theatre can generate "automatic" suspense if in fact we become all too aware of the camera-as-actor, i.e. the camera's role as an agent of capture and the montage of expressions.

Significant here is the shift in attention and dramaturgical weight executed by Mitchell's adaptation of the script for this production (dramaturgy and translations by Maja Zade), namely, moving a minor character to the center. Apparently Mitchell was drawn to what she considers "the spirit of radicalism" in Strindberg's investigation of early naturalism, paying much attention to the author's fervid responses to Zola's essay on "Naturalism in Theatre" (1888) in his own ideas on staging, set design, lighting and acting. She quotes some of Strindberg's stage directions for *Fröken Julie* (written in 1888) in the Program Notes as examples of remarkable formal innovations which one would have to imagine as radical in their historical context (if it were possible to do so after a century of mainstream Anglo-American theatre culture accustomed to the new realism Strindberg's play was advocating). Mitchell admits being attracted to the idea that a theatrical revolution could be "contained within a lifelike rendition of banal and simple actions described in a stage direction" rather than dramatic dialog.

Fräulein Julie's setting is the kitchen of a country mansion, where on Midsummer's Eve the daughter of the house, Miss Julie, has a sexual encounter with a

servant, Jean, who is her cook's fiancé. Mitchell takes as her starting-point the stage direction for Kristin, the cook, who was the least significant of his three characters according to Strindberg, described by him as a subordinate figure "without individuality." Left in her kitchen while her fiancé, and their mistress, Fräulein Julie, join the midsummer's night festivities, Kristin must perform, Strindberg directed, "as if the actress were really alone in the place":
  A schottische tune played on a violin is heard faintly in the distance. While

> humming the tune, Kristin clears the table after Jean, washes the plate at the kitchen table, wipes it, and puts it away in a cupboard. Then she takes off her apron, pulls out a small mirror from one of the table-drawers and leans it against the flower jar on the table; lights a tallow candle and heats a hairpin, which she uses to curl her front hair. Then she goes to the door and stands there listening. Returns to the table. Discovers the handkerchief which Miss Julie has left behind, picks it up, and smells it, spreads it out absentmindedly and begins to stretch it, smooth it, fold it up, and so forth.
>
> (Fräulein Julie, stage direction, reprinted in Program Notes,
> Barbican Theatre)

Mitchell's adaptation uses Zade's translation which substantially reworks Strindberg's text changing dialog, in some instances, into inner monologs (read out as filmic voice over by actors positioned inside small sound recording booths on either side of the stage). This technique is in keeping with Mitchell's earlier filmic productions such as *The Waves* (adapted from Virgina Woolf) or the 2012 Schauspielhaus Cologne staging of *Reise durch die Nacht* (adapted from Friederike Mayröcker's prose novella), which features the main character as a narrator of internal monolog. Mitchell thus turns Kristin's solitary reverie into the main filter through which the action of the play is mediated and a particular emotional mood or affective angle (Kristin's un/conscious and reflective gestures) is created. Kristin's stretching out, smoothing and folding of the handkerchief is foregrounded. Julie and Jean become the subordinate figures in this triangle.
  The set construction for this filtering is crucial, as the staging attempts to design a mixed reality performance that allows the overlaying of the physical environment with the filmed screen-projections. The most fascinating aspect of this framework for an "augmented reality"[16] is the simultaneity and overall synchronicity of filmic and theatrical spaces, and the traversable relationships between film and physical stage reality which the audience can observe at all times. Mitchell and her co-director, Leo Warner, situate the drama in three conjoined yet separate realms: an enclosed kitchen mid-stage left (adjacent to a small sitting room, in the center, and a bedroom mid-stage right), where the audience has a partial view of the figures within; at the front of the stage, where Kristin's double appears, and on a screen above the set, where the drama is shown in live close-up, filmed by the ensemble themselves (Figure 8.4).

FIGURE 8.4 Jule Böwe (on screen above the stage/film set with camera operators) in *Fräulein Julie*, Barbican 2013. Photo: Thomas Aurin.

Alex Eales' set construction of the interior rooms – there is also an "exterior" space stage left used only sparsely, for example, for the dance – is simple and under-stated, while the sepia-toned costumes aspire to historical accuracy and much attention is given to numerous props, mostly kitchen utensils and household items as mentioned in the stage direction above, which the actors take from a cupboard placed off-center right. In the opening scene, the screen hangs in front of the interior rooms before it is raised up above the set, and we watch a quiet pastoral scene of a landscape with flowers, accompanied by the cello. As the screen is raised up, our attention is drawn to the creation of the landscape: a camera has been filming a miniature "landscape set" built onto a small cardboard platform, the flowers on it are lit by a small stage lamp and we see the cameraman filming this set, behind him the cellist in the sitting room playing her instrument. Our attention is also directed to the two sound-generating members of the ensemble who had entered stage left to position themselves at a table with a vast array of tiny wooden, glass and metallic objects, jars, paper, cloth, matches, etc., which they use during the course of the performance to create the live sound for their microphones. They can watch the projected film on a small control monitor facing them on their table; there are additional closed circuit monitors in the interior set visible only to the camera operators. Like in the radio dramas of old, the Foley artists create the sync-sound effects for the action we see on screen, and this begins, perplexingly, with Laura Sundermann tapping her left foot onto a floor board (equipped with contact mic) to simulate Kristin's walk around the interior kitchen set, "alone in the place" while our imagination has already been taken far afield into the midsummer eve.

This impressionist play on nature and naturalism sets the tone for the production, and the presence of the cello adds to the elegiac mood evoked in my mind once the softly spoken German passages, which the London audience can read in English surtitles, begin to voice or translate Kristin's internal reflections. The cook's unseen labor is her poetry. The frequent shots of Jule Böwe looking into a mirror intensify the almost Bergmanesque atmosphere Mitchell achieves in her careful emphasis on the character's probing if resigned self-reflexiveness. It is this self-reflexiveness which ultimately haunts her, when she slowly realizes her entrapment and the betrayal by the callous fiancé.

It is tempting, then, to ask how the performance can maintain its psychological atmosphere of suspense, and the continuity of the implicit drama of class conflict between a mismatched couple, through a self-consciously filmic exploration of gazes (the intrigue of overhearing, overlooking, eavesdropping) and the montage of multiple angles which we in the audience experience as a paradoxical realism of emotional affect – paradoxical in the sense that we are not immersed into the make-belief of theatrical drama, action, and characterization, but asked to witness at all times the construction of a film, i.e. the artificial constructedness of each gesture, facial expression, unspoken word, glance, reaction, and anticipation in the relationships between the three characters? The complete engineering, in other words, of the atmospheric address. The emotional atmosphere is technically conditioned and crafted right in front of us.

One could argue there is perhaps an inevitable Brechtian *Verfremdungseffekt* built into Mitchell's production as it demonstrates, and thus makes explicit, the *gestus* of Kristin's (in)action, filmed from every angle with multiple cameras (high and low angle, over the shoulder shot, shot-reverse shot, etc.), as well as doubling the actions, say of her hands, by a second actress whom we have to observe re-enacting the *gestus* on another side of the stage (simultaneously while watching Jule Böwe, the camera operators, sound makers, and the resultant film scene on screen and the *découpage* – the edit cuts). This multi-perspectivally edited scenario of continuous montage is demanding. No illusion of realism can be sustained in this way, except if the audience were willing to immerse themselves in a virtual and poetic world not reliant on dramatic realism but a congenial acceptance of this continuously performed passage between actual, tangible and yet constructed filmic (augmented reality) space.

Strindberg's play is transformed, in this particular live film theatre version, into a constructivist set (and Popova's design for Meyerhold's *The Magnanimous Cuckold* comes to mind again). One wonders how Mayakovski or Brecht (and Benjamin whose writings on the apparatus of technical reproducibility might be pertinent here)[17] would have considered both the productivist and the distanciation techniques at work here, especially since Mitchell seems rather less interested in the political dimension of theatre than in the cinemato-poetic logic of her camera and framing devices and the "perception-images," as they pass from the objective to the subjective.[18] What do these constructed and complementary images (shot-reverse shot/observer-observed) makes us think or feel?

As we are not invited to identify with the characters or the fictional drama, our attention is shifted to a level of aesthetic neo-realism dependent on the production's camera (and sound effect) self-consciousness. For example, I feel an almost kinetic, choreographic delight in becoming aware of Sundermann's leg movement, as the taps with her left foot generate the sound for the walk of the actress we see on screen and below in the interior kitchen. Sundermann raises her left leg in a succinct slow motion manner, then fluidly taps down her foot and swiftly raises her leg again. She is actually expressly enacting this movement for the sound, as if framing it for an *optical sound test*, in the sense in which Benjamin explains the function of the camera and the audience's identification with the camera, except that this movement is not filmed but generates sound. The screen images often focus on close-ups of facial expression or the actions done with Kristin's hands – but the illusion effect of *synchresis*[19] is perfectly maintained throughout most of the evening, every sound on film is created by Sundermann and Nagel, with the exception of the spoken words that are captured by the five cameras or, more emphatically during the poetic inner monologs, recited by the actors in the sound booths (they wear headphones as if they were in a recording session). The production thus plays deliberately on the pretense, on the sense of ostentation or affection of an emotion. It also, in my view, refers us slyly to the characteristic procedures of radio plays or sound recordings generally, and Benjamin already in 1932 notes that the (epic) theatre "is utterly matter-of-fact, not least in its attitude towards technology," adding that "its discovery and construction of *gestus* is nothing but a retranslation of the methods of montage – so crucial in radio and film – from a technological process to a human one" (Benjamin 1999b: 584). Benjamin's theory implied, back then, that radio represented an advanced technical stage: it made its technology more evident and was able to "take up" older cultural productions such as theatre (by means of adaptation), replacing it so-to-speak with a "training of critical judgement." While I do not see Mitchell's staging intending to use the montage principle of epic theatre (which is based on interruption), it does to some extent resemble what Benjamin calls the "dramatic laboratory" – focusing on the theatre's unique opportunity to "construct" human behavior and action (Benjamin 1999b: 585). It is this sense of constructionism that appealed to my critical judgment while watching the technical process.

We observe, again and again, the intricate passing from objective to subjective perception in the stage film composition (while we note the camera operators scurrying about the exterior/interior set with the tripods making sure they have set up their camera angle quickly to be on time for the montage, dragging their cables around with great dexterity). I will use two examples here. In the opening sequences, this play seems to become a film about a lonely servant, whose life is a series of repeated domestic tasks and whose existential thoughts are bound by quotidian work. Kristin prepares a meal of kidney for Jean; she winds up the clock; she cleans the vegetables; she tests the sharpness of a knife, and then cuts the kidneys. It is women's house work, unglamorous, often tedious, and as Strindberg noted in his stage directions, the actress need not look in the direction of the audience, she can turn her back to the public. The cameras do the work for us, they act as point of view (POV). We see what Kristin sees. The subjective image is, according to Deleuze, the thing seen by someone qualified or interior to the set: we see the knife through the eyes of Kristin. We observe the sensory, active and affective factors of her testing the knife and cutting the meat.

Unobtrusively, perhaps barely noticed by the audience, Kristin's double is enacting the same gestures downstage right at the table where the close-up of the cutting is shot, now we see the small lamp being switched on, now the camera being focused on the table top, then the cutting – in this scene the double does not even need a camera operator, she becomes her own operator, focusing the lens on the small lit area on the table top, then proceeding to do the cut which we then see edited into screen image sequence. The camera-actor is left on its own. In this case, Deleuze suggests, we need to think of the image as objective, "when the thing or the set are seen from the viewpoint of someone who remains external to that set." But he then goes on to say the observing and the observed can be reversible or complementary, undecidedly, for "what is to tell us that what we initially think external to a set may not turn out to belong to it?" (Deleuze 1976: 73–74). Ironically, Mitchell of course is using a double set, so to speak, downstage and center stage, and the double actresses perform the same action for us to experience as montage on screen. In regard to the close-up of the affection-image – and Deleuze immediately mentions Bergman's *Persona* as an exemplary, complex manifestation of the close-up *being* the face (turning it into a phantom – "both face and effacement") – we only see Jule Böwe's melancholic face, her silent expression (as her double, in the sound booth, reads a series of poetic lines about "Das Nachleuchten der Einsamkeit, die Spur gibt es, und die Nachwelt gibt es ...") when she looks at herself in the mirror. The double is effaced but lends Böwe's character her voice, perfectly "mismatching" (as Chion would say) the individual's voice and face, the unity of sound and image, the subjective and objective. It matches, but it is a false match. For this context of matching/synchronizing or defacing image and voice, image and soundtrack, Chion has much to say about the various possibilities of modulation between the real and the rendered, the so-called "phantom audio-visions," contradicting the "naturalist" perspective of an illusion of unity, a perspective that postulates a natural harmony between sounds and images, disturbed through almost inevitable technical falsifications in the filmmaking process that make postproduction rendering necessary and desirable, such as the addition of sound effects, etc. (Chion 1994: 95–96).

In Mitchell's production, the Foley artists seek to stay close to a phantom "naturalism" in many of the sound effects they produce (some match precisely the action we see being created in the kitchen or the sitting room, such as the lighting of the candles Miss Julie enacts near the end, or Kristin's dressing up after the bedroom scene), but we also discover, in some moments, an even more intriguing asynchronicity (in the bedroom scene) or negative sound, when the image calls for a sound but the film does not produce it for us to hear. Similarly, the textual speech of the voice over, in the poetic inner monologs, does not necessarily correspond to the diegetic narrative of the images created by the actors nor reflect Strindberg's naturalism. But Mitchell directs a powerful audio-visual scene late in the performance (the bedroom scene), when Kristin wakes up at night and hears Julie's and Jean's voices in the room below. We see her in bed on screen (her body doubled by Cathlen Gawlich downstage right), then moving to the table, taking a glass and holding it to her ear, amplifying the sound of the voices (performed by Strauss and Wolfram in the sound booths), and their muffled dialog suddenly becomes clearer. The modulation of the sound here is accomplished by the sound actors who use digital effects to achieve this image of reverberating conductivity which reflects the claustrophobic atmosphere of the triangular relationships with a rare power (Figure 8.5).

The camera work of the ensemble is outstanding throughout, the timing and precise cueing of each new camera angle and transition perfectly enacted, revealing the ease and professional comfort with which the Schaubühne ensemble approaches the multimedia *mise en scène* and subdued on-camera acting, perhaps not surprising given the interest German theatres have invested in deconstructivist dramaturgical visions (Frank Castorf at the Berlin Volksbühne has worked

FIGURE 8.5 Luise Wolfram, lighting candles, and camera and camera operator on the stage/film set in *Fräulein Julie*. Barbican, 2013 © Thomas Aurin.

consistently with camera teams on stage to disturb any central perspective and illusionism; Chris Kondek has created eccentric video-theatre works at the Munich Kammerspiele and Hamburg Schauspielhaus; Hotel Modern, Rimini Protokoll and other independent companies have used intermedial staging techniques for some years on the continent).[20]

What remains to be interrogated, then, is the overall impact that augmented theatre has on our imagination: the question of how the construction of live film on stage reflects an atmosphere of a turbulent play of gazes instrumental for its psychological dramatization of the character's relationships of entrapment, and how the use of digital modulation in real-time sustains or enhances the perceptional complexity posited by the multiple camera angles. The proscenium stage conventionally does not offer a multi-perspectival experience as it can be achieved in installations or processional/site-specific performances enabling audiences to choose variable points of attention. The Barbican stage, where I witnessed Mitchell's production, is a proscenium stage; thus the fragmentation of viewpoints has to be produced through the complex *mise en scène* of multiple simultaneous spaces and the shifting roles of the inter-actors who fluidly exchange activities (character, camera person, sound maker, voice-over reciter, prop person, lighting technician, grip, etc.). This mosaic quality offers a rich palette of modulations affecting the entire spatio-temporal organization of live film in the theatre. Mitchell stays with a cinematographic framework not attempting more advanced computational interactivity as we have seen it in dance. Clearly, contemporary dance/technology productions use the apparatus differently, focusing more intrinsically on movement and the many possibilities of motion graphics and digital animation, thus remaining indebted – if we follow Kostanić's analysis – to the sensory-motor gesturality of the movement-image and its accelerations and decelerations. Chunky Move's *Glow* (on tour between 2006 and 2008) and *Mortal Engine* (shown at Southbank Center, 2012), or Wayne McGregor/Random Dance's *Atomos* (shown at Sadler's Wells, 2013), and his later *Autobiography* (Sadler's Wells, 2017) are good examples of this. On the kinetic level, *Fräulein Julie* offers no suspense at all. In light of my references to dance, she also does not seem interested in exploring the encounter between image and body and how movement or behavior might be affected by the technological agent.

Mitchell's take on Strindberg's early naturalism transposes *Fräulein Julie* into a form of poetic neo-realism reliant on "purely optical and sound situations" which Deleuze associates with the transition from early cinema (movement-image based) to neo-realist cinema (time-image based).[21] Deleuze describes a scene from De Sica's *Umberto D* that exemplifies a focus on the eyes and the gaze, mentioning the young maid in the film going into the kitchen, making a series of mechanical, weary gestures, cleaning a bit, driving the ants away from the water fountain, picking up the coffee grinder, stretching out her foot close to the door with her toe; then her eyes meet her pregnant belly. Deleuze proceeds to argue that De Sica, Rosselini, Antonioni, Visconti, and later Truffaut and Hitchcock, were moving beyond the action-image of the old realism, focusing instead on optical situations of tactile images and sound, images of time and contemplation. With Hitchcock and later filmmakers, the emphasis shifted increasingly toward the "indiscernibility of the actual and the virtual," the real and the imaginary – the physical and the mental folded into each other (Deleuze 1989: 85).

While the visionary aesthetics of Visconti or Tarkovsky are incontrovertibly powerful instances of the history of film, and Hitchcock's mastery of psychological suspense crystallizes the mirror effect of cinema – the double of the mental image including the spectator into the vertigo of the film (the viewer, as the protagonist of *Rear Window*, is immobilized in the gaze) – the theatre audience is not glued into the screen in the same way. The audience is not fully in-cluded or enveloped, so to speak, in the thought-montage but reminded, at all times, of the construction carried out, the cut, the positioning of the camera angle right there in front of us, on this darkened stage, so that we can see Kristin's gaze and her growing suspicion of Jean and Julie on screen as we look up. We look up and down, down and up. We are neither immobilized nor trained to be critical and pressed into seeing our political concerns (Benjamin). Nor do we care about the frequent gazes into the mirror, since we hardly ever find out anything about Julie's or Jean's desires or motivations. We are a little bit lost in translation. The energy dissipates via the surtitles, too. One does not listen to the voice of surtitles. We are aware of the differentials between stage and screen and also their intermediate flow, and while this is hardly Strindberg's idea of naturalism, what distinguishes Mitchell's work is her insistence on a steady, quiet and consistent aesthetic of poetic realism that will be intriguing to some, and impressive to most (who cannot avoid admiring the stage craft). But it may surely be off-putting to others who recognize the repression and effacement of physical, energetic and political theatre; and who apprehend the hypermediated contrivance of a multi-perspectival scenography that feels emotionally cold, removed, inconsequential. These acting bodies do not resonate. There is no tonality that vibrates in my body.

While Mitchell's lack of concern for acting might be acceptable or even commonplace in the context of postdramatic theatre, it appears contradictory given her expressed interest in psychological realism and human behavior, and her hope that the images "crystallize the inner landscape of the characters' feelings" (as she states in the interviews). Her choice of adaptation also remains a puzzle, as her complicated staging of the gaze, with the filmic techniques she involves so comprehensively and competently, does not quite succeed in raising the Strindberg narrative or the performance to the level of cinematic suspense or complexity audiences are accustomed to from Buñuel, Chabrol, Truffaut or Scorsese or any of the contemporary filmmakers (e.g. Wes Anderson, David Lynch, but also Lars von Trier and Dogme95-influenced handheld cinematography) following in the footsteps of Hitchcock's realization of deep and irrational disturbances of the unconscious. And unlike, say, Marguerite Duras (in *India Song*), Mitchell never really bothers to interrogate the potential disconnection between sound and image, the irrational cut between the two.

Thus her live theatre film weakens, even decarnates, the theatricality of the physical body without inventing, in her modulations of the mixed reality of augmented theatre, a critical exploration of the construction of sound framing, and alignment of her images. She claims that there "always will be errors since it is all live," as she says in the interview with Mark Kermode, but I did not notice any glitches, slippages and breaks in the continuity to stimulate my imagination and illuminate the pressing question of how images can be actionable and disrupt the habituation of the sensory apparatus to technological artifice. How images may be excessive of the atmospheric properties, point beyond psychological realism and the scientific or materialist gaze. How can our imagination drift beyond the seen or heard? What happens if the screens go blank?

Even more noticeable, perhaps, is the paradoxically damaging effect of Mitchell's intermedial strategy: by dispossessing the live theatre of some of its particular strengths of physical, emotional, kinetic energies and resonances, at the same time she cannot achieve the uniquely "focussed" body (Kostanić) of cinematography as we are not watching the screenic image alone, in this constructivist production, but the projection of its weakened, eviscerated double on a cluttered proscenium stage. If she wanted to make the chaos of the clutter evocative of a certain incompatibility between technology and physical behavior, I would have liked to experience it in ways that are not at all familiar to me from the movies, nor from what remains from the theatre.

Notes

Fräulein Julie (after August Strindberg), Schaubühne Berlin production directed by Katie Mitchell and Leo Warner, was presented at the Barbican, London, April 30– May 4, 2013.

1 The tradition of *Regietheater* (director's theatre) in Germany has often been debated quite virulently by theatre critics and scholars, as there have of course been competitive and often very contrasting stage interpretations of dramatic works created by artistic directors of ensembles at the larger state theatres. An exhibition at the Deutsche Theatermuseum München (July 2020–April 2021) tried to delineate historical interconnections in *Regietheater: Eine deutsch-österreichische Geschichte,* tracking the various lines of critical-creative directing and including such artists as Max Reinhardt, Leopold Jessner, Fritz Kortner, Gustaf Gründgens, Peter Stein, Peter Zadek and Claus Peymann (a very male lineage, unfortunately, reflecting male privilege that more recently has been altered to some extent; the impact of female choreographers in Europe, North and South America over the past 100 years tells a very different story).

2 As mentioned earlier, the most comprehensive study of new technological performance is found in Dixon and Smith 2007. See also Birringer 2009, Kwastek 2013, and Salter 2015.

3 A critical discussion of process and generative concepts in performance is long overdue; an awareness of the significant role of controllers and computational parameters for live synthesis (in software programming) is more prevalent in music technology and electronic arts contexts. Cf. the exhibition *Process as Paradigm – art in development, flux and change,* curated by Susanne Jaschko and Lucas Evers, for LABoral Centro de Arte y Creacion Industrial in Gijon, Spain (April 23–August 30, 2010). LABoral's website: http://www.laboralcentrodearte.org/en/exposiciones/el-proceso-como-paradigma-23.04-30.08.2010. The exhibition catalog can be found here: http://www.romankirschner.net/portfolio/process-as-paradigm/.

4 *Sisyphus of the Ear* premiered in Ufa and Moscow (Russia) in October 2016, and was created after I experienced hearing loss during the summer. Fabrizio Manco had completed a PhD thesis on *Ear Bodies: Acoustic Ecologies in Site-Contingent Performance* (University of Roehampton 2016) and gave me valuable feedback. His inquiry into *ear body* – a bodied experience of sound and listening where the whole body becomes an ear – derives from his own experience of chronic tinnitus and provocatively addresses hearing/listening in contexts of performance practice where sound and body move and perform by relating to the constantly changing acoustic environment. *Sisyphus of the Ear* is such a contingent performance: it takes in the sound of the quarry. The other film, *a thousand machines,* was conceived as a screening with live music (percussion and electronics), the music also created by Brazilian composer Paulo C. Chagas. Both films toured widely and have also been exhibited online, most recently at VERY LARGE WORKS (https://thewrong.org/).

5 Ackerman and Pucher 2006: 9. See especially Charlie Keil, "'All the Frame's a Stage': (Anti)Theatricality and Cinematic Modernism," pp. 76–91.

6 Cf. Phelan 1993; Auslander 2008. A careful summary of the antagonistic debate is offered in Dixon and Smith's *Digital Performance: A History of New Media in Theater, Dance, Performance Art, and Installation,* pp. 115–34. The gradual take-over of the digital (video production) and its conflicted relationship to photography/analog filmmaking is a subject that has earned much less attention in performance studies, but remains controversial in debates on film aesthetics and among camera artists who defend the distinct quality of film stock and film projection. Without wanting to enter this debate, it should be admitted here that I have used the term *projection* without addressing the difference between the flickering images coming from a film projector in a cinema (an opto-mechanical device for displaying motion picture film) and the pixelated electronic image reaching screen or projection surface from a video beamer

("beamer" is the word generally used for projectors in Germany, to distinguish the current type of HDMI image projection from the older cinematic 35 mm or 16 mm projectors).

7 Artaud 2010: 47. Unlike M.C. Richards' first translation for Grove Press (1958), Corti uses "wholly substantiated attraction," which of course resonates with Tretyakov's "theatre of attractions" and Eisenstein's "montage of attractions." For a fascinating commentary on the Russian theatre/film avant-garde, which has inspired some of the DAP-Lab's dance rehearsals, see Gerald Raunig, *A Thousand Machines* (2010). Raunig's title riffs on Deleuze, and I then riffed on Raunig's in my *Metropolis* remix.

8 See his chapter "On the Balinese Theatre," esp. pp. 40–41. While working on revisions for this essay, I came across an inspiring evocation of Artaud's writings by the late Herbert Blau: "Performing in the Chaosmos: Farts, Follicles, Mathematics and Delirium in Deleuze," in Cull 2009: 22–34.

9 Artaud's notion of a "raw cinema," according to Stephen Barber, concerns the violent and disruptive unleashing of the spectator's senses experienced in the body's projection; see his *The Screaming Body: Antonin Artaud – Film Projects, Drawings and Sound Recordings*, Creation Books, 1999, p. 26.

10 Cf. Birringer 1989. In my 1998 *Media and Performance: Along the Border* the emphasis of my investigations had shifted more emphatically to technological performance, video and new media art practice. I considered these shifts to have important political and transcultural implications and therefore published a more ethnographic performance study of my field work around the same time (*Performance on the Edge: Transformations of Culture*, London: Athlone Press, 2000).

11 See Katie Mitchell's interviews with Mark Kermode on the BBC Culture Show (July 29, 2008), with Dan Rabellato at the public platform show (National Theatre, August 12, 2008), and her comments on several short films by Pinny Grylls, for example, http://www.nationaltheatre.org.uk/video/katie-mitchell-on-pains-of-youth; http://www.nationaltheatre.org.uk/video/cinematography-in-a-multimedia-production; http://www.nationaltheatre.org.uk/video/exploring-multimedia-in-katie-mitchells-productions. I should add that obviously in the current context the references to cinematography are no longer to celluloid film but digital moving-image technologies.

12 Cf. Soloski 2014. Jay Scheib's performance, in which he operates one of the cameras himself, opened at The Kitchen on January 8, 2014, and was broadcast live at the ACM Empire 25 Cinema (as a simulcast). This live performance integrates multiple video screens and, as in his previous *Simulated Cities/Simulated Systems* trilogy, interrogates the impact of counterfeits and simulacra on our perception of the live/real.

13 My underlying allusion is to the title of Mitchell's last production I saw, *Alles Weitere kennen Sie aus dem Kino* (based on a script by Douglas Crimp adapted from Euripides' *The Phoenician Women*), created for Schauspielhaus Hamburg in November 2013.

14 This description is by Goebbels himself (from the program issued at the work's presentation at Ambika P3, London, 2012). See my "Choreographic Objects: *Stifters Dinge*," *Body, Space and Technology* 11:02: http://people.brunel.ac.uk/bst/vol1102/. The musicalization of scenography/choreography is a fascinating subject that could be traced from Wagner to the futurists, constructivists and later experiments in performance art following Cage, Fluxus, the Judson Dance Theatre and E.A.T. (Rauschenberg's and Klüver's *9 Evenings of Theatre and Engineering*, 1966). The term "composed theatre" has been put to interesting use in Matthias Rebstock's and David Roesner's book, *Composed Theatre: Aesthetics, Practices, Processes*, Bristol: Intellect, 2012.

15 Marko Kostanić, "The Choreographic Unconscious: Dance and Suspense," in *Semi-Interpretations, or How to Explain Contemporary Dance to an Undead Hare*, BADco. program, Zagreb 2012. BADco.'s political and dramaturgical interventions into the relations of performance and image are of significance here; during a recent workshop in Zagreb (2012), we had a prolonged discussion about the "responsibility for things seen" and the "actionable image," i.e. the question whether images can have

agency. I am redirecting the question here to the role of agency of the camera in generative live processing of images.

16 See, for example, the helpful introduction to the nature of augmented theatre and various interleaved trajectories through hybrid structures of real/virtual space, time, interfaces and roles in Steve Benford and Gabriella Giannachi, *Performing Mixed Reality*, Cambridge, MA: MIT Press, 2011, pp. 2–3. See also, Kaisu Koski, *Augmented Theatre*, Entschede: Ipskamp, 2007. Mitchell's book, *The Director's Craft: A Handbook for the Theatre*, London: Routledge, 2009, contains only three pages (pp. 90–92) on the use of video in theatre, claiming, astonishingly, that the use of video in mainstream theatre is still in its infancy.

17 Walter Benjamin's essay, "The Work of Art in the Age of Technological Reproducibility" (1939) is of course very well-known; I am referring to the third version here, and am particularly thinking of the passages (Section VIII) in which Benjamin discusses the actor being subjected to "optical tests" and the audience becoming critics of these optical tests:

> The recording apparatus that brings film actor's performance to the public need not respect the performance as an integral whole. Guided by the cameraman, the camera continually changes its position with respect to the performance. The sequence of positional views which the editor composes from the material supplied him constitutes the completed film. It comprises a certain number of movements, of various kinds and duration, which must be apprehended as such through the camera, not to mention special camera angles, close-ups, and so on. Hence, the performance of the actor is subjected to a series of optical tests. This is the first consequence of the fact that the actor's performance is presented by means of a camera. The second consequence is that the film actor lacks the opportunity of the stage actor to adjust to the audience during his performance, since he does not present his performance to the audience in person. This permits the audience to take the position of a critic, without experiencing any personal contact with the actor. *The audience's empathy with the actor is really an empathy with the camera. Consequently, the audience takes the position of the camera; its approach is that of testing.*
> (pp. 259–60)

18 Cf. Deleuze 1986: 73–75. Deleuze points out that a film is never made up of a single kind of image but of combinations, of the inter-assemblage of movement-images, perception-images, affection-images, and action-images; his philosophy of the principal types of image movement (expanded in the second volume, *Cinema 2: The Time Image*) is too complex to be summarized here but individual sections in *Cinema 1* on montage and especially on facial close-ups provide a thought-provoking context for a reading of Mitchell's live film staging.

19 Michel Chion not only defines *synchresis* as the "spontaneous and irresistible weld produced between a particular auditory phenomenon and visual phenomenon when they occur at the same time," but also addresses the many instances of elasticity in film when patterns of synchronization or negative and offstage sounds – and the whole acousmatic dimension of non-diegetic sound effects and layerings, are not accompanied by the sight of a sound's source of cause (1994: 63–64).

20 For further context on intermedial productions in Germany, see the special sections on "Theater und Video" in *Theater Heute* (April 2004), pp. 18–31, and on "Kunst, Bühne & Videotapes" in *Theater Heute* (August/September 2007), pp. 4–31, the latter dedicated to Kondek and the late Christoph Schlingensief.

21 Cf. Deleuze's opening chapter "Beyond the Movement-Image" in *Cinema 2: The Time Image* (1989: 1–23).

9

KINETIC ATMOSPHERES

Sounding off

We move in slow-motion,
creeping inch-by-inch through the milk-light
beneath brown bulbs of the tamarind.
We raft over the
crack
gap
gulf
of millennia
Clinging to weed and log
We, the dauntless
We, the night-wanderers
We, the moonlit acrobats.
        ("Red Ghosts/Shadows of the Dawn," *kimosphere no. 4*)[1]

In this brief postscript, I return for a moment to the Red Ghosts that we evoked in one of our *kimospheres* – the virtual lemurs that jumped through the VR world we invited our audience to visit and travel through. They also, as a word poem, moved in slow motion across the screen of a video game that visitors could play. Moving the cursor enabled jumping the lines. Visitors are inter-actors who enter to play, to listen and move around, with shifting proximities to what they lean into. As I described in Chapter 4, the kinetic atmospheres as an immersive world or living system actualize various phenomena and sensory circumstances, occurrences, objects, screens, fabrics, lighting moods, projections, instruments and resonating sounds that become intertwined. Intertwinement is not Wonderland, as if visitors had fallen "through the screen" together, and were now on the other side, in a sphere in which everything real is unreal, immanent in an artificial way.

DOI: 10.4324/9781003114710-9

Rather, kinaesthetic circumstances stimulate feeling the body move – move inside, alongside, outside. Such movement experiences invite each other as they accumulate, carry over, and carry on, like little time machines. Resonances configure the presences in this world. They are also distances. There are no virtual lemurs, they are real and imagined. In this sense, a costume also can become environment, because it resonates being worn, it absorbs influences and gives something out, in the same way in which sound objects, in *kimosphere no. 3*, were handed between visitors or played together: these sounding objects live there and are beholden. Like pebble stones you pick up and throw onto the surface of a pond to see how they skip across the water.

One way to imagine kinetic atmospheres, then, is to imagine being these pebbles within the skipping movement and sensation. We are implicated in atmospheric spaciousness, in waves and surfaces, as we are unsustained. We are implicated but our bodily perceptions can hardly be sustained indefinitely since we always move on, and our lives are like our sonorous existence (perhaps this is an embryological heritage, as Katsura Isobe suggested to me). The first sensation, when entering the installation, would be hearing sound as you adjust your eyes to the dim light and the red apparitional mosquito nets suspended from the ceiling. It is the lighting that makes them appear red, and as you move closer you sense their fabric qualities, lightness and porosity. The lighting also makes these nettings stand out within the atmospheric milieu, the blackness and darkness of the enveloping volumetric space of the warehouse. These mosquito nets which hug the multi-channel loudspeakers are also "red ghosts," soft agents of camouflage. One or two of our dancers on occasion become visible underneath, standing motionless under the nets, wearing them as masks, *personae*. The Red Ghosts video game is set up as but one interface in a larger theatrical architecture in which the real and the virtual merge, with the virtual complementing the real in a tangible way as these realities are layered on top of and within each other. You hear the clicking of keys, before you might see the typewriter and the typist, behind one of the nets. The typist, on the other side of the video screen, writes fragments from "Red Ghosts." To a certain extent, both the poetry of the game and the loop of the sound artist Sara Belle typing are allegories, evoking the feeling of slow time or slow space, with refrain, return beat – a wop bop a loo bop a lop bam boom. This was pertinent for the temporally extenuated experience we had devised for *kimosphere no. 4*.

Slow motion, distended time and sonic resonances that imply oscillation. A rebound, as Jean-Luc Nancy suggests, is "a gap, a distance, or a lapse that makes possible a pulsation or is the result of such a pulsation – or both" (Nancy 2020: 13). Pulsating sounds and recorded voices intermingle with all the other visual and non-visual propagations in such a kinetic atmosphere that is potentially inexhaustible, as all light and sound waves, and all tactile and olfactory affects spread out and appear in different ways of appeal, in different (dis)harmonies. I would liken the experience of such kinetic atmospheres to a kind of night wandering where our senses are heightened, in the dark, as we focus on all the stimuli we receive and process to keep navigating, sharpening our reactions to a diffuse and volatile environment. Our hair bristles.

Composing kinetic atmospheres, thus, is like designing a complex instrument, as Ellen Fullman would argue after having worked for many years building and performing her *Long String Instrument* (an installation of over 40 strings spanning 70 feet in length). Fullman says she "bows the strings lengthwise" (walking along them and rubbing them with her hands), and the strings vibrate in multiple modes, with "spectral patterns of harmonics" stretched to each string length (Fullman 2020: 82). Most interestingly, Fullman's installation of course takes up a very large space in buildings she chooses; as she invites audience inside the actual resonating atmosphere, the architecture as a whole is transformed into a musical instrument with varying timbres and frequencies in the harmonics. Fullman also suggests that the physics of string vibration are so astonishing to her that she does not want to "impose any unnecessary expression or gestures upon it; by refining my performance technique I wish to reveal and share what I discover as I listen to the spectrum of harmonics unfold" (81).

Revealing. Such atmospheric conditions in the sculptural sonic architecture are kinetic to me in the sense in which they move and change, remain mobile and unpredictable; the oscillations are permanent and thus *resonance* is the crucial paradigmatic phenomenon of the many cases I have reviewed in this book. Fullman's modesty, on wanting to reveal a spectrum of harmonics rather than force it through unnecessary gestures, is beautiful. I strive for that too. Sound in particular, as Nancy points out, "conjugates with space, it crosses space, and the duration of this crossing is a part of the sound, an intrinsic property… it configures a presence in the world" (Nancy 2020: 15). This presence is also the receiver's presence to the world, since the sound flows through all flesh and bones. I would like to imagine that the various artworks, performances and installations discussed here point to a concept of kinetic atmospheres that allows a very wide range of reception experiences yet stimulates above all a "deep listening" in the sense in which Pauline Oliveros understood "deep" to imply complexity and the blurring of edges (beyond habitual understandings) – heightened attention and expanded perception of sonic, visual and tactile matters or appearances.[2] In Chapter 8 I concentrated on a kind of deep listening to the live "camera show" Mitchell and her ensemble generated on a multimedia sound stage, with attention to some extent focused on the imaging of the action and the kinetic camera techniques, although the role of the sound actors was equally significant. The world is listening too, and so one always has to remember mutualities, blurring of edges over materials and other organisms. In other chapters I looked at lighting, design, *mise en scène*, choreography, wearables and materials as constitutive of such resonant and relational entanglement of bodies and environments, articulating the terms of engagement I have witnessed in the performing and media arts over the last decades – engagement with technological media infrastructures enabling new techniques of atmospheric composition.

Technical capacities were always a concern in my writing, but above all I wanted to share a deep curiosity with the aesthetics of kinetic atmospheres and how the new somatechnics affect the work of actors, dancers, choreographers, directors, designers and composers, and what we can learn from an ecophilosophical embrace of environments – the arts as a part of our earthly elemental forces and perceptional experience. In Chapter 6 I also began to reflect on the viral conditions of the inter-pandemic era we have come to face. The "elemental media," as Durham Peters describes them so vividly (in *The Marvelous Clouds*), now include the corona virus we are exposed to, and the many atmospheric publics and contested political spheres challenging all of our aesthetic practices and the infrastructural ethics of being alive in the cosmos. Leaning on the past, we wait for competent shamans and try to hang on to a narrative of redemption.[3]

We move in slow motion as we now come to my last examples, again evoking perhaps a sense of forest knowledge, as suggested in my introduction. Such knowledge will be equally useful for built environments in dense urban contexts where immersive performance – in the times of pandemic safety restrictions – is not likely to happen again soon, unless you imagine yourself wandering around getting lost in teamLab's *Borderless* or in the vast simulation universes, such as Yuxi Cao and Lau Hiu Kong's *Dimensional Sampling*, projected in the Nxt Museum. In Berlin, club culture and raves had been a staple of contemporary night life, but it all stopped in mid-March 2020 with the universal lock-down. Months later, in late July, a small queue formed in front of the fabled Berghain techno club. People have lined up carefully to gain entry, but nothing is really the same as there will be no dancing to techno beats, no ecstatic rave with sweaty and twitching bodies, no flickering strobe lights.

On offer in the large former boiler hall adjacent to the club is an electro-acoustic installation, *Eleven Songs,* by tamtam (the Austrian sound artists Sam Auinger and Hannes Strobl). Visitors can stroll through it with a mask in front of their faces, at the appointed time, distanced from one another. The dance floors in Berghain are still closed and will remain so for an indefinite period of time; but during the summer, the operators of the club opened the "Halle" to us. We become melancholic flâneurs who can listen at least, if not dance. The artist duo fills the vast hall with dark sounds: the sonic experience takes hold of my body, with a densely sensual interweaving of abstracted sounds and textures (as one might hear them in György Ligeti's *Atmosphères,* or post-metal ambient music like decarnation's *Apocalypse*). Every bone seems to vibrate, the nostrils begin to flutter, my hair bristles, vibrations creep up from the floor to the head and then back down again, a clot of sound in the belly. The sounds surround the listener as they become the inner core of a large instrument or a fire storm, waves flow through all the pores.[4] The physical impulse of course is to dance, to sway with the sounds, to follow. The artists have opened the windows of the boiler hall so that light penetrates the otherwise dark room; it is not soundproof but noises from the street mix with *Eleven Songs* again and again, reminding us of the weather world of urban ambience.

... and suddenly it all blossoms

As with Fullman's *Long String Instrument,* the Austrian artists speak of activating the hall as if it were an instrument, making us hear the reflections of sound from the walls, and never from loudspeakers. And thus the resonances are transfers of energy and transformative magic. We are earbodies and these earbodies dance.

Opening the windows of the old industrial boiler hall is a good idea. Quite a few of the DAP-Lab's kinetic atmospheres were staged at the Artaud Center for Performance (London), also a former boiler house. All windows to the outside are sealed and painted black. However, we tried to bring materials inside from the outside, and recycle or upcycle some of the materials, observing how they behaved and aged, fragile as they were, or how they grew stronger through performance, and how they might even enjoy an unfinished design agency, in terms of their emergence and persistence.[5] The tree trunks I hauled into *Mourning for a dead moon* (performed in London) were the actual old fir trees that had to be cut down in my garden in Germany, so the old partial trees traveled across the borders to be resurrected, the countryside returning to the metropolis. Irritatingly, architect Rem Koolhaas thinks of countryside as our great "ignored realm," now gloriously rediscovered by planners and architects against their own older ideologies of "total urbanization" (AMO and Koolhaas 2020: 2).[6]

And thus our question of sustainability, from Chapter 2, cannot be answered. Physical time is real time, and perhaps it all can blossom again. Light falls through leaves and branches, conjuring a stage or a precipice, a relic of convictions and then one hears the percussion, distant beats, intoned by shamans. They call, and fading up are colors of cymbals and gongs, and passageways, the dance always moving as if movement were captured in the lesser moments of disrupted dreams, the screen dances of our unconscious. The dreams can act as prophecies too. This book was dedicated to movement, movement through life and away from it too, into small distances, aimed at perceiving and listening to more and more landscapes and terrains of nature (real and designed) as bearers of intelligence, shared commons. In reverse, as you tried to catch the moonlit acrobats looking into dark goggles of virtual realities, hopefully you caught glimpses of sensual worlds where you touch others without fear.

Notes

1 Core elements of *kimosphere no. 4* include 8-channel sound design by Sara S. Belle; "Red Ghosts" and "Horlà" cut up poetry by Emma Filtness; "Shadows of the Dawn" cut up by Johannes Birringer; performance by Yoko Ishiguro, Helenna Ren, Haein Song, and Sara S. Belle; biosignal interface by Claudia Robles-Ángel; *Horlà* 3D film by Paul Moody; "Red Ghosts" interactive game design by Ashley Rezvani; and coral reef projections by Chris Bishop and Johannes Birringer. Project direction and editing of postproduction by Johannes Birringer, with additional film by Martina Reynolds. For film excerpts, see: https://www.youtube.com/watch?v=0aIW6Klfm1g.

2 Meeting Ellen Fullman and Pauline Oliveros in Texas was one the most stimulating experiences in my life. Oliveros' "deep listening" idea is reputedly owed to her 1988

descent into an underground cistern to make a recording. Her approach to teaching it was meant to inspire an art of listening and responding to environmental conditions – as a heightened sonic awareness combining two ways of processing information, "attention and awareness," of focal attention combined with global attention to the surroundings. In 2005 Oliveros published her *Deep Listening: A Composer's Sound Practice,* New York: iUniverse, Inc., Listen to Oliveros/Dempster/Panaiotis: https://www.youtube.com/watch?v=U__lpPDTUS4. Fullman's Long String Instrument, which I last witnessed in her performance at the Houston CounterCurrent festival in April 2019, has been very evocative for my thinking on embodied psychoacoustic experiences.

3 For a fascinating discussion of the shamanic tradition and a hopeful, different kind of ethno-futurism, less resigned than Frank Wilderson's Afropessimism, see Kreuger 2017. For an article on Senegalese *maraboutage,* see Angela Kökritz, "Wie verhext," *Die ZEIT,* no. 16, 8 April 2020. The most significant research on *maraboutage* and the role of the imaginary in Senegalese life is available in Ibrahima Sow, *La maraboutage au Sénégal,* Dakar: editions IFAN Cg. A. Dio, 2013.

4 For a trailer introducing the installation of *Eleven Songs* at the Halle, Berghain, and comments by the sound artists Sam Auinger and Hannes Strobl, see: https://www.youtube.com/watch?v=0X61A55sMSw&feature=emb_logo. Evoking the idea of a *Raumpartitur* or spatial score that I mentioned in conjunction with designer Penelope Wehrli, Auinger argues that what interested them most was how a spatial architecture brings certain characteristics of materiality, volume and form with it, and thus also has its own *Klangmaterial* (sound material). Auinger also suggests that after adding their sonic compositions to the space, it invites "thinking with your ears," generating new images: "The room becomes psychotropic."

5 *Mourning for a dead moon* was premiered at the Artaud Center, and for Scene 3 Michèle Danjoux had designed a prototype (PlasticsDress) for dancer Zhi Xu which consisted of discarded plastic materials saved in her household over a period of time. The entangling plastic construction offered the stimulus for Xu's choreographic responses. See Danjoux' presentation "Dance, Costume, Climate and Contamination," for the 2020 Critical Costume conference in Oslo: https://www.youtube.com/watch?time_continue=6&v=0P3XLY0CuD8&feature=emb_logo. The tree trunks were from my garden in Germany, where they belong to the Tree Resurrection Inc. Gallery.

6 See AMO and Koolhaas and the *Countryside in Your Pocket* book they published after the New York City Guggenheim Museum exhibition *Countryside, The Future* (February 20–August 2020), curated by Richard Armstrong and Troy Conrad Therrien. The exhibition was interrupted by the corona virus, see https://www.guggenheim.org/exhibition/countryside. In the pocket book, Therrien enthusiastically and uncritically describes the utopian shift OMA, Koolhaas' earlier Office for Metropolitan Architecture, has undertaken to become AMO and dedicate its efforts to rediscovering "archetypal agriculture" and "countrysiding as a healthy flattening of consciousness," responding to "volatile, uncertain, ambiguous social scenarios" with its own kind of climate activism and post-human architecture (15–16). The exhibition is a vast bulletin board of texts, photographs and cut outs plastered on the Guggenheim Museum's curved wall surfaces.

BIBLIOGRAPHY

Ackerman, Alan and Pucher, Martin, eds. 2006. *Against Theatre: Creative Destructions on the Modernist Stage*, Basingstoke: Palgrave Macmillan.

Agamben, Giorgio. 2004. *The Open. Man and Animal*, trans. Kevin Attell, Stanford, CA: Stanford University Press.

Alston, Adam. 2016a. *Beyond Immersive Theatre*, Basingstoke: Palgrave Macmillan.

Alston, Adam. 2016b. "Making Mistakes in Immersive Theatre: Spectatorship and Errant Immersion," *Journal of Contemporary Drama in English* 4(1): 61–73.

Amerika, Mark. 2007. *Meta/Data: A Digital Poetics*, Cambridge, MA: MIT Press.

AMO and Koolhaas, Rem. 2020. *Countryside: A Report*, Cologne: Taschen.

Aronson, Arnold. 2005. *Looking into the Abyss: Essays on Scenography*, Ann Arbor: University of Michigan Press.

Artaud, Antonin. 2010. *The Theatre and Its Double*, trans. Victor Corti, Richmond: One World Classics.

Assheuer, Thomas. 2016. "Epidemie des Argwohns: Der Terrorismus ist keine mediale Erfindung. Er zeigt Wirkung. Der Verdacht dominiert neuerdings unsere Wahrnehmung des Alltags," *Die ZEIT* 32, 30 July.

Attali, Jacques. 1985. *Noise: The Political Economy of Music*, trans. Brian Massumi, Minneapolis: University of Minnesota Press.

Auslander, Philip. 2008. *Liveness: Performance in a Mediatized Culture*, 2nd ed., London: Routledge.

Avanessian, Armen. 2017. *Overwrite: Ethics of Knowledge — Poetics of Existence*, trans. Nils Schott, Berlin: Sternberg Press.

Bablet, Denis and Bablet, Marie-Louise. 1982. *Adolphe Appia: 1862–1928 Actor-Space-Light*, London: John Calder Ltd.

Barad, Karen. 2007. *Meeting the Universe Halfway: Quantum Physics and the Entanglement of Matter and Meaning*, Durham, NC: Duke University Press.

Bartaku. 2019. "Seed Scarification: Serious Taking." In Johannes Birringer and Josephine Fenger, eds., *Tanz der Dinge/Things That Dance*, Bielefeld: transcript, pp. 193–198.

Basbaum, Sérgio Roclaw. 2020. "Sense and Meaning: Body Senses and Language in a Synesthetic Key," *Proceedings VI International Congress Synaesthesia, Science & Art*, Granada: Fundación Internacional Artecittá, pp. 97–101.

Baudry, Jean-Louis. 1978. *L'Effet cinema*, Paris: Albatros.

Bauhaus: Art as Life. An exhibition at the Barbican Art Gallery, London, May 3- August 12, 2012.

Beacham, Richard. 2011. *Adolphe Appia: Texts on Theatre*, 2nd ed., London: Routledge.

Beacham, Richard. 2006. "'Bearers of the Flame': Music, Dance, Design, and Lighting, Real and Virtual – the Enlightened and Still Luminous Legacies of Hellerau and Dartington," *Performance Research* 11(4): 81–94.

Beacham, Richard C. 1993. *Adolphe Appia: Texts on Theatre*, London: Routledge.

Benford, Steve and Giannachi, Gabriella. 2011. *Performing Mixed Reality*, Cambridge, MA: MIT Press.

Benjamin, Walter. 2003. "The Work of Art in the Age of Technological Reproducibility (1939)." In Howard Eiland and Michael W. Jennings, eds., *Selected Writings, Volume 4, 1938–1940*, Cambridge, MA: Belknap Press of Harvard University Press, pp. 251–83.

Benjamin, Walter. 1999a. *Illuminations*, ed. Hannah Arendt, London: Random House.

Benjamin, Walter. 1999b. "Theatre and Radio (1932)." In Michael W. Jennings, Howard Eiland, and Gary Smith, eds., *Selected Writings, Volume 2, 1927–1934*, Cambridge, MA: Belknap Press of Harvard University Press, pp. 583–85.

Benjamin, Walter. 1999c. "The Work of Art in the Age of Mechanical Reproduction." In Hannah Arendt ed., *Illuminations*, London: Random House, pp. 211–44.

Bennett, Jean. 2010. *Vibrant Matter: A Political Ecology of Things*, Durham, NC: Duke University Press.

Bergson, Henri. 1991. *Matter and Memory*, trans. N.M. Paul and W.A. Palmer, New York: Zone Books.

Bergson, Henri. 1896/1988. "Of the Selection of Images for Conscious Presentation. What Our Body Means and Does," In N.M. Paul and W.S. Palmer trans., *Matter and Memory*, New York: Zone Books, pp. 17–76.

Birringer, Johannes. 2020. "Queering the Bauhaus," *PAJ: A Journal of Performance and Art* 125: 120–124.

Birringer, Johannes. 2019a. "Black Dada Overtones: Robert Hodge's Visible Music" (unpublished manuscript).

Birringer, Johannes. 2019b. "Low End Resilience Theory," *PAJ: A Journal of Performance and Art* 123: 28–43.

Birringer, Johannes. 2019c. "*Sous les pavés, la plage.*" In Jelena Novak and John Richardson, eds., *Einstein on the Beach: Opera Beyond Drama*, London: Routledge, pp. 49–65.

Birringer, Johannes. 2018a. "Becoming-Atmosphere," *Performance Research* 23(4/5): 162–68.

Birringer, Johannes. 2018b. "Sisyphus of the Ear," *The Polyglot Magazine of Poetry and Art* 2 (fall): 18–19.

Birringer, Johannes. 2017a. "Audible and Inaudible Choreography: Atmospheres of Choreographic Design." In Sabine Kaross & Stephanie Schroedter, eds., *Klänge in Bewegung: Spurensuche in Choreografie und Performance*, Bielefeld: transcript Verlag, pp. 121–42.

Birringer, Johannes. 2017b. "Metakimospheres." In Susan Broadhurst and Sara Price, eds., *Digital Bodies: Creativity and Technology in the Arts and Humanities*, London: Palgrave Macmillan, pp. 27–48.

Birringer, Johannes, 2016, "Kimospheres, or Shamans in the Blind Country," *Performance Paradigm* 12: http://performanceparadigm.net/index.php/journal/article/view/176.

Birringer, Johannes. 2015. "Gestural Materialities and the Worn *Dispositif*." In Nicolás Salazar Sutil and Sita Popat, eds., *Digital Movement: Essays in Motion Technology and Performance*, Basingstoke: Palgrave Macmillan, pp. 162–85.

Birringer, Johannes. 2013a. "Bauhaus, Constructivism, Performance," *PAJ: A Journal of Performance and Art* 104: 39–52.

Birringer, Johannes. 2013b. "Retro-engineering: The Sound of Wearables." In Kara Reilly, ed., *Theatre, Performance, and Analogue Technology: Historical Interfaces and Intermedialities*, Palgrave Macmillan, pp. 133–58.

Birringer, Johannes, 2012a. "Choreographic Objects: *Stifters Dinge*," *Body, Space and Technology* 11:02, available online: http://people.brunel.ac.uk/bst/vol1102/.

Birringer, Johannes. 2012b. "Gesture and Politics," *VLAK: Contemporary Poetics & the Arts*, 380–88.

Birringer, Johannes, 2010, "Moveable Worlds/Digital Scenographies," *International Journal of Performance Arts and Digital Media* 6(1): 89–107.

Birringer, Johannes. 2008a. *Performance, Technology and Science*, New York: PAJ Publications.

Birringer, Johannes. 2008b. "After Choreography," *Performance Research* 13(1): 118–22.

Birringer, Johannes. 2007. "Bodies of Color," *PAJ: A Journal of Performance and Art* 87: 35–46.

Birringer, Johannes. 2002. "Algorithms for Movement," *PAJ: A Journal of Performance and Art* 71: 115–19.

Birringer, Johannes. 2001. "Suite Fantastique: Architecture and the Digital," *Body, Space & Technology* 2(1). DOI: http://doi.org/10.16995/bst.254.

Birringer, Johannes. 2000. *Performance on the Edge: Cultural Transformations*, London: Continuum.

Birringer, Johannes. 1998. *Media and Performance: Along the Border*, Baltimore, MD: Johns Hopkins University.

Birringer, Johannes. 1991. *Theatre, Theory, Postmodernism*, Bloomington: Indiana University Press.

Birringer, Johannes and Danjoux, Michèle. 2021. "Wearable Technology for the Performing Arts." In David Bryson and Jane McCann, eds., *Smart Clothes and Wearable Technology*, Amsterdam: Elsevier.

Birringer, Johannes and Danjoux, Michèle. 2013. "The Sound of Movement Wearables," *Leonardo* 46(3): 233–40.

Birringer, Johannes and Danjoux, Michèle. 2009. "Wearables in Performance." In Jane McCann and David Bryson, eds., *Smart Clothes and Wearable Technology*, Cambridge: Woodhead Publishing, pp. 388–419.

Birringer, Johannes and Danjoux, Michèle. 2008. "Wearable Performance," *Digital Creativity* 20(1–2): 95–113.

Blanke, Olaf. 2012. "Multisensory Brain Mechanisms of Bodily Self-Consciousness," *Nature Reviews Neuroscience* 13: 556–71.

Blau, Herbert. 1990. *The Audience*, Baltimore, MD: Johns Hopkins Press.

Blau, Herbert. 1987. *The Eye of Prey: Subversions of the Postmodern*, Bloomington: Indiana University Press.

Blau, Herbert. 1982a. *Take Up the Bodies: Theater at the Vanishing Point*, Urbana: University of Illinois Press.

Blau, Herbert. 1982b. *Blooded Thought: Occasions of Theatre*, New York: PAJ Publications.

Blume, Torsten. 2014. "An Undertaking Contrary to Nature for Purposes of Order," *Bauhaus: Zeitschrift der Stiftung Bauhaus Dessau* 6: 4–13.

Bleeker, Maaike, ed. 2017. *Transmission in Motion: The Technologizing of Dance*, New York: Routledge.

Bogart, Anne. 2007. *And Then, You Act: Making Art in an Unpredictable World*, New York: Routledge.

Bogart, Anne. 2006. *The Viewpoints Book: A Practical Guide to Viewpoints and Composition*, New York: TCG.

Böhme, Gernot. 2017. *The Aesthetics of Atmospheres: Ambiences, Atmospheres and Sensory Experiences of Space*, trans. Jean-Paul Thibaud, London: Routledge.

Böhme, Gernot. 1995. *Atmosphäre: Essays zur neuen Ästhetik*, 7th ed., Frankfurt: Suhrkamp.

Böhme, Gernot. 1993. "Atmosphere as the Fundamental Concept of a New Aesthetics," *Thesis Eleven* 36: 113–26.

Bois, Yve-Alain and Krauss, Rosalind E. 1997. *Formless*, New York: Zone Books.

Bolter, Jay David and Gromola, Diane. 2003. *Windows and Mirrors. Interaction Design, Digital Art, and the Myth of Transparency*, Cambridge, MA: MIT Press.

Bolter, Jay David and Grusin, Richard.1999. *Remediation: Understanding New Media*, Cambridge, MA: MIT Press.

Bonnet, François and Sanson, Bartolomé, eds. 2020. *Spectres II: Resonances*, Rennes: Shelter Press.

Borch, Christian, ed. 2014. *Architectural Atmospheres: On the Experience and Politics of Architecture*, Basel: Birkhäuser.

Brandstetter, Gabriele. 2007. "Schwarm und Schwärmer. Übertragungen in/als Choreographie." In Gabriele Brandstetter, Bettina Brandl-Risi, and Kai van Eikels, eds., *Schwarm E/motion: Bewegung zwischen Affekt und Masse*, Freiburg: Rombach, pp. 65–91.

Brandstetter, Gabriele, Brandl-Risi, Bettina, and van Eikels, Kai, eds. 2007. *Schwarm E/motion: Bewegung zwischen Affekt und Masse*, Freiburg: Rombach.

Brandstetter, Gabriele and Wiens, Birgit, eds. 2010. *Theater ohne Fluchtpunkt/Theatre without Vanishing Point*, Berlin: Alexander Verlag.

Brecht, Bertolt. 1964 [1929]. "Der Dreigroschenprozess." In John Willett, ed., *Brecht on Theatre*, London: Methuen, pp. 156–79.

Broadhurst, Susan and Machon, Jo, eds. 2006. *Performance and Technology: Practices of Virtual Embodiment and Interactivity*, Basingstoke: Palgrave Macmillan.

Brouwer, Joke, with Arjen Mulder, Anne Nigten, Laura Martz, eds., 2005, *aRt&D: Artistic Research and Development*, Rotterdam: V2_Publishing/NAi Publishers.

Burt, Ramsay. 2006. *Judson Dance Theater: Performative Traces*, London: Routledge.

Buskirk, Martha, 2006. *The Contingent Object of Contemporary Art*, Cambridge, MA: MIT Press.

Calza, Gina Carlo, ed. 2005. *Ukiyo-e*, London: Phaidon Press.

Camurri, Antonio and Volpe, Gualtiero, eds. 2004. *Gesture-Based Communication in Human-Computer Interaction*, Berlin: Springer.

Canetti, Elias 1984 [1960]. *Crowds and Power*, trans. Carol Stewart, New York: Farrar, Straus and Giroux.

Carver, Gavin and Beardon, Colin, eds. 2004. *New Visions in Performance: The Impact of Digital Technologies*, Lisse: Svets & Zeitlinger.

Chapple, Fred and Kattenbelt, Chiel, eds. 2006. *Intermediality in Theatre and Performance*, Amsterdam: Rodopi.

Chatzichristodoulou, Maria, Jeffries, Janis and Zerihan, Rachel, eds. 2009. *Interfaces and Performance*, London: Ashgate Publishing.

Chion, Michel. 1994. *Audio-Vision*, trans. Claudia Gorbman, New York: Columbia University Press.

Chun, Wendy Hui Kyong and Thomas Keenan. 2006. *New Media, Old Media: A History and Theory Reader*, London: Routledge.

Connor, Steven. 2004. *The Book of Skin*, Ithaca, NY: Cornell University Press.

Coole, Diana, and Frost, Samantha, eds. 2010. *New Materialisms: Ontology, Agency, and Politics*, Durham, NC: Duke University Press.

Corin, Florence, ed. 2004. *Interagir avec les Technologies Numériques*, special issue of *Nouvelles de Danse* 52.

Cormier, Zoe. 2017. *"Blue Planet II*: Fish Are the Sex-Switching Masters of the Animal Kingdom," Available online: https://ourblueplanet.bbcearth.com/blog/? article=incredible-sex-changing-fish-from-blue-planet.

Couchot, Edmond and Hillaire, Norbert. 2006. *L'Art Numerique: Comment la technologie vient au monde de l'art*, Paris: Flammarion.

Cramer, Franz Anton. 2014. "In the Here and Now of Geometry," *Bauhaus: Zeitschrift der Stiftung Bauhaus Dessau* 6: 19–29.

Curtin, Adrian and Roesner, David. 2015. "Sounding out 'the Scenographic Turn': Eight Position Statements," *Theatre & Performance Design* 1(1): 107–25.

Cull, Laura, ed. 2009. *Deleuze and Performance*, Edinburgh: Edinburgh University Press.

Danjoux, Michèle. 2017. *Design-in-Motion: Choreosonic Wearables in Performance*, PhD Thesis, London College of Fashion, University of the Arts London.

Danjoux, Michèle, 2014. "Choreography and Sounding Wearables," *Scene* 2(1–2): 197–220.

Deleuze, Gilles. 1992a. "Postscript on the Societies of Control," *October* 59 (Winter): 3–7.

Deleuze, Gilles. 1992b. "What Is a *Dispositif*?" In Timothy J. Armstrong, ed. *Michel Foucault Philosopher*, New York: Routledge, pp. 159–68.

Deleuze, Gilles. 1986. *Cinema 1: The Movement-Image*, trans. Hugh Tomlinson and Barbara Habberjam, Minneapolis: University of Minnesota Press.

Deleuze, Gilles. 1989. *Cinema 2: The Time Image*, trans. Hugh Tomlinson and Robert Galeta, London: The Athlone Press.

D'Evie, Fayen, 2017. "Orienting through Blindness: Blundering, Be-Holding, and Wayfinding as Artistic and Curatorial Methods," *Performance Paradigm* 13: 42–72.

Dinkla, Söke and Leeker, Martina, eds. 2002. *Dance and Technology/Tanz und Technologie: Moving towards Media Productions - Auf dem Weg zu medialen Inszenierungen*, Berlin: Alexander Verlag.

Dixon, Steve, with Smith, Barry. 2007. *Digital Performance: A History of New Media in Theater, Dance, Performance Art and Installation*, Cambridge, MA: MIT Press.

Doruff, Sher. 2003. "Collaborative Culture," *Making Art of Databases*, Rotterdam: V2_ Publishing/NAI Publishers.

Doruff, Sher. 2009. "The Tendency to 'Trans-': The Political Aesthetics of the Biogrammatic Zone." In Maria Chatzichristodoulou, Janis Jeffries and Rachel Zerihan, eds., *Interfaces and Performance*, London: Routledge, pp. 121–40.

Douglas, Stan and Eamon, Christoph, eds. 2009. *The Art of Projection*, Ostfildern: Hatje Cantz Verlag.

Durham Peters, John. 2015. *The Marvelous Clouds. Toward a Philosophy of Elemental Media*, Chicago, IL: University of Chicago Press.

Eisner, Lotte H. 2008 [1969]. *The Haunted Screen: Expressionism in the German Cinema and the Influence of Max Reinhardt*, Oakland: University of California Press.

Eliasson, Olafur, 2014, "Atmospheres, Art, Architecture." In Christian Borch, ed., *Architectural Atmospheres: On the Experience and Politics of Architecture*, Basel: Birkhäuser Verlag, pp. 90–107.

Esposito, Roberto. 2008. *Bios: Biopolitics and Philosophy*, trans. Timothy Campbell, Minneapolis: University of Minnesota Press.

Flusser, Vilém. 1999. *Towards A Philosophy of Photography*, London: Reaktion Books.

Forsythe, William. 2008. *Suspense, Exhibition Catalogue*, ed. Markus Weisbeck, Zürich: Ursula Blickle Foundation.

Foster, Hal. 2020. *What Comes after Farce?* London: Verso.

Fullman, Ellen. 2020. "Instrument is Composition, Resonance is Harmony." In François Bonnet and Bartolomé Sanson, eds., *Spectres II: Resonances*, Rennes: Shelter Press, pp. 81–84.

Fülöp-Miller, René. 1926. *Geist und Gesicht des Bolschewismus: Darstellung und Kritik des kulturellen Lebens in Sowjet-Russland*, Zürich: Amalthea.

Gaensheimer, Susanne and Kramer, Mario, eds. 2016. *William Forsythe: The Fact of Matter*, Bielefeld: Kerber Verlag.

Gaillot, Michel. 1999. *Multiple Meaning: Techno - An Artistic and Political Laboratory of the Present*, Paris: Disvoir.

Gaines, Malik. 2017. *Black Performance on the Outskirts of the Left: A History of the Impossible*, New York: New York University Press.

Garelick, Rhonda. K. 2007. *Electric Salome: Loïe Fuller's Performance of Modernism*, Princeton, NJ: Princeton University Press.

Garwood, Deborah. 2007. "The Future of an Idea: *9 Evenings* – Forty Years Later," *PAJ: A Journal of Performance and Art* 85: 36–48.

Gethmann, Daniel, ed. 2010. *Klangmaschinen zwischen Experiment und Medientechnik*, Bielefeld: transcript Verlag.

Giannachi, Gabriella. 2004. *Virtual Theatres: An Introduction*, London: Routledge.

Giesekam, Greg. 2007. *Staging the Screen: The Use of Film and Video in Theatre*, Basingstoke: Palgrave Macmillan.

Gil, José. 2006. "The Paradoxical Body," *TDR: The Drama Review* 50(4): 21–35.

Goodman, Steve, Heys, Toby, and Ikoniadou, Eleni, eds. 2019. *AUDINT: Unsound: Undead*, Falmouth: Urbanomic Media Ltd.

Graham, Beryl and Cook, Sarah. 2010. *Rethinking Curating Art after New Media*, Cambridge, MA: MIT Press.

Grau, Oliver. 2003. *Virtual Art: From Illusion to Immersion*, Cambridge, MA: MIT Press.

Greenblatt, Stephen. 1980. *Renaissance Self-Fashioning: From More to Shakespeare*, Chicago, IL: University of Chicago Press.

Groys, Boris. 1992. *Gesamtkunstwerk Stalin*, trans. Charles Rougle, Princeton, NJ: Princeton University Press.

Hainge, Greg. 2013. *Noise Matters: Towards an Ontology of Noise*, New York: Bloomsbury.

Halligan, Fionnuala. 2017. "*Dragonfly Eyes*: Locarno Review," *Screen*, August 10. Available online: https://www.screendaily.com/reviews/dragonfly-eyes-locarno-review/5120610.article.

Hansen, Mark B.N. 2006. *Bodies in Code: Interfaces with Digital Media*, New York: Routledge.

Hansen, Mark B.N. 2004. *New Philosophy for New Media*, Cambridge, MA: MIT Press.

Halprin, Anna. 1995. *Moving Towards Life: Five Decades of Transformational Dance*, Hanover, NH: Wesleyan University Press.

Haraway, Donna. 2016. *Staying with the Trouble: Making Kin in the Chthulucene*, Durham, NC: Duke University Press.

Harney, Stefano and Moten, Fred. 2013. *Undercommons: Fugitive Planning & Black Study*, Wivenhoe: Minor Compositions.

Hartman, Saidiya. 2019. *Wayward Lives, Beautiful Experiments Intimate Histories of Social Upheaval*, New York: W.W. Norton & Co.

Hayles, N. Katherine. 1999. *How We Became Posthuman: Virtual Bodies in Cybernetics, Literature, and Informatics*, Chicago, IL: University of Chicago Press.

Henriques, Julian F. 2011. *Sonic Bodies: Reggae Sound Systems, Performance Techniques and Ways of Knowing*, London: Continuum.

Hindson, Catherine. 2015. "Dancing on Top of the World: A Serpentine through Late Nineteenth-Century Entertainment, Fashion and Film." In Marketa Uhlirova, ed., *Birds of Paradise: Costume as Cinematic Spectacle*, Köln: König Books, pp. 65–77.

Hookway, Branden. 1999. *Pandemonium: The Rise of Predatory Locales in the Postwar World*, New York: Princeton Architectural Press.

Hsieh, Tehching and Heathfield, Adrian. 2008. *Out of Now: The Lifeworks of Tehching Hsieh*, London: Live Art Development Agency.

Hünnekens, Annette. 1997. *Der bewegte Betrachter: Theorien der Interaktiven Medienkunst*, Köln: Wienand.

Ihde, Don. 2002. *Bodies in Technology*, Minneapolis: University of Minnesota Press.

Ingold, Tim. 2011. *Being Alive: Essays on Movement, Knowledge and Description*, London: Routledge.

Iimura, Takahiko. 2021. "New Media Curating Digest April 1," *New Media Curating*, listserv discussion on curating digital art.

Jablonski, Nina G. 2006. *Skin: A Natural History*, Berkeley: University of California Press.

Jackson, Shannon. 2011. *Social Works: Performing Art, Supporting Publics*, London: Routledge.

Jackson, Shannon and Weems, Marianne. 2015. *The Builders Association: Performance and Media in Contemporary Theatre*, Cambridge, MA: MIT Press.

Jäger, Joachim. 2007. "Learning Embodiment from Film and Video Art: Interview with Joachim Jäger." Available online: http://www.artificial.dk/articles/jager.htm.

Joseph, Branden W. 2009. "'My Mind Split Open': Andy Warhol's Exploding Plastic Inevitable." In Stan Douglas and Christoph Eamon, eds., *The Art of Projection*, Ostfildern: Hatje Cantz Verlag, pp. 71–92.

Junko, Muto. 2005. "Enjoying Actor Prints: Imagining the Voices of Actors and Music." In Gina Carlo Calza, ed., *Ukiyo-e*, London: Phaidon Press, pp. 10–11.

Keil, Charlie. 2006. "'All the Frame's a Stage': (Anti-)Theatricality and Cinematic Modernism." In Alan Ackerman and Martin Pucher, eds., *Against Theatre: Creative Destructions on the Modernist Stage*, Basingstoke: Palgrave Macmillan, pp. 76–91.

Kelly, Caleb. 2017. *Gallery Sound*, New York: Bloomsbury.

Kelly, Caleb. 2009. *Cracked Media: The Sound of Malfunction*, Cambridge, MA: MIT Press.

Kleber, Pia and Trojanowska, Tamara. 2019. "Performing the Digital and AI: In Conversation with Antje Budde and David Rokeby," *TDR* 63(4): 99–112.

Kluitenberg, Erik. 2008. *Delusive Spaces: Essays on Culture, Media and Technology*, Amsterdam: NAi Publishers & Institute of Network Culture.

Koski, Kaisu. 2007. *Augmenting Theatre: Engaging with the Content of Performances and Installations on Intermedial Stages*, Entschede: PrintPartners Ipskamp.

Kozel, Susan. 2008. *Closer: Performance, Technologies, Phenomenology*, Cambridge, MA: MIT Press.

Kracht, Christian. 2008. *Ich werde hier sein im Sonnenschein und im Schatten*, Cologne: Kiepenheuer & Witsch.

Krauss, Rosalind. [1978] 1986. "Sculpture in the Expanded Field." In *The Originality of the Avant-Garde and Other Modernist Myths*, Cambridge, MA: MIT Press, pp. 276–90.

Kreuger, Anders. 2017. "Ethno-Futurism: Leaning on the Past, Working for the Future," *Afterall* 43: 117–33.

Kuppers, Petra. 2004. *Disability and Contemporary Performance: Bodies on Edge*, New York: Routledge.

Kurzweil, Ray. 1999. *The Age of Spiritual Machines: When Computers Exceed Human Intelligence*, Cambridge, MA: MIT Press.

Kwastek, Katja. 2013. *Aesthetics of Interaction in Digital Art*, Cambridge, MA: MIT Press.

LaBelle, Brandon. 2018. *Sonic Agency: Sound and Emergent Forms of Resistance*, London: Goldsmiths Press.

Lamarre, Thomas. 2009. *The Anime Machine: A Media Theory of Animation*, Minneapolis: University of Minnesota Press.

Lanier, Jaron. 2006. *Homuncular Flexibility*, Seattle: Edge Foundation, Inc.

Lasky, Kathryn. 1998. *Shadows in the Dawn: The Lemurs of Madagascar*, San Diego: Gulliver Green.

Latour, Bruno. 2011. "Some Experiments in Art and Politics," *e-flux* 23(3): 1–11. Available online: http://www.e-flux.com/journal/some-experiments-in-art-and-politics/.

Latour, Bruno. 2005. *Reassembling the Social. An Introduction to Actor-Network Theory*, Oxford: Oxford University Press.

Lee, Sook-Kyung and Frieling, Rudolf, eds. 2019. *Nam June Paik*, London: Tate Publishing.

Leeker, Martina, ed. 2001. *Maschinen, Medien, Performances: Theater an der Schnittstelle zu digitalen Medien*, Berlin: Alexander Verlag.

Lehmann, Hans Thies. [1999] 2006. *Postdramatisches Theater*, Frankfurt/M: Verlag der Autoren. Trans. Karen Jürs-Munby, *Postdramatic Theatre*, London: Routledge.

Lestienne, Francis. 2008. "Les sciences du mouvement: art & handicap," *Bains numériques* 2: 81–87.

Leuenberger, Eva Maria. 2019. *dekarnation*, Graz: Droschl.

Lockwood, Patricia (2021) "I Hate Nadia Beyond Reason," *London Review of Books* 43(4): 19–22.

Louppe, Laurence. 2010. *Poetics of Contemporary Dance*, trans. Sally Gardner, Alton: Dance Books Ltd.

Louppe, Laurence, ed. 1994. *Traces of Dance*, trans. Brian Holmes, Paris: Éditions Dis Voir.

Lozano-Hemmer, Rafael, ed. 2000. *Alzado Vectorial/Vectorial Elevation: Relational Architecture No.4*, Mexico City: Conaculta.

Lucie, Sarah. 2019. "Posthuman Visions," *PAJ: A Journal of Performance and Art* 122: 75–79.

Mackenzie, Adrian. 2002. *Transductions: Bodies and Machines at Speed*, London: Continuum.

Malik, Suhail, with Avanessian, Armen. 2016. "The Time-Complex. Postcontemporary," Available online: http://dismagazine.com/discussion/81924/the-time-complex-postcontemporary/.

Magruder, Michael Takeo. 2010. "*The Vitruvian World*: A Case Study in Creative Hybridisation of Virtual, Physical and Networked Space." In Franziska Schroeder, ed., *Performing Technology: User Content and the New Digital Media*, Newcastle: Cambridge Scholars Publishing, pp. 51–71.

Manning, Erin. 2009. *Relationscapes: Movement, Art, Philosophy*, Cambridge, MA: MIT Press.

Manovich, Lev. 2001. *The Language of New Media*, Cambridge, MA: MIT Press.

Marranca, Bonnie. 2008. *Performing Histories*, New York: PAJ Publications.

Marzano, Stefano, Green, Josephine, van Heerden, Clive, Mama, Jack, and Eves, David. 2001. *New Nomads: An Exploration of Wearable Electronics*, Rotterdam: o10 Publishers.

Matt, Gerald and Miessgang, Thomas, eds. 2002. *Deanimated: The Invisible Ghost (Martin Arnold)*, Wien: Springer Verlag.

McCarren, Felicia. 2003. *Dancing Machines: Choreographies of the Age of Mechanical Reproduction*, Palo Alto: Stanford University Press.

McCormack, Derek P. 2018. *Atmospheric Things: On the Allure of Elemental Envelopment*, Durham, NC: Duke University Press.

McCormack, Derek P. 2014. *Refrains for Moving Bodies in Affective Spaces*, Durham, NC: Duke University Press.

McCullough, Malcolm. 2004. *Digital Ground: Architecture, Pervasive Computing, and Environmental Knowing*, Cambridge, MA: MIT Press.

Medeiros, Maria Beatriz de. 2006. *Corpos Informáticos: arte, corpo, tecnologia*, Brasilia: FAC.

Medeiros, Maria Beatriz de. 2005. *Aisthesis: estética, educação e comunidades*, Chapecó: Argos.

Meisenheimer, Wolfgang. 2007. *Choreography of the Architectural Space/Choreografie des Architektonischen Raumes*, Paju Book City: Dongnyok Publishers.

Merleau-Ponty, Maurice. 2003. *Phenomenology of Perception*, trans. Donald A. Landes, London: Routledge.

Milgram, Paul and Kishino, Fumio. 1994. "A Taxonomy of Mixed Reality Visual Displays," *EICE Transactions on Information Systems* E77-D(12). Available online: http://etclab.mie.utoronto.ca/people/paul_dir/IEICE94/ieice.html.

Mitchell, Katie, 2009. *The Director's Craft: A Handbook for the Theatre*, London: Routledge.

Mitchell, Thomas, Hyde, Joseph, Tew, Philip, and Glowacki, David R. 2016. "Danceroom Spectroscopy: At the Frontiers of Physics, Performance, Interactive Art and Technology," *Leonardo* 49(1): 138–47.

Mitra, Royona. 2015. *Akram Khan: Dancing New Interculturalism*, London: Palgrave Macmillan.

Morey, Miguel and Pardo, Carmen. 2002. *Robert Wilson*, Barcelona: Ediciones Polígrafa.

Morton, Timothy. 2016. *Dark Ecology: For a Logic of Future Coexistence*, New York: Columbia University Press.

Moser, Mary Anne and MacLeod, Douglas. 1996. *Immersed in Technology: Art and Virtual Environments*, Cambridge, MA: MIT Press.

Moten, Fred. 2003. *In the Break: The Aesthetics of the Black Radical Tradition*, Minneapolis: University of Minnesota Press.

Munster, Anna. 2006. *Materializing New Media: Embodiment in Information Aesthetics*, Hanover: Dartmouth College Press/University of New England.

Nancy, Jean-Luc. 2020. "Resonance of Sense." In François Bonnet and Bartolomé Sanson, eds., *Spectres II: Resonances*, Rennes: Shelter Press, pp. 13–16.

Nancy, Jean-Luc. 2007. *Listening*, trans. Charlotte Mandell, New York: Fordham University Press.

Newling, John. 2007. *An Essential Disorientation*, Warsaw: SARP.

Nichols, Bill, ed. 2001. *Maya Deren and the American Avant-Garde*, Berkeley: University of California Press.

Noble, Margaret. 2010. "The Art of Projection," Available online: http://margaretnoble.net/educator/the-art-of-projection.

Obrist, Hans Ulrich. 2014. *Ways of Curating*, London: Penguin.

Oddey, Alison and White, Christine, eds. 2006. *The Potentials of Space: The Theory and Practice of Scenography and Performance*, Bristol: Intellect.

Pallasmaa, Juhani. 2014. "Space, Place and Atmosphere: Peripheral Perception in Existential Experience." In Christian Borch, ed., *Architectural Atmospheres: On the Experience and Politics of Architecture*, Basel: Birkhäuser Verlag, pp. 18–41.

Parikka, Jussi. 2015. *A Geology of Media*, Minneapolis: University of Minnesota Press.

Paul, Christiane. 2003. *Digital Art*, London: Thames and Hudson.

Paulsen, Kris. 2017. *Here/ There: Telepresence, Touch and the Art of the Interface*, Cambridge, MA: MIT Press.

Pavitt, Jane. 2008. *Fear and Fashion in the Cold War*, London: V & A Publishing.

Pfeifer, Rolf and Bongard, Josh. 2007. *How the Body Shapes the Way We Think: A New View of Intelligence*, Cambridge, MA: MIT Press.

Phelan, Peggy. 1993. *Unmarked: The Politics of Performance*, London: Routledge.

Philippopoulos-Mihalopoulos, Andreas. 2015. *Spatial Justice: Body, Lawscape, Atmosphere*, London: Routledge.

Picchi, Giuliano. 2017. "Sabine Theunissen, Lulu," *Scenography Today*, May 24. Available online: https://www.scenographytoday.com/sabine-theunissen-lulu/.

Pilikian, Vaughan. 2021. "Parallel Seizure: Art and Culture at the End of Days," *PAJ: A Journal of Performance and Art* 43(1): 3–18.

Polidorou, Doros. 2021. "The Tamarind Forest: An Augmented Virtuality Experience," *Digital Creativity* 32(1): 71–77.

Popper, Frank. 2007. *From Technological to Virtual Art*, Cambridge, MA: MIT Press.

Povall, Richard. 2016. "Branching Out: Creative Collaborations with Trees," SCUDD listserv (email posting, 25 January).

Quinz, Emanuele, ed., 2004. *Interfaces*, special issue of *Anomalie digital_arts* 3.

Quinz, Emanuele, ed. 2004. *Digital Performance*, special issue of Anomalie digital_arts 2.

Rainer, Cosima, Rollig, Stella, Daniels, Dieter, and Ammer, Manuela, eds. 2009. *See this Sound: Versprechungen von Bild und Ton*, exhibition catalogue. Cologne: Verlag Walther König.

Raunig, Gerald. 2016. *Dividuum: Machinic Capitalism and Molecular Revolution*, trans. Aileen Derieg, Pasadena: Semiotext(e).

Raunig, Gerald. 2010. *A Thousand Machines*, trans. Aileen Derieg, Cambridge, MA: MIT Press.

Read, Alan. 2020. *The Dark Theatre: A Book about Loss*, Abingdon: Routledge.

Read, Alan. 2013. *Theatre in the Expanded Field: Seven Approaches to Performance*, London: Bloomsbury.

Rebstock, Matthias and Roesner, David, eds. 2012. *Composed Theatre: Aesthetics, Practices, Processes*, Bristol: Intellect.

Rokeby, David. 2019. "Perspectives on Algorithmic Performance through the Lens of Interactive Art," *TDR* 63(4): 88–98.

Salazar Sutil, Nicolás. 2018. *Matter: Transmission: Mediation in a Paleocyber Age*, London: Bloomsbury Publishing.

Salazar Sutil, Nicolás. 2015. *Motion and Representation: The Language of Human Movement*, Cambridge, MA: MIT Press.

Salazar Sutil, Nicolás and Popat, Sita, eds. 2015. *Digital Movement: Essays in Motion Technology and Performance*, London: Palgrave Macmillan.

Saletnik, Jeffery and Schuldenfrei, Robin, eds. 2009. *Bauhaus Construct: Fashioning Identity, Discourse and Modernism*, New York: Routledge.

Salter, Chris. 2015. *Alien Agency: Experimental Encounters with Art in the Making*, Cambridge, MA: MIT Press.

Salter, Chris. 2010. *Entangled: Technology and the Transformation of Performance*, Cambridge, MA: MIT Press.

Salvesen, Brett, ed. 2018. *3D: Double Vision*, Munich: Prestel.

Scheper, Dirk. 1988. *Oskar Schlemmer Das Triadische Ballett und die Bauhausbühne*, Berlin: Akademie der Künste.

Simondon, Gilbert. 2012. *Die Existenzweise technischer Objekte*, trans. Michael Cuntz, Zürich: diaphanes.

Simondon, Gilbert. 1969. *Du mode d'existence des objets techniques*, Paris: Aubier Montagne.

Slater, Mel, Spanlang, Bernhard, Sanchez-Vives, Maria V., and Blanke, Olaf. 2010. "First Person Experience of Body Transfer in Virtual Reality," *PLoS One* 5(5): e10564.

Sloterdijk, Peter. 2004. *Sphären III – Schäume*, Frankfurt: Suhrkamp [partial translation: *Terror from the Air*, trans. Amy Patton & Steve Corcoran, Los Angeles: Semiotext(e), 2009].

Soloski, Alexis. 2014. "A Double Dose of Chekhov," *The New York Times*, 5 January. Available online: http://www.nytimes.com/2014/01/06/theater/jay-scheib-stages-platonov-or-the-disinherited.html?_r=0.

Sondheim, Alan. 2009. *The Accidental Artist: Phenomenology of the Virtual*, Second Life Exhibition at Odyssey, June 2008-January 2009. Available online: http://www.alansondheim.org/sltheory.txt.

Spuybroek, Lars. 2008. *The Architecture of Continuity, Essays and Conversations*, Rotterdam: V2_Publishing/NAi Publishers.

Spuybroek, Lars. 2004. *NOX: Machining Architecture*, London: Thames & Hudson.

Stewart, Kathleen. 2011. "Atmospheric Attunements," *Society and Space* 29(3): 445–53.

Steyerl, Hito. 2012. *The Wretched of the Screen*, Berlin: Sternberg Press.

Stoppiello, Dawn with Mark Coniglio. 2003. "Fleshmotor." In Judy Malloy, ed., *Women, Art, and Technology*, Cambridge, MA: MIT Press, pp. 440–50.

Suzuki, Tadashi. 1987. *The Way of Acting: The Theatre Writings of Tadashi Suzuki*, trans. J. Thomas Rimer, New York: Theatre Communications.

Thibaud, Jean-Paul. 2011. "The Sensory Fabric of Urban Ambiances," *The Senses and Society* 6(2): 203–15.

Thomas, Lisa May and Glowacki, David R. 2018. "Seeing and Feeling in VR: Bodily Perception in the Gaps between Layered Realities," *International Journal of Performance Arts and Digital Media* 14(2): 145–68.

Tomkins, Calvin. 2005. *Off the Wall: a Portrait of Robert Rauschenberg*, New York: Picador.

Tribe, Mark and Jana, Reena. 2007. *New Media Art*, Cologne: Taschen.

Trimingham, Melissa. 2011. *The Theatre of the Bauhaus*, London: Routledge.

Tsing, Lowenhaupt Anna. 2015. *The Mushroom at the End of the World: On the Possibility of Life in Capitalist Ruins*, Princeton, NJ: Princeton University Press.

Uexküll, Jakob von. 2010 [1934]. A *Foray into the World of Animals and Humans*, trans. Joseph D. O'Neil. Minneapolis: University of Minnesota Press.

Uhlirova, Marketa, ed. 2015. *Birds of Paradise: Costume as Cinematic Spectacle*, Cologne: König Books.

Valentini, Valentina. 2009. "On the Dramaturgical Aspects of Bill Viola's Multi-media Installations," *Performance Research* 14(3): 54–64.

Verstraete, Pieter. 2010. "The Listener's Response," *Performance Research* 15(3): 88–94.

Viola, Bill. 2010. "Artist to Artist: Peter Campus Image and Self," *Art in America*, (February).

Whitelaw, Mitchell. 2001. "Inframedia Audio: Glitches and Tape Hiss," *Artlink* 21(3): 49–52.

Wiberg, Mikael. 2017. *The Materiality of Interaction: Notes on the Materials of Interaction Design*, Cambridge, MA: MIT Press.

Wilcox, Emily. 2018. *Revolutionary Bodies: Chinese Dance and the Socialist Legacy*, Oakland: University of California Press.

Wilderson, Frank B. III. 2020. *Afropessimism*, New York: Liveright Publishing Corp.

William Forsythe: Suspense. An Exhibition at Ursula Blickle Stiftung, Kraichtal, May 17 – June 29, 2008. In collaboration with Markus Weisbeck.

William Forsythe: The Fact of Matter. An Exhibition at MMK Museum für Moderne Kunst Frankfurt, 17 October 2015 – 13 March 2016.

Wilson-Bokowiec, Julie and Bokowiec, Mark. 2009. "Sense & Sensation: The Act of Mediation and Its Effects," *Intermedialites: History and Theory of the Arts, Literature and Techniques* 12: 129–42.

Winkler, Todd. 2000. "Audience Participation and Response in Movement-Sensing Installations," paper presented at ISEA, Paris.

Won, Andrea Stevenson, Bailenson, Jeremy, Lee, Jimmy, and Lanier, Jaron. 2015. "Homuncular Flexibility in Virtual Reality," *Journal of Computer-Mediated Communication* 20(3): 241–59.

Worthen, William B. 2008. "Antigone's Bones," *TDR* 52(3): 10–33.

Wright, Charles. 2019. *Oblivion Banjo*, New York: Farrar, Straus, and Giroux.

Xu, Zhi. 2021. *Techno-Choreography and the Embodiment of Chineseness*, PhD Thesis, Brunel University London.

Zarrilli, Phillip B. 2010. "Oral, Ritual, and Shamanic Performance." In Gary Jay Williams, ed., *Theatre Histories: An Introduction*, London: Routledge, pp. 37–38.

Zielinski, Siegfried. 2002 [2006]. *Archäologie der Medien. Zur Tiefenzeit des technischen Hörens und Sehens*, Reinbek: Rowohlt [trans. Gloria Custance, *Deep Time of the Media: Toward an Archaeology of Hearing and Seeing by Technical Means.* Cambridge, MA: MIT Press.]

Zielinski, Siegfried. 1994. "Historic Modes of the Audiovisual Apparatus," *Iris* 17: 7–24.

Zumthor, Peter. 2006. *Atmospheres: Architectural Environments – Surrounding Objects*, Basel: Birkhäuser Verlag.

INDEX

9 780367 632618